T0260563

Coastal and Estuarine Studies

Coastal and Estuarine Studies

61

Jonathan T. Phinney, Ove Hoegh-Guldberg,
Joanie Kleypas, William Skirving, and
Al Strong (Eds.)

Coral Reefs and Climate Change: Science and Management

American Geophysical Union
Washington, DC

Published under the aegis of the AGU Books Board

Jean-Louis Bougeret, Chair; Gray E. Bebout, Carl T. Friedrichs, James L. Horwitz, Lisa A. Levin, W. Berry Lyons, Kenneth R. Minschwaner, Andy Nyblade, Darrell Strobel, and William R. Young, members.

Library of Congress Cataloging-in-Publication Data

Coral reefs and climate change : science and management.
p. cm. – (Coastaland estuarine studies ; 61)
Includes bibliographical references.
ISBN-13: 978-0-87590-359-0 (alk. paper)
ISBN-10: 0-87590-359-2 (alk. paper)
1. Coral reefs and islands–Environmental aspects. 2. Climatic changes–Environmental aspects. I. American Geophysical Union.

QH95.8.C67 2006
577.7'89–dc22

2006035298

ISBN-13: 978-0-87590-359-0
ISBN-10: 0-87590-359-2

ISSN 0733-9569

Copyright 2006 by the American Geophysical Union, 2000 Florida Ave., NW, Washington, DC 20009, USA.

Figures, tables, and short excerpts may be reprinted in scientific books and journals if the source is properly cited.

Authorization to photocopy items for internal or personal use, or the internal or personal use of specific clients, is granted by the American Geophysical Union for libraries and other users registered with the Copyright Clearance Center (CCC) Transactional Reporting Service, provided that the base fee of $1.50 per copy plus $0.35 per page is paid directly to CCC, 222 Rosewood Dr., Danvers, MA 01923. 0733-9569/06/$1.50+0.35.

This consent does not extend to other kinds of copying, such as copying for creating new collective works or for resale. The reproduction of multiple copies and the use of full articles or the use of extracts, including figures and tables, for commercial purposes requires permission from the American Geophysical Union.

Printed in the United States of America

CONTENTS

Preface
Jonathan T. Phinney, Ove Hoegh-Guldberg, Joanie Kleypas, William Skirving and Allan E. Strong vii

1 **Corals and Climate Change: An Introduction**
John E. N. Veron and Jonathan Phinney 1

2 **Tropical Coastal Ecosystems and Climate Change Prediction: Global and Local Risks**
Terry Done and Roger Jones 5

3 **Coral Reef Records of Past Climatic Change**
C. Mark Eakin and Andréa G. Grottoli 33

4 **The Cell Physiology of Coral Bleaching**
Sophie G. Dove and Ove Hoegh-Guldberg 55

5 **Coral Reefs and Changing Seawater Carbonate Chemistry**
Joan A. Kleypas and Chris Langdon 73

6 **Analyzing the Relationship Between Ocean Temperature Anomalies and Coral Disease Outbreaks at Broad Spatial Scales**
Elizabeth R. Selig, C. Drew Harvell, John F. Bruno, Bette L. Willis, Cathie A. Page, Kenneth S. Casey, and Hugh Sweatman 111

7 **A Coral Population Response (CPR) Model for Thermal Stress**
R. van Woesik and S. Koksal 129

8 **The Hydrodynamics of a Bleaching Event: Implications for Management and Monitoring**
William Skirving, Mal Heron, and Scott Heron 145

9 **Identifying Coral Bleaching Remotely via Coral Reef Watch – Improved Integration and Implications for Changing Climate**
A. E. Strong, F. Arzayus, W. Skirving, and S. F. Heron 163

10 **Management Response to a Bleaching Event**
David Obura, Billy Causey, and Julie Church 181

11 **Marine Protected Area Planning in a Changing Climate**
Rodney V. Salm, Terry Done, and Elizabeth McLeod 207

12 **Adapting Coral Reef Management in the Face of Climate Change**
Paul Marshall and Heidi Schuttenberg 223

List of Contributors 243

PREFACE

The effects of increased atmospheric carbon dioxide and related climate change on shallow coral reefs are gaining considerable attention for scientific and economic reasons worldwide. Although increased scientific research has improved our understanding of the response of coral reefs to climate change, we still lack key information that can help guide reef management. Research and monitoring of coral reef ecosystems over the past few decades have documented two major threats related to increasing concentrations of atmospheric CO_2: (1) increased sea surface temperatures and (2) increased seawater acidity (lower pH). Higher atmospheric CO_2 levels have resulted in rising sea surface temperatures and proven to be an acute threat to corals and other reef-dwelling organisms. Short periods (days) of elevated sea surface temperatures by as little as 1-2°C above the normal maximum temperature has led to more frequent and more widespread episodes of coral bleaching–the expulsion of symbiotic algae. A more chronic consequence of increasing atmospheric CO_2 is the lowering of pH of surface waters, which affects the rate at which corals and other reef organisms secrete and build their calcium carbonate skeletons. Average pH of the surface ocean has already decreased by an estimated 0.1 unit since preindustrial times, and will continue to decline in concert with rising atmospheric CO_2. These climate-related stressors combined with other direct anthropogenic assaults, such as overfishing and pollution, weaken reef organisms and increase their susceptibility to disease.

The economic ramifications of reduced or disappearing coral reefs are staggering: between 100 and 500 million people from 50 nations directly depend on coral reefs for food; while tourism and fishing are multibillion dollar industries. Indirect services from coral reefs, such as protective barriers from wave action and tsunamis, are also important, especially since sea level is expected to rise quicker in response to climate change.

Given the bleak outlook for shallow coral reefs, what can natural resource managers do? This book attempts to bridge the science and management of coral reefs in the face of a changing climate. It provides an overview and background of the scientific issues as well as present management strategies to limit the effects of climate change. The science of coral reefs and climate change, and resulting management actions, will continue to evolve. The information in this book portrays the urgency to improve the science and management of coral reefs in a changing climate as well as the necessary background information to do so.

The editors thank the authors who graciously accepted our invitation and submitted the chapters comprising this publication. They are also indebted to the many reviewers from the coral reef scientific and management community – a small, but dedicated lot – whose critiques and comments greatly improved the accuracy and clarity of the book. The financial support from the NOAA Coral Program to offset publication and distribution costs is gratefully acknowledged. Finally, the editors wish to acknowledge our AGU acquisitions

Coral Reefs and Climate Change: Science and Management
Coastal and Estuarine Studies 61
Copyright 2006 by the American Geophysical Union.
10.1029/61CE01

editor, Allan Graubard, and staff member, Dawn Seigler, who patiently steered this publication towards completion.

Jonathan T. Phinney, Ove Hoegh-Guldberg,
Joanie Kleypas, William Skirving
and Allan E. Strong

1

Corals and Climate Change: An Introduction

John E. N. Veron and Jonathan Phinney

Shallow water coral reefs are widely considered the pinnacle of nature's achievement in the Ocean realm, for not only are they among the most beautiful places on Earth, they are the nexus of marine biodiversity. Reefs are more than biological entities, however; they are geological structures as well – in fact, the biggest and most enduring structures ever made by life on Earth. This unique combination of beauty, biology, and geology makes them immensely valuable. They head the list of natural tourist destinations; provide a livelihood for millions of people spread around coasts of the tropical world; and are a rich source of pharmaceutically active compounds.

This is quite a list of superlatives, and there are others. Being both biological and geological entities, coral reefs leave exceptional records of ancient marine environments, and of the life-forms that once occupied them. These records are set in limestone, a component of the carbon cycle – arguably the most important of all the Earth's great biogeochemical cycles. Carbon dioxide (CO_2) plays a critical role in the global carbon cycle because it is a key compound in photosynthesis, in the carbonate buffering system of the oceans, and in the Earth's radiative balance of the atmosphere. Not surprisingly, changes in the amount of carbon dioxide in our atmosphere have many different and far-reaching consequences for all reefs – past, present, and future.

Coral reefs exist at the interface of the ocean, land and atmosphere. This dynamic realm leaves reefs exceptionally vulnerable to environmental change. Corals build reefs that in turn create the habitats that allow them to proliferate during sea level fluctuations. To build these massive reefs in nutrient-poor (oligotrophic) ocean realms, corals maximize their energy sources by maintaining a symbiotic relationship with algae – zooxanthellae – that inhabit their body tissues and provide most of the energy for coral growth. This symbiotic relationship requires a commitment to live near the surface of warm tropical oceans, which are the most likely areas to be affected by atmospheric changes, including changes in temperature and carbon dioxide concentration.

A central concern – and the subject of this volume – is what happens when the environment becomes less than optimal for coral reef ecosystems, as is now happening because of increases in atmospheric carbon dioxide. We know that the present increases are due to anthropogenic activities, but we also know that increased CO_2 levels have occurred in the geological past, so a logical question is: What happened to reefs during other high-CO_2 periods?

Before turning to this question, we must add one final superlative to our list: reefs, alone of all major ecosystems, have been decimated by all five of the great mass extinctions of

Coral Reefs and Climate Change: Science and Management
Coastal and Estuarine Studies 61
Copyright 2006 by the American Geophysical Union.
10.1029/61CE02

Earth history. In fact this is about the only thing all five extinctions have in common. The time reefs have taken to recover from mass extinctions is certainly real: 6 to as many as 13 million years; time enough to build the Great Barrier Reef ecosystem several times over. These immense time intervals, nicknamed "reef gaps," appear to be associated with high levels of atmospheric carbon dioxide in all but one (still unclear) case. This is not to say that carbon dioxide was the cause of past mass extinctions, but the association deserves further study. In the past, several gases including carbon dioxide have come from geological sources such as volcanism – primarily the outcome of sea-floor spreading – and exploding asteroid impacts. The carbon cycle has also been influenced by variations in solar irradiance due to changes in solar output as well as in the Earth's orbital geometry around the sun, and by sudden releases of methane from (methane) hydrates stored in the deep ocean as ice slurries.

Are such drastic changes about to happen again? Given the rarity of mass extinctions, this seems unlikely. But biological change is also unpredictable, particularly because the responses of living organisms to large environmental changes are generally nonlinear and likely to occur as a series of threshold events. Corals are tolerant to both changes in heat and water chemistry – up to a point. And it is possible to understand and predict climate change within definable bounds of uncertainty. Unfortunately, determining these bounds of uncertainty is largely derived from the ongoing global "experiment" of rising atmospheric CO_2 since our understanding of past climate changes on corals remains so incomplete. A looming concern for scientists and managers is whether or not we will understand and manage these uncertainties before most of the present shallow reefs are gone.

After the end of the Cretaceous mass extinction 65 million years ago and the beginning of our era, the Cenozoic, there were no living reefs. High levels of carbon dioxide and methane persisted, then peaked at the end of the Paleocene (55 million years ago) in an event known as the Late Palaeocene Thermal Maximum. At this time, for reasons unknown, over 1,000 gigatonnes of carbon, probably as methane, were released into the atmosphere in 1,000 years or less. This amounts to 25-50% of the entire CO_2 release now anticipated from human activities. The abrupt release is associated with a 5-7°C rise in deep ocean temperatures (as recorded in sediments) and a dramatic shoaling of the calcium carbonate compensation depth. If this was a rehearsal of what is happening today, it is an interesting one, for peak levels of carbon dioxide remained in the atmosphere for over 100,000 years.

Global temperatures reached a maximum during Early Eocene (50 million years ago), then started to decline. By Mid Eocene, carbon dioxide levels were roughly similar to those of today. By Late Eocene, the first cooling cycles had become clearly established: glaciers had formed on Antarctica. And after a gap of many millions of years, reefs once again proliferated around the world. The environmental trail fades at this time but becomes clear again in the Early Miocene (20 million years ago). Since then carbon dioxide levels have been uniformly low – below 360 ppm – and so have most measures of ocean temperature except for occasional short-term excursions, mostly from unknown causes. The Ice Age fluctuations in atmospheric CO_2 that occurred throughout the Pleistocene and into the Holocene, roughly between 200 and 300 ppm, are far less than projected CO_2 increases for this century (present-day concentration is already greater than 380 ppm, and is projected to at least reach 560 by the middle of this century).

Mass coral bleaching – the most immediate biological consequence of global warming – has caused widespread degradation of reefs around the tropical world. Yet this extraordinary phenomenon has only been known for two decades. Nothing is comparable to these worldwide massive bleaching events in human history, at least not in the ocean realm and not as a result of unintentional human cause. Already the reasons behind the bleaching are well known: increases in maximum sea surface temperatures are pushing corals beyond their thermal tolerances and are particularly widespread during warm phases of the El Niño-Southern Oscillation. The resulting bleaching, caused by a breakdown in coral-algal symbiosis, is now a regular event. In addition, ocean acidification from changing oceanic buffering capacity has been shown to slow coral calcification. The consequences of this phenomenon are still unclear, but it is likely to affect not only corals and other calcifying reef organisms but reef-building altogether.

By any standards, the plight of reefs comes from a strange chain of circumstances. Increases in atmospheric carbon dioxide composition cause the Earth's atmosphere to warm, which in turn warms the upper ocean. This warm water causes a breakdown in coral-algal symbiosis, and the breakdown leads to mass bleaching. Many of the linkages between carbon dioxide and coral bleaching seem improbable at first. In this chain of events occur environmental or physiological changes that are either invisible (e.g., changes in atmospheric composition) or barely measurable (e.g., biochemical changes). Yet the effects are real, and now there are sound explanations for them.

Not surprisingly, mass bleaching events have set off numerous alarm bells for coral reef managers and scientists. Fifty years from now, for example, the Earth's population will have increased approximately by another 3 billion. And even if major third world countries have not adapted first world living standards, greenhouse gas emissions will almost certainly be greater than they now are. Fifty years from now we will also have a much better understanding of the response of coral reefs to climate change. Corals will likely survive on the geologic timescale, as they have in the past, but based on current trends it is unlikely that coral reef ecosystems will improve, at least on time-scales relevant to humans.

This book thus provides a brief background on the current science and management options for shallow coral reefs in a changing climate. There are two major threats to shallow water corals from increased atmospheric carbon dioxide concentrations: increased sea surface temperatures and lowered pH (increased acidification). Chapters one through seven provide the scientific underpinnings of corals and climate change. Chapter one situates reefs within the larger context of tropical ecosystems, including mangroves and seagrass, with comments on the general deterioration of ecosystem resilience. Chapter two provides an overview of paleoclimatology, and the methodologies used when describing chronologies of past climate change in fossil corals. Chapters three, four and five provide overviews of the physiological and environmental effects of increased sea surface temperatures (chapter three) and lower pH (chapter four); and how increased temperature may make corals more susceptible to water borne pathogens and disease (chapter five). Modelling is a critical tool for managers and scientists toward understanding, and possibly forecasting, coral vulnerability to increased temperatures; thus, chapters six and seven provide a physiological model and a hydrodynamic model, respectively, as a means to predict the vagrancies of thermal stress on surface corals. Chapters eight through eleven move the reader into the management realm, and offer a "toolbox" of management options. Chapter eight provides an overview of remote sensing techniques to predict possible areas of elevated sea surface temperatures where bleaching alerts can be issued. Chapter nine provides case studies from two distinct tropical regions, Kenya and the Caribbean, on how to manage a bleaching event once it occurs. Chapter ten and eleven summarize more long range planning options for managers in the face of climate change, including establishing marine protected areas to protect biodiversity and manage for ecosystem resilience.

The science and management of coral reefs within a changing climate are relatively young but fast-growing disciplines that befit the seriousness and urgency of the issue. Applying techniques from other scientific disciplines, such as genetics and remote sensing, are well-established in coral reef science and offer incredibly important insights. However, the application of technology requires training scientists and managers to utilize it. This highlights a serious gap in coral reef science and management, in that most coral reefs are in "developing" countries while most technology development is in "developed" countries. Increasing collaboration between scientists and managers, as well as between developed and developing countries, is a simple and proven method to transfer technology and increase capacity. Here, the issue is one of resources, both money and expertise, to address the problem. In the meantime, coral reef scientists and managers need to broaden their own expertise to include social science, economics, and an understanding of how to best inform policy making from local to national levels. The next generation of coral reef scientists and managers will likely be more astute in these fields, but it is uncertain whether there is enough time to reverse the downward trajectory of coral reefs and to ensure that future generations can experience a thriving coral reef ecosystem. This statement is a difficult way to start a book, but given the present state and future projections of coral reefs, it is a necessary one.

2

Tropical Coastal Ecosystems and Climate Change Prediction: Global and Local Risks

Terry Done and Roger Jones

1. Introduction

Coastal marine ecosystems occupy shallow coastal waters and extend up rivers and across coastal lands to the normal inland influence of tidal seawater intrusion [Allee et al., 2000]. In the tropics (Figure 1), coastal ecosystems are characterized more than anything else by mangroves, seagrass meadows and tropical calcareous reefs (i.e., coral reefs and coralline algae ridges) – the rich, diverse and interconnected engines of marine productivity for tropical islands, coasts and continental shelves. Our goal is to consider the prospects for these habitats in the relatively short term of a few decades- as their local environmental settings change with the changing global climate. We focus on the resilience of their defining and essential attributes, and their values to humans [Holling, 1973; McClanahan et al., 2002; Nyström et al., 2000]. These habitats are being exposed to seas that are projected to become gradually warmer (Figure 2A), deeper (Figure 2B); less alkaline (Figure 2C) and more productive [Sarmiento et al., 2004] than their present states [Kennedy et al., 2002]. The extent to which the ecosystems can accommodate these gradual changes in average conditions is of great importance. So too are the implications of changes in the intensity and frequency of extreme events, such as storms and precipitation. The Intergovernmental Panel for Climate Change [IPCC, 2001b] called for improved understanding of regional differences in the physical drivers and change in ecological function of coastal systems as the basis for developing appropriate adaptive responses, from management of coastal fisheries and shoreline stabilization [Burke et al., 2000] to the design of networks of marine protected areas [Salm et al., 2001; West and Salm, 2003].

1.1. Mangrove Forests, Seagrass Meadows, Coral Reefs and Coralline Algae Ridges

In geological terms, individual mangroves forests, seagrass meadows, coral reefs and coralline algae ridges have generally had very short histories of occupancy at their present sites, (given sea level was 120 m lower only 20,000 years ago – Peltier [2002]). On any given shore, the current relative importance, extent and juxtaposition of these habitats is largely a product of the particularities of their Holocene histories (last 10,000 years) and the influence they themselves have on local environmental settings. In the Lesser Antilles,

Coral Reefs and Climate Change: Science and Management
Coastal and Estuarine Studies 61
Copyright 2006 by the American Geophysical Union.
10.1029/61CE03

A

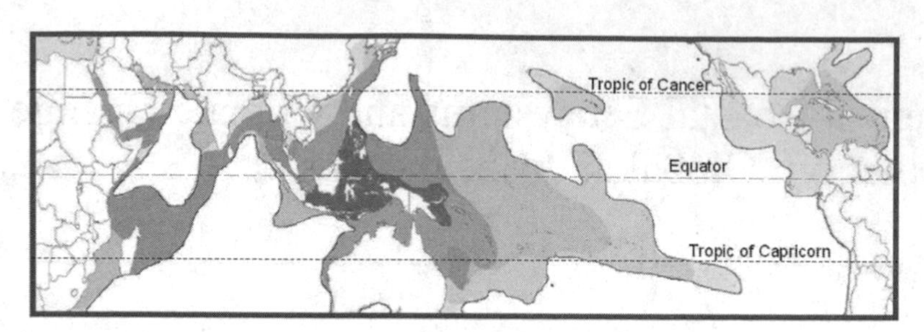

B

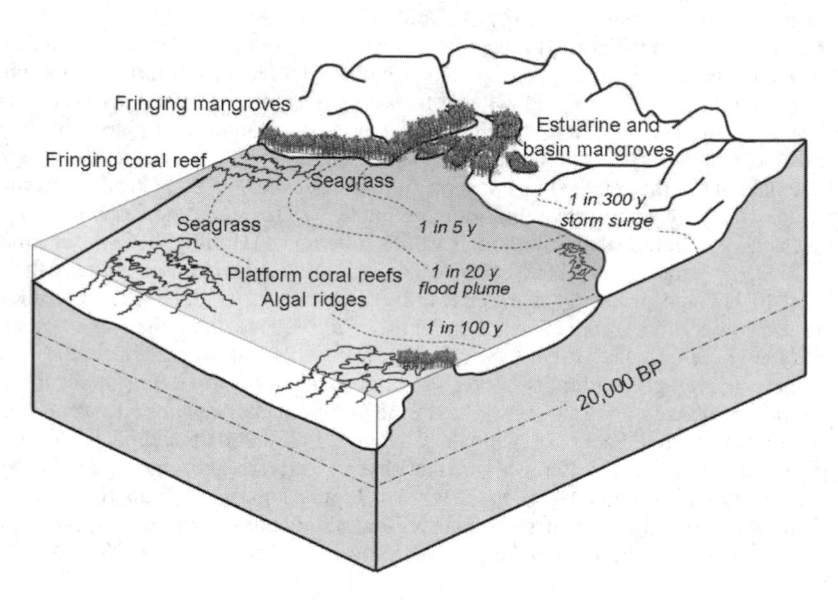

Figure 1. Tropical marine ecosystems. A. Global distribution of coral species diversity, from >500 (darkest) to <50 species (lighest), after Veron 2000. Tropical mangrove and seagrass meadows have a similar overall distribution pattern but much lower species diversity (mangroves ~40 species, seagrasses ~60 species.) B. Local variety in types and settings of mangroves, seagrass meadows and coral reefs. Also shown are sea level at 20,000 years before present (–120 m), and the reach of 1 in 5, 20 and 100 y flood plumes and the 1 in 300 y storm surge. The frequency and reach of flood plumes and storm surges that exceed specific damage thresholds will change as climate changes.

for example, Adey and Burke [1977] and Macintyre et al. [2001] paint pictures of the current state and juxtaposition of coral reefs and algal ridges as dynamic ongoing processes with a strong Holocene legacy. What is now coral reef may have once been algal ridge, and vice versa. In some settings, Adey and Burke suggested that insufficient Holocene time has elapsed to allow reefs and ridges to grow into shallow waters, while in others, it had.

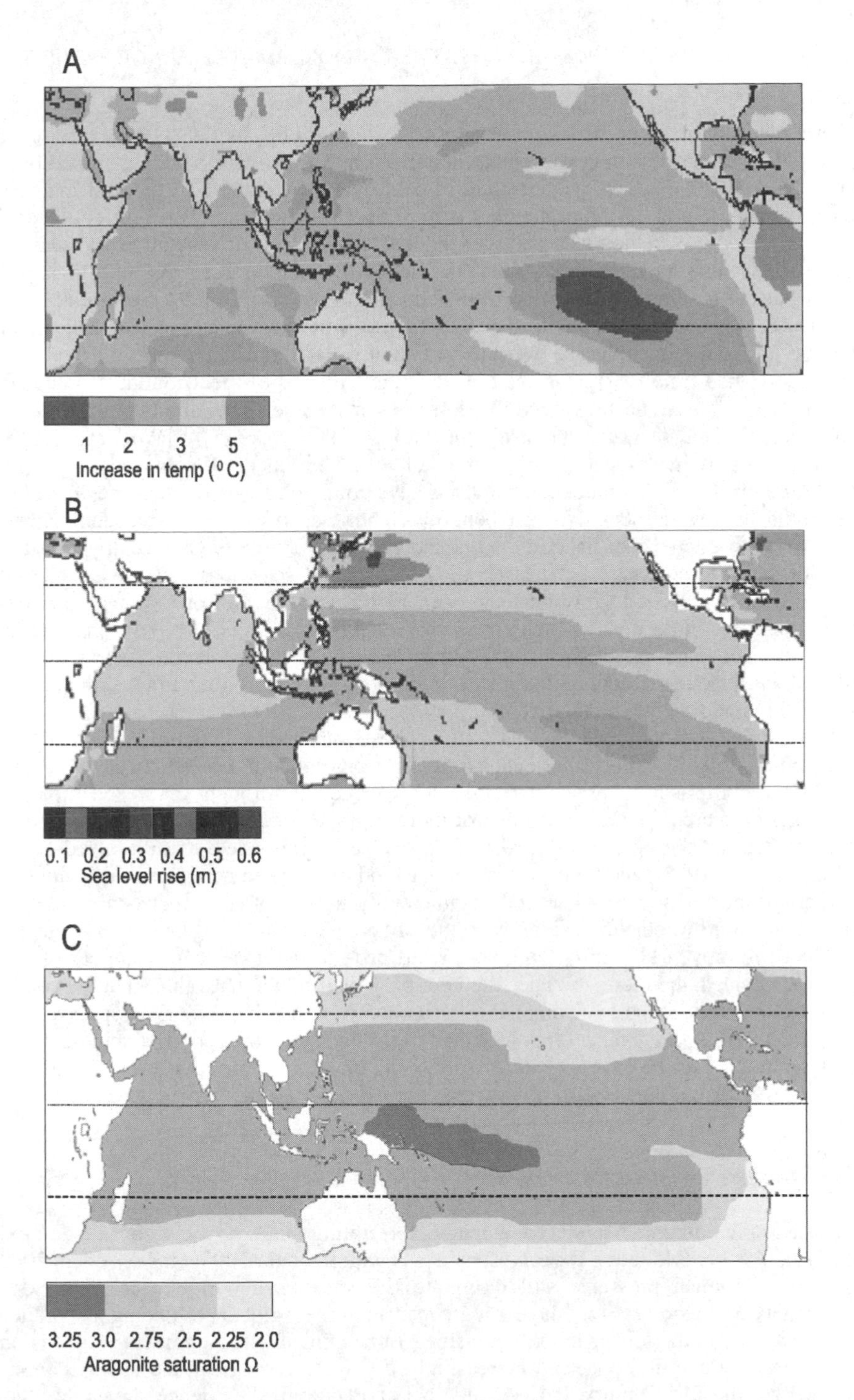

Figure 2. Tropical marine ecosystems. A. Projected increase in temperature by 2100 (business as usual Had 3). B. Projected sea-level rise by 2100 (business as usual Had 3). C. Projected Aragonite saturation state by 2070. After Gattuso et al., 1998.

Where relatively younger reefs have developed to seaward of older algal ridges that developed in their preferred strong wave areas in earlier times, the wave blocking caused by the corals leaves the ridge now degenerating in calm waters. Likewise in the vast mangrove expanses of northern Australia, the only way to deeply understand their current location, extent and species zonation is to consider them as the legacies of vegetation and sedimentary processes that have taken place over dramatically changing sea-level and fluvial-tidal energy balance histories of the Holocene [Chappell and Woodroffe, 1994; Semeniuk, 1994].

But as ecological entities that must confront rapid, anthropogenic climate change, it may be more important to emphasise two things: 1) that the biological assemblages in all these system have had time for the turnover of many generations of their dominant organisms, and 2), that they have been exposed to relatively constant sea level and benign conditions of temperature and seawater chemistry for the last 6,000 years or so [Woodroffe, 1992]. The systems may have 'been there' in terms of some aspects of climate in the geological past [Pandolfi, 1999], but human populations have come to rely on them as they have been during the last few decades. We have benefited from their often not appreciated contributions to the productivity of fisheries and to shoreline protection. We have enjoyed them as some of nature's most scenic and fascinating landscapes, seascapes and playgrounds. We have unwittingly come to rely on their primary biogenic framework (above and below ground biomass in the case of mangroves and seagrass meadows; the skeletons of corals and other calcifying organisms in reefs and algal ridges): it is these frameworks that provide habitat for our seafood, and protect our shores by accumulating and stabilizing sediments within and below the tidal zone.

Biological connectivity within and among these systems is important. Each of them occurs in discrete littoral units, and each unit relies on critical linkages between units of its same type, comprising sources and sinks for each other through larval and biological exchanges of shared species. When two or more types occur contiguous with one another, as they often do, they process shared waters, exchange particulate and dissolved organic matter [Alongi, 1998], and their plants, animals and mobile sediments can intermingle. In the Caribbean, the biomass of several commercially important coral reef fish species was found to be more when adult habitat were closely connected to mangroves than when they were not [Mumby et al., 2004]. To manage and protect the values and resources of these systems, we need, therefore, to understand not only their individual values, but the connectivity and interdependencies among them. As a contribution to building a synoptic view, we have structured this chapter by first looking at two unifying concepts (risk and resilience) and then looking at the effects of different climate change variables (sea level, CO_2, temperature, pH) across the systems. Finally, we focus briefly on the systems one at a time.

1.2. Threats to System Resilience

These tropical marine habitats have been deteriorating under pressures of human modification and overexploitation for centuries [e.g., Pandolfi et al., 2003; Hughes et al., 2003]. Such direct human pressures still dominate the short-term outlook for their extent, productivity and biodiversity – three key indicators of their global well being and value to human society. In the longer term, increasing global climate change effects seem to compound direct human impacts, and increasingly dominate ecological and socio-economic change over the 21st Century. Paleo-ecological and ecological evidence shows that these ecosystems have persisted in place as recognisable entities though past climate change [Pandolfi, 1999]. The threat that human activities pose to ecosystems is in large part reduction of the ecological resilience that underpins their persistence as productive, diverse and useful assemblages [Loreau et al., 2001; Bellwood et al., 2004].

For each system, a key aspect of their ecological resilience is the capacity for the structurally dominant organism – be it a stand of mangrove, seagrass or coral – to repeatedly reinstate itself as the dominant structural form in the face of repeated disturbance. This is the basis for each particular place to retain its identity and value to humans in the face of damage, destruction, erosion and resource extraction that are, or have become, a normal part of their existence. It is the capacity for individual places to restore their prior attributes of productivity and architectural structure in periods of years to decades between major disturbances that has made these places so valuable as habitats for fisheries and wildlife, protectors of shorelines, builders of beaches and islands, places of great biodiversity and beauty, and assets for tropical tourism [McManus and Polsenberg, 2004].

For mangroves, seagrass meadows, coral and coralline algae reefs, sea-level rise and increases in ocean temperatures will potentially affect local survivorship, extent of individual formations and regional distributions. Increasing concentrations of atmospheric CO_2 will directly affect plant productivity and indirectly affect growth rates in calcareous organisms such as corals, clams and calcareous macroalgae, potentially impeding a reef's ability to grow vertically in pace with sea-level rise. The greater the climate change, the greater the potential detrimental effect on humans, through the loss in area of productive habitat, and loss of ecosystem goods and services in areas that remain. Summaries of potential hazards, consequences and risk modifying factors are provided in Tables 1, 2 and 3.

There are two main purposes for this chapter. The first is to describe potential climate change effects, and the second is to address their likelihood of occurrence at local scales. IPCC [2001a] provides a series of regional assessments, but in relation to coral reefs and related systems, each tropical region's chapter more or less reiterates the same critical outcomes. We describe a process for assessing the likelihood of exceeding critical thresholds at the local scale, by taking into account a) how physical hazards may change under climate change, and b) which properties of an ecological system and its local geographic setting exacerbate or ameliorate its vulnerability to those hazards.

2. Risk Assessment and Risk Management for Coastal Systems

The impacts of global change will vary greatly from place to place, requiring locally appropriate adaptive responses by planners and policymakers. A focus on local scales- say a bay, beach or province over which a resource manager or politician has influence – brings into play major issues of uncertainty about the potential impact and the adaptive response required. As in weather forecasting, precise medium to long-term predictions for particular times and places are unworkable. However projections made within a risk assessment and risk management framework can guide human adaptations to limit harm in an environment of uncertainty. Risk analysis is the process of assessing the likelihood of a harmful outcome, while risk management aims to reduce likelihood, consequences, or both. For example, mitigation of greenhouse gases reduces the likelihood of a given set of a rise in sea level, while building a sea wall reduces its consequences. This risk framework can help society understand and make decisions about the trade-off between the relative costs and benefits of adapting versus the costs and benefits of radical changes in the way we generate energy and use fossil fuels.

Climate change risks to tropical coastal ecosystems can be considered in terms of the probability of exceeding critical thresholds [after Pittock and Jones, 2000]. This is a more tenable objective at present trying to assign probabilities to particular model predictions. Our framework for risk analysis is presented in Figure 3 where a cumulative probability curve connects the 100% certainty of exceeding 'no change' and the 0% likelihood of exceeding the 'highest change' scenario. In our hypothetical example, there are two rates

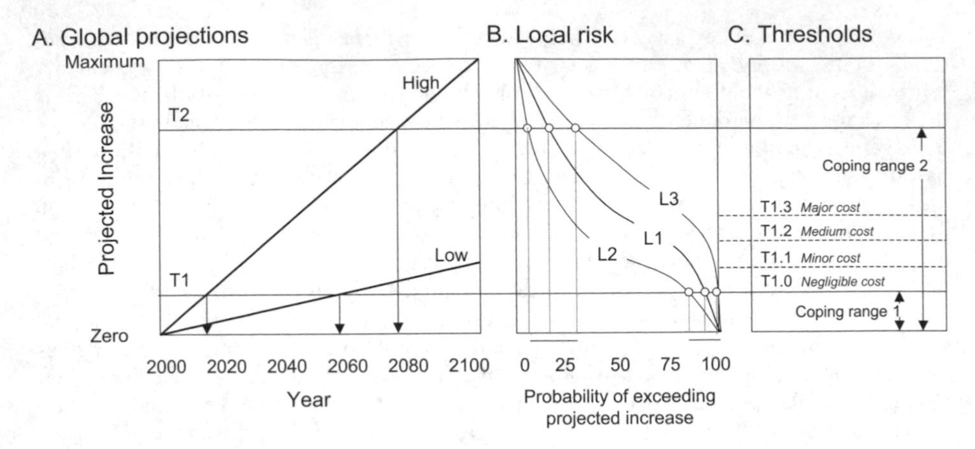

Figure 3. Relating threshold exceedance to the likelihood of climate change. A. Ranges of increase of hypothetical climate variables such as CO_2, sea-level or sea temperature, showing two critical thresholds for onset of effects detailed in C. B. Cumulative probability distributions cast as likelihood of exceedance of any projected level of increase in part A. L1 represents the global mean probability, L2 a locale in which local geography ameliorates change in the climate variable of interest, and L3, a locale in which it exacerbates global change. These probability distributions – that combine two ranges of uncertainty, randomly sampled and multiplied in a manner consistent with Schneider [2001], are typical of those with two or more component ranges of uncertainty.

of change and two critical thresholds (Figure 3A). The cumulative probability of exposure is given in Figure 3B. A critical threshold is defined as the level exceeding a tolerable limit of harm. In the example, there is an 80-100% probability that the low threshold (T1) will be exceeded. At the higher rate of global change, it will occur by 2015, and at the lower rate, by 2060. For the high threshold (T2), the odds are lower (5-30%) and they come into play around 2080. In each case, the range in the odds is largely a function of local circumstances (Figure 3B) that ameliorate (L2) or exacerbate (L3) the hazard compared to the global mean hazard (L1).

Application of this approach to any real system requires that we address four main issues: uncertainties in climate change at global and regional levels (encompassing global climate modelling and downscaling issues); the formulation of critical thresholds; attributes of the local physiographic and environmental settings that modify the likelihood of critical thresholds being exceeded; and intrinsic attributes of the local ecosystem that determine the extent to which it accommodates, is damaged by, or benefits from, the changed conditions.

2.1. *Uncertainties in Global Climate Predictions*

The uncertainty in global climate predictions is caused almost equally by uncertainty about future greenhouse gas emissions and uncertain climate science [Pittock et al., 2001]. This uncertainty, embodied in the familiar Special Report on Emission Scenarios (SRES) envelopes of diverging 21st Century trajectories for rises in CO_2, temperature and sea level, is caricatured in Figure 3A. The IPCC [IPCC, 2001b] explicitly does not assign a relative likelihood to the different emission scenarios [Nakiçenovic and Swart, 2000; Schneider, 2001]. Despite continual improvements in global climate models, their assumptions and predictions are generally not valid at regional scales [Murphy et al., 2004]. Nevertheless, as Pittock et al. [2001] observed, we can be sure that the cumulative probability of exceeding any particular level does decline with increasing magnitude (Figure 3B).

2.2. Critical Thresholds

Critical thresholds are delimited at their upper and lower limits by the coping range and ecological collapse, respectively. The coping range of a system (Figure 3C) represents a degree of change so slight that it imposes no diminution on ecosystem functionality, and no cost to the local dependant human population. Coping ranges 1 and 2 can simply represent ecological/human systems with smaller and larger capacity to accommodate environmental change before non-negligible costs are inflicted. (Alternatively, a coping range could represent the loss that an authority is prepared to accept, before reacting). Coping range 1 could be a change that falls within normal environmental and ecological variability at that site, but that may have previously been exceeded under the influence of extreme natural events (e.g., past floods, droughts, heatwaves, storms). If global climate change increases the frequency of exposure to such hazards, Coping range 1 is breached more often and by a greater amount and greater ecological costs are imposed (T1.1 to T1.3). An extreme event such as a 1 in 1000 y storm could precipitate ecological collapse in a single step, or there may be a stepwise breakdown of homeostatic mechanisms of ecological resistance, tolerance and resilience (Table 1).

TABLE 1. Stages of ecological deterioration: resistance, tolerance, resilience and critical failure. This terminology is intended for the ecological assemblage, not a dependent human population.

- Resistance (Tn.0) – the upper limit to the coping range of the ecological system- is characterized by absence of physiological symptoms or of change in life expectancy for existing individuals and new cohorts;
- Tolerance (Tn.1) is the accommodation of minor changes to the vigour or life expectancies of key functional species (such as dominant species of mangroves, seagrasses and corals) without loss of extent, productivity or biological diversity of the area. Biological adaptations developed as responses to these changes may increase Tn.1 relative to climate hazards, whereas repeated and/or cumulative stresses may decrease it.
- Resilience (Tn.2- whose distinction from 'tolerance' is somewhat fuzzy) is the capacity for the ecological assemblage to recover expeditiously, should the system be damaged by an extreme event. It is the property that ensures that, in spite of a temporary setback, there are no lasting changes in function, composition and/or amenity to the human population. ('Expeditious' and 'lasting' relate to 'reasonable expectations' for an ecological system of a particular type in a particular place). Such resilience is characterized by speed, strength and reliability in the processes of repair, recolonisation and new growth. It is as much a property of a damaged site's connectivity to, or isolation from sources of supply of larvae of key functional groups and resource species, as it is of the local environmental conditions.
- Critical failure (Tn.3) signifies loss of resilience, where the ecosystem loses its fundamental function and attributes (such as reef-building capacity in coral reefs; physical structure and habitat roles of mangroves and sea-grass meadows). It could also occur directly as a single catastrophic event, such as physical removal of an underlying sedimentary deposit by a storm or a flood.

2.3. Regional Variability and Local Modification of Environmental Change

Global climate models (e.g., Figure 2A-C) show that odds of a region exceeding a particular threshold will not necessarily be the same as the global odds (L1 in Figure 3B).This regional variability is averaged out in the global cumulative probability of exceedance curve (L1 in Figure 3A). Going down scale, the local cumulative probability curves (L2 and L3) are a function of local oceanography and local physiography of coastal land- and seascapes. For example, 'local oceanography' could on the one hand, be a reliable upwelling system that prevents local reef waters from heating by a particular amount – say 1°C (L2), or on the other, a calm shallow area where warming is exacerbated (L3). 'Local physiography of the coast' could be a particular combination of embayment and bathymetry that greatly lessens (L2) or increases (L3) the odds of exposure to a 0.5 m sea-level rise or 1°C heating.

2.4. Intrinsic Attributes of the Local Coastal System

T1 and T2 (Figure 3A) indicate local differences in the amount of change (e.g., degrees of heating or centimetres of rise in mean sea level) that a coastal system can accommodate without any detrimental effect. Examples could include the coral bleaching onset threshold (in degrees above 'normal') in a poorly adapted (T1) or better adapted (T2) coral community, or the sea-level rise that will inundate lower (T1) or higher (T2) coastal infrastructure. In other words, particular types of mangrove, seagrass or coral assemblages may have different coping thresholds (Tn.0) and cost thresholds (Tn.0 toTn.3) that would have been determined by the recent environment at the site. For example, T1 may characterize a system exposed historically to a relatively narrow range of temperature and tide, and T2, a place historically exposed to much greater variability in temperature and tide.

2.5. Local Risk Analysis – a Coral Reef Example

Coral bleaching (discussed in more detail below) is a potentially fatal stress response that begins as a sub-lethal paling and whitening of corals that occurs when local summer sea surface temperature rises to local threshold T1.0, which is typically 1-2°C above the local long-term mean summer maximum [Goreau and Hayes, 1994]. The driver for coral bleaching is not the mean temperature *per se*, but rather runs of extremely hot days [Berkelmans, 2002]. Local temperature thresholds can nevertheless be represented in relation to projected increase in *global means* (G in Figure 4A) assuming there is a proportional increase in *regional means* and *local extremes*. For a nearshore reef in the central Great Barrier Reef (GBR), Jones [2003] obtained the projected *regional mean* out to the year 2050 by extracting the central GBR trajectory (the average of eight GCMs). He then superimposed *local extremes* on that trajectory, based on an instrumental record of daily temperature variability covering the decade of the 1990s, and identified three thresholds, based on observations and measurements from bleaching events on the Great Barrier Reef in 1998 and 2002: T1 (non-lethal bleaching); T1 + 0.5° C which kills the most heat-sensitive corals; and T1 + 2.5°C , which kills the most heat-resistant corals. By examining the intersections between the threshold lines (Figure 4C), the regional warming curve excised from a GCM, (Figure 4A), and the regional likelihood curve (R in Figure 4B), he concluded that this regional average trajectory has a 95% probability of exceeding the threshold that kills heat-sensitive corals by 2015, and an 80% probability of killing heat-resistant corals by 2040. However in these same years, the predicted local warming at L1 was only 70% to 80%

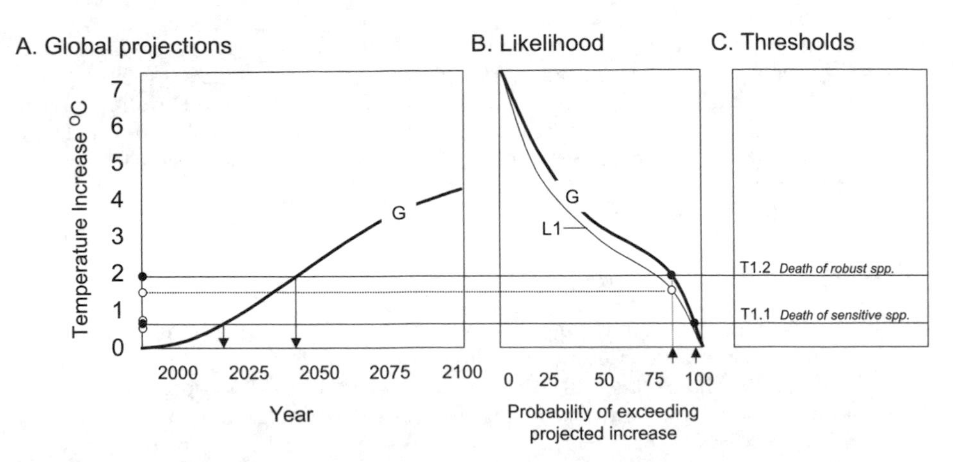

Figure 4. Application of risk assessment framework to a coral reef. A. Most likely global warming curve (mean of 8 models). B. Threshold exceedence probability for global warming curves (G) and for a coral reef at Magnetic Island, Great Barrier Reef, in which rate of warming is slower than global rate due to local geographic setting. C. Illustrative thresholds for coral mortality.

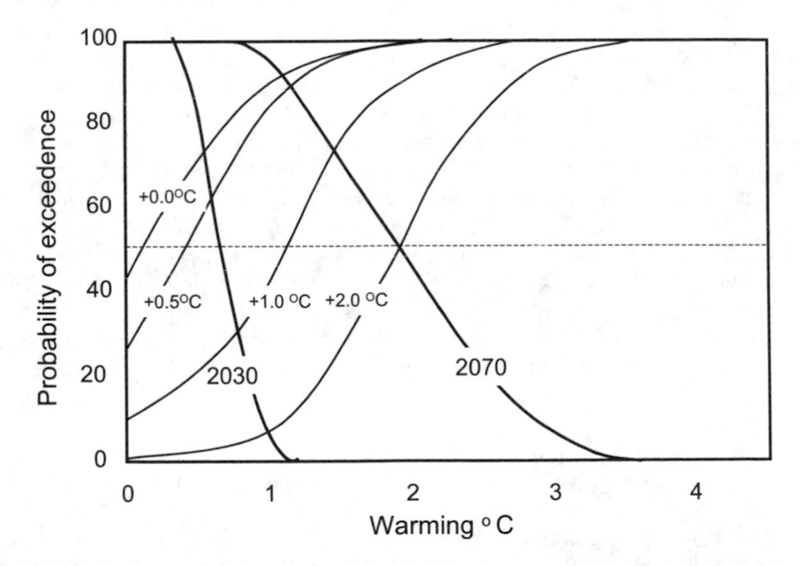

Figure 5. Probabilities of local warming in 2030 and 2070 superimposed on bleaching and mortality relationships for Magnetic Island, Great Barrier Reef. Probability for exceedance of bleaching thresholds is expressed as the annual risk at a given temperature above a 1990s baseline.

of predicted regional warming. This means that the local probabilities are somewhat lower, and the year of exceedance, somewhat later.

An alternate presentation of these relationships (Figure 5) predicted bleaching levels that were consistent with observations in recent years. At current temperatures (i.e., warming = 0°C in Figure 5), the annual probability of a bleaching event is just over 40% (or 4 events per decade). The annual probability of bleaching +0.5°C is ~30% (3 per decade) and bleaching +1.0°C is <10% (<1 event per decade). With a 1°C warming (70% likely by 2070), these frequencies increase to 9, 6 and 1 per decade. Site L1 projected temperatures are 70-80% of global projections, so to keep local warming at site L1 below the critical 0.5°C threshold, global warming would need to be kept below 0.6°C.

3. Coastal Marine Systems and Their Environment

With the risk model of Figure 3 in mind, we now consider potential impacts and interactions of a number of aspects of global change, potentially leading to the exceedance of critical damage thresholds (e.g., T1.3 in Figure 2, where resilience capacity collapses). Later in this chapter, we consider local attributes of sites, settings and habitats that may ameliorate or exacerbate local hazards compared to the global average for a series of ecosystems.

3.1. Rising Sea Level

A mean global sea level rise of between 90 and 880 mm (about 1 to 9 $mm.y^{-1}$) is projected for the 21st Century, the huge range reflecting both the uncertainties about future emissions and the inexact science [IPCC, 2001]. Rising sea level is clearly of major concern in coastal systems, and the commonest predictions are of flooding and coastal erosion [Carter and Woodroffe, 1994]. Sediment erosion from one place and deposition in another is the essence of shoreline morphodynamics, and, irrespective of climate change, a matter of critical importance in relation to both human inhabitation of coastal areas, and the integrity and value of ecological assemblages that occupy those shores. With climate change, regional sea level (Figure 2B) will likely continue to rise for centuries due to the thermal inertia in the oceans. Locally, the key variables will be mean sea level rise relative to the land surface, the sea floor and/or the reef top, and the heights of wave set-up and storm surges.

For mangroves, rising sea level will potentially affect their capacity to retain their substratum at a rate that keeps pace with sea level rise [Twilley et al., 1996]. Woodroffe [1992] considers that lower rates of rise (<1 $mm.y^{-1}$) would have little impact, but that higher rates (5-8 $mm.y^{-1}$) will probably be sustainable only in settings with ample autochthonous sediment. For some settings and species assemblages, there is the prospect of mangroves advancing across lowlands, while in others, they will disappear as substrata are eroded away [Ellison and Farnsworth, 1996; Bacon, 1994; also, see below]. Where the sediment remains in place but the edaphic conditions and immersion regime change, infaunal and microbial communities will change with consequences for their roles in the processes of decomposition, recycling and sequestration of organic matter [Alongi, 1998]. Snedaker [1995] stresses the importance of local rainfall regime in the likelihood of subsidence. Subsidence occurs when rates of decomposition of sediment organic matter exceed rates of production. This is likely with reduced rainfall and runoff, because it causes higher salinity, greater seawater-sulfate exposure and decreased production. Sediment elevations would be maintained in places with higher rainfall and runoff, which would result in reduced salinity, reduced exposure to sulfate, and increased rates of delivery of terrigenous nutrients.

Likewise for coral reefs it is possible to envisage net gains in some situations. Even slow growing individual corals in shallow reef areas can grow vertically faster than the faster predicted rate of sea level rise. For example, vast beds of shallow water *Goniopora* corals were easily able keep up with a sea level rise at a rate near the maximum predicted for the 21st Century [6.6 ± 2.8 $mm.y^{-1}$ from 7.5 to 7 thousand years ago; Yu et al., 2004]. Thus, where shallow coral populations are dense and can survive increased temperatures (but see below), prolific reef areas could keep up with any projected rate of sea level rise.

With changes in water depth, shallow benthic communities will be exposed to changed tides, currents, salinity and light regimes [Short and Neckles, 1999]. In seagrass meadows and coral reefs, for example, a rising sea level could raise the deeper margin through light attenuation effects, and the shallow margin through upward displacement of the intertidal zone.

A higher sea level will exacerbate the impacts of tropical cyclones and the frequency of extreme storms. For example, a 20 cm sea-level rise could add 50 cm onto the 1 in 100 year storm surge (from 2.3 to 2.8 m – McInnes et al. [2003]) at mangrove-fringed Cairns adjacent to the Great Barrier Reef. This would double the area of the city that would be inundated by such a storm-surge compared to its effect at the sea level of 1990. In 1990, a 2.8 m storm surge was a 1 in 300 year inundation event: at +20 cm, it will become a 1 in 100 y inundation event, with implications for the long term stability of substrata for mangroves, beaches and reef-top sediments (see below).

3.2. Rising Temperature

Marine air temperatures are generally expected to warm more slowly than the global mean [IPCC, 2001b], notably near areas of ocean upwelling, and where cold currents affect the local air mass. Rising air temperatures will influence the survival, phenology and net primary productivity of intertidal ecosystems such as mangroves [Clough, 1992] and seagrasses [Short and Neckles, 1999; Duarte, 2002], interacting with changing levels of atmospheric CO_2, soil salinity, solar radiation flux density, changes in vapour pressure differences.

Sea temperatures will also, in most cases, rise at less than the global average air temperature because of the thermal inertia of the oceans. By the end of the 21st Century, most Pacific coral reef areas now considered 'optimal' are predicted to become 'marginal' [Guinotte et al., 2003]. Some habitats that are currently too cool for reefs will become warm enough, but few will satisfy all habitat requirements (notably temperature plus substratum depth and stability) for latitudinal expansion of substantial mangrove forests or reefs, as opposed to establishment of mere clumps of trees or corals [Guinotte et al., 2003; Buddemeier et al., 2004; Twilley et al., 1996].

Coral reefs are extremely sensitive to rising sea temperatures. One major negative impact is the increased virulence of diseases of coral reef biota [Harvell et al., 2002; Rosenberg and Ben-Haim, 2002] that has decimated reef populations already, particularly in the Caribbean. There are also negative effects on coral metabolism (e.g., Fitt et al. [2001]), coral reproduction [Szmant and Gassman, 1990] and larval settlement [Jokiel and Guinther, 1978]. One putative positive impact has been hotly disputed. Lough and Barnes [2000] recorded an increase in the rate of calcification in specimens of massive colonies that lived through ~0.25°C warming of mean sea temperatures in the 20th Century. This growth outpaced projected negative effects of reduced alkalinity (see below). McNeil et al. [2004] went on to suggest that climate change will increase the net rate of coral reef calcification by the year 2100. However far more significant is the catastrophic heat-induced bleaching and mortality of corals taking place in the platy, branching and bushy corals living alongside those same robust massive corals [Brown, 1997; Hoegh-Guldberg, 1999;

Kleypas et al., 2005]. Coral bleaching and mortality have been exceptionally widespread and intense in the last decades of the 20th Century, and the first years of the 21st [Wilkinson, 2002; Buddemeier et al., 2004].

In terms of Figure 3C, the onset of non-lethal coral bleaching of the most sensitive types of corals is an example of a Tn.0 threshold in Figure 3. With prolonged and severe heat stress, these can be killed by the heat (Tn.1, Tn.2). Where new corals cannot become re-established on damaged sites within a reasonable time frame (a few years) a Tn.3 threshold (collapse of resilience) is indicated. But impact and recovery potential are patchy phenomena, both at local scales [Fisk and Done, 1985; Loya et al., 2001; Marshall and Baird, 2000] and across oceans and archipelagos [Wilkinson, 2002; Sheppard, 2003; Berkelmans et al., 2004; Wooldridge and Done, 2004]. Later, we further consider local risk in terms of the concepts introduced in Figure 3.

3.3. Rising Atmospheric CO_2 - Seawater Chemistry Effects

Rising concentrations of atmospheric CO_2 lead to increased concentrations of CO_2 dissolved in surface waters of the oceans, causing increases in the relative proportion of CO_2 to HCO_3^- [Short and Neckles, 1999], and decreasing the concentration of CO_3^{2-} [Buddemeier et al., 2004]. The waters are becoming more acidic and less saturated with respect to the precursors of aragonite, which is the strong reef-building form of calcium carbonate. For today's reefs – lying mostly between the tropics of Cancer (24°N) and Capricorn (24°S), coral reef waters have an aragonite saturation state of >4, which is optimal for reef calcification and accretion [Gattuso et al., 1998; Guinotte et al., 2003]. However by 2070, this entire latitudinal range is projected to be 'marginal', at 3.0-3.5 as a direct consequence of increased atmospheric CO_2 increase and global warming (Figure 2C). In relation to this aspect of climate change, the lines L1, L2 and L3 in Figure 3B represent probabilities of such detrimental changes in water chemistry – changes that will lead to measurable declines in reef calcification, hence accretion of reef framework and production of rubble and sands. However some recent reports [Guinotte et al., 2003; Kleypas and Langdon, 2002] mention a potential buffering process based on dissolution of some types of carbonate. This is considered in more detail below.

3.4. Rising Atmospheric CO_2 - Vegetation Effects

The increased CO_2 in the atmosphere also potentially accelerates photosynthesis. Farnsworth et al. [1996] found that *Rhizophora mangle* seedlings grown for a year at 2 × CO_2 exhibited significantly increased biomass, total stem length, branching activity, and total leaf area, and also greatly accelerated maturation (earlier reproduction, lignification of the main stem and production of aerial roots). These attributes in themselves would appear to be beneficial in an era of rapid sea level rise, and the lines in Figure 3B could be considered lines of their likelihood of occurrence. However there are other physiological changes that may significantly affect survival in a changed climate. For example, Farnsworth et al. [1996] also found changes in epidermal cell sizes, density of stomata, and lowering of foliar chlorophyll, nitrogen, and sodium concentrations in the 2 × CO_2 mangroves. In the mangrove, sea grass and coral reef systems of tropical coasts, any decline in plant nutritional value may have implications for the quantity and quality of plant matter entering consumer and detrital pathways, and hence their fishery support value. For mangroves, Twilley et al. [1996] note how nitrogen cycling in the canopy and nutrient dynamics in the soil are coupled, the strength and nature of the linkage being species- and location-dependant. Also location-dependent is the potential for anaerobic mangrove sediments to provide long-term storage for the carbon fixed in mangrove leaves [Clough, 1998; Matsui, 1998]. Presumably, this capacity, modest though it is on a global scale [Ayukai, 1998], would be further reduced by any increase of erosion or aeration of such deposits, through climate related changes in storm impacts.

3.5. Rising Ultra-Violet B Radiation

Shallow tropical waters are being exposed to increased fluxes of UV-B radiation due to thinning of the ozone layer, and increased exposures are predicted through the 21st Century [IPCC, 2001b]. Penetration of UV-B to submerged habitats is affected by concentrations of UV-absorbing substances that run off from the land, by clarification of the water under doldrum conditions (predicted to increase in global change scenarios), and changes in plankton composition and productivity [Zepp, 2003]. For littoral plants and zooxanthellate corals, the increased fluxes are predicted to lead to inhibition of photosynthesis and increased metabolic costs of tissue repair or production of UV-B blocking compounds [Short and Neckles, 1999]. The tolerance to UV radiation varies among species of phytoplankton, seagrass, mangroves, zooxanthellae and macro-algae, suggesting that the productivity of some species will be favoured over others. In coral reefs, for example, the greatest impacts are expected in shallow water corals, with possibly a tendency to limit the minimum depth of occurrence of some species, and potential for selection among coral and zooxanthellae for more UV resistant species and genotypes [Shick et al., 1996]. Zepp [2003] noted that extensive stratification in El Niño conditions may be greatly increasing exposure of Florida's reefs to UV and PAR and thus exacerbating coral bleaching. Decomposing phytoplankton detritus and decaying litter from seagrasses and mangroves appear to be the major sources of UV-absorbing substances, and the damage to corals more sensitive to changes in their concentrations than to changes in atmospheric ozone per se.

4. Local Modifiers and System Properties

For the mangroves, reefs and seagrasses of any specific location, the probability of exceeding various damage thresholds (T1 – T3) will be ameliorated (L2) or exacerbated (L3) by local circumstances (Figure 3). Not all sites will be equally vulnerable, and thus, there will be greater or lesser imperative for adaptive actions by humans, and greater or lesser selective pressures on the coastal organisms and systems themselves. Here, we look in more detail at local environmental factors in relation to features and potential coping strategies of each habitat type.

4.1. Mangroves

Mangroves occur on about 25% of tropical coastlines [Spalding et al., 1997], but this figure is declining as a result of extensive draining and reclamation, cutting for fuel and conversion to aquaculture [Burke et al., 2000]. Mangrove forests can dominate tropical intertidal zones with major inputs of either terrigenous sediments (river deltas, lagoons and estuaries), or marine sediments, notably carbonate shores and reefs where the only significant sediment inputs are the product of local calcareous production and sedimentary processes [Semeniuk, 1994; Twilley et al., 1996]. A forest's current location, extent and zonation are often legacies of vegetation and sedimentary processes that have taken place over the Holocene period, an environment of dramatically changing sea-level and fluvial-tidal energy balance [Chappell and Woodroffe, 1994]. These authors evoke likely future scenarios for mangrove habitats, as they speak of 'the big swamp' phase of rising sea level, and of transitions to and from freshwater plains that accompany changes in the relative dominance of ebb and flood tides and fluctuating sea levels. In some settings, severe droughts can lower a mangrove forest's water table, increase its salinity, and drive saline flows towards adjacent freshwater systems [Drexler and Ewel, 2001].

A simplified view of potential sea-level effects has relative vulnerability controlled by 1) the tidal range and 2) the elevation of the substratum surface, relative to mean sea level [Clough, 1992]. Ellison [2003] suggested that the strongest predictor of vulnerability to sea level rise might be substratum elevation at the seaward margin. For mangroves, the issue has always been one of balance between how much sediment and vegetation advance in low to moderate conditions between extreme events, and how much is physically removed by extreme events [Chappell and Woodroffe, 1994]. Changes in the regimes of extreme cyclones as discussed by McInnes et al. [2003] will clearly be important drivers for cycles of damage and rejuvenation, particularly in low-lying, unsheltered mangrove fringes.

Bacon [1994] predicted a variety of responses to rising sea level in a sample of Caribbean mangrove systems of several types: 'fringe', 'riverine', 'basin', 'scrub or dwarf' and 'overwash' (after Lugo and Snedaker [1974] – cited in Twilley et al. [1996]). His analysis suggested that about 20% of his sample sites would lose area, and about 70% would have no net area loss, (although there may be species changes or migration inland). The remainder were likely to be rejuvenated and/or increased in area and forest development. Semeniuk [1994], working on mangrove forests in northwest Australia, similarly stressed that responses to sea level rise will be variable, depending on five major factors: climate (arid or humid); Holocene sea-level history (and whether presently rising or falling); coastal stability (currently eroding or prograding); tidal range; geomorphological setting (degree of heterogeneity). Ellison and Farnsworth [1996] point out that climate change effects on mangroves (sea-level and CO_2 fertilization) compound other important anthropogenic impacts (extraction, pollution and land reclamation), and that any benefits from CO_2 will be very unlikely to compensate for effects of sea-level rise, especially in the highly vulnerable 'fringe' settings [Farnsworth et al., 1996].

In terms of our conceptualisation of risk (Figure 3), low-lying mangrove stands – be they 'riverine' or 'fringe'- may have a lower onset threshold for any sea-level effect (e.g., T1) than 'scrub' or 'basin' stands established further inland and at a slightly higher elevation (T2). For sites with projected detrimental or beneficial impacts in relation to fisheries and wildlife values, a series of sea-level rise costs or benefits above baselines of T1 or T2 could be conceived (e.g., fisheries benefits of extended mangroves due to increased mean sea-level, versus increased cost preventing or repairing damage to property as storm surge frequency increases). Bacon [1994] considers the best predictors of site-specific responses to sea level change to be site physiography and relation to neighbouring coastal systems, plus the immediate hinterland, and recent changes to wetlands resulting from human intervention. Riverine mangrove substrata may be best able to keep up with sea level rise, because of major external inputs of sediment from rivers [Ellison, 2003]. By contrast, the most vulnerable mangrove systems may be those of low relief islands and atolls in carbonate settings without rivers. Theirs are comparatively sediment-poor environments with poor capacity for vertical accretion.

Other key local environmental drivers that will change with global climate change are highly variable and location-dependent, including soil salinity, solar radiation flux density and vapour pressure differences [Clough, 1992; Semeniuk, 1994]. All of these are driven by local temperature, rainfall and river flow regimes that will change differently across

TABLE 2. Mangroves. Putative consequences of projected changes in climate variables and local factors that will ameliorate or exacerbate the hazard and/or the consequence.

Hazard	Consequence	Ameliorating/Exacerbating factors
Rising atmospheric CO_2	Nil or increased rates of photosynthesis and plant, depending on salinity.	Vapour pressure deficit, in as much as it affects utilization of CO_2 and water balance homeostasis.
	Advance in the timing of flowering [He and Bazzaz, 2003] with potential to de-synchronise with pollinators.	Access to sufficient fresh water, N, P and trace elements.
Rising air/ seawater temperature	Reduced survival as aridity increases.	Topographic and meteorological conditions that affect local weather, in particular, relative humidity effects on vapour pressure deficit.
	Expansion of latitudinal range as minimum water temperature rises in adjacent temperate waters.	
Rising sea level	Keep up, migrate landward or contract, depending on local tidal range, forest type, topography and duration of anomalous periods of inundation by seawater.	Flatness of terrain, its height at seaward margin and tidal range.
		Vertical growth rate of mangroves and vertical and horizontal accretion rates of sediments.
Increased storm frequency or intensity	Extreme storm surges destroy habitats on open coasts – vegetation and mud-flats.	Degree of shelter from wind and storm surge provided by location.
	On sheltered coasts, increased supply of sediments and nutrients.	Prior storm exposure regime: effects will be greater, the greater the disparity.

zones of large mangrove forests. Different mangroves species display a great range of tolerances to variability in salinity, aridity and low temperature [Clough, 1992], and differences in their relationships among net photosynthesis, soil salinity and vapour pressure differences [Clough, 1992]. Elevated atmospheric CO_2 increases primary production in individual plants, but it also decreases leaf nutritional value and increases phenolic compounds, traits that have a major effect on plant/herbivore interactions [Coley et al., 2002].

Clearly, there will be a range of effects on the interrelated properties of forest structure, zonation, production, foliage food webs, nutrient dynamics, and the distribution and composition of forest soil, including its infauna, and microbial communities. An integrated approach to assessing the hazards, their potential individual consequences (Table 2) and their interactions, will be required to recognise and assess these subtle but potentially critical determinants of the likely fates of particular mangrove areas.

4.2. Seagrass Meadows

Seagrasses – like mangroves and land grasses- are rooted seed-bearing plants that flourish as meadows in soft sediments at and below the intertidal zone [Alongi, 1998]. Meadows can be covered with seagrasses such as *Zoestera* or *Posidonia* that can be as dense and tall as ungrazed pastures, or by sparse, wispy runners of *Halophila ovalis*, with oval leaves no more than a few millimetres long. They are highly productive habitats for many marine species, from shrimps and juvenile fishes that spend their adult stages in open sea or coral reefs, to dugongs and turtles. Seagrasses entrap suspended material in the water-column, and they take up nutrients from both sediments and water column [McRoy and McMillan, 1977].

'Seagrass meadows' are in many cases mixed seagrass and seaweed meadows. Brown macroalgae such as *Sargassum* spp. or green calcareous algae such as *Halimeda* spp. often occur intermingled with seaweeds. Since the benthic marine algae do not have roots and get their nutrients only from the water column, their response to changes in environment will be different to that of seagrasses). When individual plants or large areas die, dead fronds of both seagrasses and seaweeds become secondarily colonised by dense carpets of filamentous algae and other small seaweeds (epiphytic algae), as well as epiphytic animals such as hydrozoa and bryozoa. Together, the total biomass per unit area and its apportionment among live and dead seagrass, epiphytic plants and epiphytic animals, are ecological indicators of seagrass condition in any place.

Short and Neckles [1999] and Duarte [2002] have reviewed climate change effects on seagrasses, and conclude that there is potential for local improvement in some ecological indicators, as well as local declines (Table 3). The variability in potential responses is

TABLE 3. Seagrass meadows. Putative consequences of projected changes in climate variables and local factors that will ameliorate or exacerbate the hazard and/or the consequence.

Variable	Impact	Ameliorating/Exacerbating factors
Rising atmospheric CO_2	Fertilization of intertidal meadows: reduction in calcification rates of associated calcareous algae.	Access to sufficient, N, P and trace elements.
	Advance in the timing of flowering [He and Bazzaz, 2003].	Embayments and upwelling areas less N and P limited than open seas.
Rising air/ seawater temperature	In intertidal areas, periodic 'burn off' by heat and desiccation increase during aerial exposure at low tides.	Local mixing, cooling and shading; relative heat sensitivities of species present, and their acclimatization history.
	Potential for increased photosynthesis rates in shallow submerged plants.	
Rising sea level	Reduced performance or survival in those currently light saturated depths that become light limited due to increased depth and/or turbidity.	Depth of meadow relative to intertidal level.
		Light attenuation coefficient affects depth at which light becomes limiting.
Increased storm frequency or intensity	Increased frequency of physical destruction of meadows. Increased and more prolonged periods of reduced light due to sediments introduced by flooding rivers or resuspended by storms.	Degree of shelter from wind and storm surge provided by location.

attributed to physiological and natural history differences among species of seagrasses, difference between tropical and temperate areas, and differences within tropical and temperate areas. For example, in areas already around the upper limit of the thermal tolerance of their current seagrass species, further increases in water temperature will likely reduce their productivity and cause dieback, especially in intertidal areas. When this is combined with nutrient pollution, stimulation of the growth of intermingled algae and epiphytes may shade the remaining seagrass and further tend to reduce its productivity. On the other hand, in places where seagrasses are not close to their upper temperature limit, and where nutrient supply is limiting, the same incremental changes in temperature and nutrient supply may serve to increase net productivity per hectare, and hence their value as habitat, nurseries and feeding areas for benthic and demersal food webs, fisheries and wildlife. Understanding local manifestations of global climate change will require case by case analysis of the biophysical settings, the diversity of physiological vulnerabilities, and the natural history of assemblages of populations contending for the changing environmental and resource envelopes.

Similar spatial and species-specific variability in response are likely in relation to the other aspects of climate change (Table 2). In places where light is already limiting, sea-level rise *per se* will, by attenuating penetration of PAR, further reduce productivity at a given depth, and cause a shallowing of the lower depth limit. But to compensate for this loss, seagrass habitats could advance by tracking both sea-water encroachment across coastal lowland (in the absence of built infrastructure such as walls or roads), and salinity intrusions further into estuaries and rivers, replacing present benthic assemblages adapted to current lower salinity and fresh water [Short and Neckles, 1999]. The creation of sea walls to protect rural and urban property from rising sea levels will clearly limit this compensatory tendency within seagrass meadows, and hence their capacity to act as feeding and nursery areas for marine and estuarine fisheries. Conceptually, net reduction of seagrass area could be formulated as costs associated with loss of sea-food nursery function (Tn.1 – Tn.3). Individual storm and flood events can affect seagrass areas by uprooting plants, reducing water clarity, eroding beaches and banks, exporting sediments, smothering plants [Preen et al., 1995]. Any change in the frequency and intensity of high-energy storms – be it an increase or a decrease- would clearly change the ecological dynamics of seagrass systems (e.g., mean plant life expectancy; successional processes; extent of suitable substratum; continuity as nursery areas and feeding ground for fisheries and wildlife).

4.3. Tropical Calcareous Reefs- Coral Reefs and Coralline-Algae Ridges

Tropical calcareous reefs around the world include both the iconic and familiar coral reefs and less familiar coralline-algae ridges (e.g., Littler and Doty [1975]; Adey and Burke [1977]; Woodroffe et al. [1990]; Macintyre et al. [2001]). In geological parlance, they are both examples of 'Holocene bioherms' [Adey and Burke, 1977]. These are carbonate platforms often up to 15 m thick that have been built and shaped over the last 10,000 years by the interaction between biological calcification and physical forces (wave, tides, sea-level, sediment sorting and redistribution). The biological calcification and the living architecture of the iconic coral reef are dominated by hard corals, but coralline algae are also major components in coral reefs, even to the exclusion of corals in some zones. For example the massive coralline algae *Porolithon onkodes* covers around 40% of the heavily grazed seaward slope of the algal ridge on the fringing reef at Waikiki, Hawaii [Littler and Doty, 1975]. Because of its role in maintaining and providing the surf-resistant reef edge, *P. onkodes* is one of the most important reef-building organisms. However there are also extensive tropical carbonate platforms where corals are absent or insignificant contributors to the living and geologically recent architecture. Like coral reefs, they are habitats for fishes, invertebrates and marine plants, and they form barriers on which waves break. In other words, they provide a similar range of ecosystem services as do coral reefs, and they will be exposed to the same suite of environmental changes facing coral reefs.

Coral reefs are patchy and dynamic seascapes, whose state and trajectory are driven by a complex net of pressures, processes and episodic disturbances (see McManus and Polsenberg [2004] for a review). Whereas it could be argued that changes to mangroves and sea-grass meadows may be more or less balanced in terms of human amenity and ecological function, for coral reefs, most aspects of climate change are seen as unqualified threats (Table 4; Pittock [1999]; Hoegh-Guldberg [1999]; Wilkinson et al. [1999]; Buddemeier et al. [2004]). Perhaps more than any other change in recent years, mass coral bleaching events caused by scorching summer heatwaves are providing the world with its most worrying foretaste of ecological collapse. The term 'bleached reefs' is often used to describe coral reefs that are so densely covered with bleached coral they are visible as glowing white fields from aircraft [Berkelmans et al., 2004], and even detectable in the pixels of some satellite images [Elvidge et al., 2004]. When all these living parts of a reef's architecture die, perhaps becoming colonised by seaweeds, disintegrating under assault from borers and grazers, and collapsing under their own weight and the impacts of waves,

TABLE 4. Coral reefs. Putative consequences of projected changes in climate variables and local factors that will ameliorate or exacerbate the hazard and/or the consequence.

Hazard	Consequence	Ameliorating/Exacerbating factors
Rising atmospheric CO_2	Reduced availability of dissolved carbonate ion (CO_3^{2-}) causes increased sea water acidity, increase in rates of dissolution of reef carbonates, reduction in calcification rates in reef building corals and algae.	Local depth and availability of high- Mg calcite to buffer against increased acidity.
Rising air/ seawater temperature	Reduced coral survival as increases in annual maximum sea-surface temperature of 1-2°C causes coral bleaching.	Local mixing, cooling and shading; relative heat sensitivities of coral species present coral and their zooxanthellae types; acclimatization history.
	Expansion of latitudinal range as minimum water temperature rises in adjacent temperate waters.	
Rising sea level	Re-initiate vertical growth in reef-flat corals that are limited by current low tide level.	Depth of shallowest part of reef relative to intertidal level.
Increased storm frequency or intensity	Effects largely unknown; some species may decline while others may benefit.	Degree of shelter from wind and storm surge provided by location.
		Prior storm exposure regime: effects will be greater, the greater the disparity.

the reefscape loses fishery carrying capacity and aesthetic appeal [Sheppard et al., 2002; McManus and Polsenberg, 2004]. There are varied, complex and little understood effects on reef invertebrate and fish communities [e.g., Eakin, 1992; Lindahl et al., 2001; Chabanet, 2002; Spalding and Jarvis, 2002; Williams et al., 2001]. Even though death is as much a part of the cycle of existence in coral reefs as it is in any ecosystem, the widespread death of old corals over large areas caused by extreme summer heatwaves does appear to represent a genuine climate change impact [Hoegh-Guldberg, 1999].

Our descriptions of individual climate change hazards above highlighted two issues for coral reefs (see also Table 4). The first was the prospect of increased frequency and severity of heatwaves causing repeated bleaching and coral mortality, eventually overwhelming reefs' ecological resilience [Done, 1999; Hoegh-Guldberg, 1999]. The second was changes in seawater chemistry that diminish coral growth rates [Kleypas et al., 1999, 2001; Gattuso et al., 1998; Leclercq et al., 2000; Buddemeier et al., 2004]. Underpinning both is the powerful symbiosis between the coral animal host and the unicellular micro-algae (zooxanthellae) that live at densities of millions per square centimetre within their tissue. The micro-algae provide the energy rich food source that allows it to build tissues, to reproduce, and to produce strong, aragonite limestone skeletons. Bleaching is the loss of these micro-algae, the loss of the food source, and the loss of these vital functions. Too frequent bleaching across too many coral species and too many reefs means too little time for damaged reefs to recover before the next impact arrives. In analysing the risk for any particular place, therefore, it is about frequency and extent to which local environmental conditions go beyond the coping range of corals and other key reef builders. We characterise bleaching events as episodic impacts, and acidification of reef waters as a chronic stress. Here, we consider local ameliorating and exacerbating circumstances in relation chronic acidification effects on aragonite production, and second, to coral bleaching as an extreme, episodic event.

Models suggest that dissolution of shallow-water carbonates in general will not significantly buffer against potential falls in seawater pH caused by rising atmospheric CO_2 [Archer et al., 1997; Andersson et al., 2003]. However empirical observations on 'live sands' [Yates, 2000; Yates and Robbins, 2001; Halley et al., 2004] suggest there may be important exceptions to this conclusion in specific coral reef settings rich in very fine grain high magnesium calcite. Even small diurnal fluxes of CO_2 caused dissolution of the fine sediments and a measurable increase in pH. Calcite comprises around 50% of many coral reef sediments [Milliman, 1974]. This material comes from the skeletons of coralline red algae [Borowitzska, 1983; Barnes and Chalker, 1990] that are major components in many reef zones, and of soft corals, tube-worms, echinoderms and bryozoans [Milliman, 1974]. Fine particles of calcite with a potential to buffer any tendency for increased acidity are released into reef waters by processes of erosion, both physical [Davies, 1983] and biological [Hutchings, 1986]. The latter is undertaken by diverse and abundant external bioeroders (sea-urchins and reef crunching, scraping and nipping fishes that defecate their own weight in sand every day) and 'internal bioeroders' (sponges and worms that pump fine silts out through holes and burrows into the overlying waters). Reef waters are thus often milky with mixtures of fine reactive limestone particles. This raises the intriguing possibility that the dense populations of external bioeroders responsible for this milkiness may play an under-appreciated role in buffering the potentially acidifying effects of increasing atmospheric CO_2.

There thus appear to be sources and mechanisms for high magnesium calcites to buffer seawater against damaging changes in seawater chemistry [Halley and Yates, 2000; Halley et al., 2004; Kleypas and Langdon, 2002]. However the magnitude and distribution of any buffering effect that might ameliorate the tendency for acidification, and hence, for calcification slow-down is not clear. Accumulated deposits of sands and rubbles, and places with high loadings of fine suspended calcite silts, could, in principle, ameliorate local hazard, and would need to be taken into account in assessment of local risks. Kleypas and Langdon [2002], for example, raise the possibility that the buffering effect of dissolution processes could be effective in neutralizing any effect on aragonite calcification over areas as large as the Great Barrier Reef lagoon (>100,000 km^2). Patchiness in this buffering capacity is also likely among reefs and among the different zones of individual reefs, depending on their local abundance of reef limestone scrapers and defecators, and the differing propensity of different zones to accumulate fine silts and sands – a function of wave energy and water residence time [Kleypas and Langdon, 2002]. Guinotte et al. [2003] suggest that any such neutralization of negative effect on aragonite saturation will be insignificant in well-flushed areas such as reef slopes, where fine sediments do not accumulate. There are a number aspects of this phenomenon that warrant further study: Could well-flushed areas with abundant limestone scrapers and defecators – characteristic of many intact reef communities – receive an effective daily dosing of buffering materials, even if it does not accumulate there? Would a local risk assessment for effects of changed water chemistry assign high risk to well-flushed areas when limestone scrapers and/or defecators are sparse, but a low risk when they are abundant? Could sheltered areas with accumulated fine sediments that are constantly perturbed by infauna and epifauna be at a relatively lower risk of acidification? However, would such poorly mixed waters be more prone to overheat, and thus present an elevated risk of coral bleaching and death?

Local physical processes can also either ameliorate or exacerbate regional heating anomalies. Woodridge and Done [2004] referred to places that avoid damaging heat exposures by virtue of their physical setting as having 'good oceanography'. Salm et al. and Skirving and Steinberg (this issue) provide examples where vertical mixing of deeper cool waters up through the water column effectively reduces temperature-related stress to shallow water corals. However where upwelling is weakened in heatwave condition, such proximity may be ineffectual. For example, Zepp [2003] describes a mechanism whereby heatwave conditions over the Florida Keys reduces upwelling of cool more opaque waters, and increases photobleaching and microbial degradation of coloured dissolved organic matter (CDOM). The result is that waters transported over the reefs by currents are warmer and have lower concentrations of UV absorbing compounds than in normal summers; the likelihood of exposures to a damaging combination of high temperatures and high dosage of solar radiation is increased [Brown, 1997].

There are places with particular suites of coral species that are in some sense 'well adapted' to withstand a minor heat anomaly. Some coral species appear to be able to tolerate higher temperature anomalies than others by virtue of particular morphological and metabolic traits [Gates and Edmunds, 1999; Craig et al., 2001; Loya et al., 2001]. For example, in a comparison of the Great Barrier Reef and Kenya, McClanahan et al. [2004] found that *Stylophora* and *Pocillopora* were susceptible in both regions, *Cyphastrea*, *Goniopora Galaxea* and *Pavona* were resistant in both regions, *Acropora*, and branching *Porites* were moderately affected on the Great Barrier Reef but highly affected in Kenya, and the opposite was true for *Pavona*. These interspecific and inter-regional differences contribute to community-wide bleaching thresholds that vary according to the reef's location and usual summertime temperature regime [Berkelmans, 2002] and, possibly, to the presence of particular genotypes of zooxanthellae [Little et al., 2004; Rowan, 2004]. The risk analysis presented above, and similar analyses carried out elsewhere (e.g., Hoegh-Guldberg [1999] and others), shows that there is significant probability of critical thresholds being exceeded within this century and a high probability of them being exceeded by the end of the century.

The most significant unknown that remains is how rapidly corals can adapt to warmer temperatures. For an extant coral population to persist, it, or its successors at a site that becomes damaged, will have to lift its mortality threshold (e.g., its T1.1 or T1.2) by a significant proportion of the global rate of warming (0.14 to 0.58°C per decade). This raises the question as to how fast can reasonably be expected? Some modes of adaptation are recognized (e.g., Dunne and Brown [2001]; Coles and Brown [2003]; Baker et al. [2004]; Little et al. [2004]; Rowan [2004]) but their rates have not yet been quantified in terms of increased tolerance to higher temperatures or geographic spreading. Periods of new coral settlement and growth following inevitable disturbances in coming decades will be critical in determining the extent to which reef communities will accommodate warming seas. There are relatively heat resistant corals and coral-zooxanthellae partnerships present in coral reef areas [Marshall and Baird, 2000; Loya et al., 2001; McClanahan et al., 2004], and thus a potential for local increases in critical thresholds Tn.0 to Tn.3. However the speed at which they can propagate across reefs and regions is not known. Ironically, coral reefs may become better adapted by losing all their sensitive species.

Local vulnerability of coral reef biota to changing environment will depend on local reef geomorphology and sedimentary settings. For example, the height of the reef relative to mean sea level and the tidal range will together effect prospects for coral survival on reef flats; the nature of sea-floor sediments and the degree of shelter afforded by land or reef masses will determine the extent to which resuspended sea floor sediments become a threat in a stormier environment; a greater frequency of floods or damaging storms shortens the period for recovery of heavily damaged populations; increased sub-lethal exposure to lowered salinity may increase coral vulnerability to heatwaves; increasing temperatures may make pathogens more virulent [Harvell et al., 2002] and increase their contribution to chronic coral mortality; the state of regional populations of coral reef species will affect the size of the regional pool of larvae available for recovery of damaged reefs. Some seemingly minor quirks of natural history can exert major influences on ecological resilience: if the daily rate of scraping of the substratum by fishes and invertebrates is too low, algal carpets may limit opportunities for settlement of coral larvae (e.g., Sammarco et al. [1974]; Sammarco [1980]; McClannahan [1997]; Hatcher and Larkum [1983]; McCook [1999]; Steneck [1994]); if it is too high, tiny proto-corals will be scraped off before they establish as juvenile corals; active 'farming' of reef areas by some territorial damselfish can largely exclude coral establishment from extensive areas of the reef [Done et al., 1991; Ceccarelli et al., 2001].

To conclude, in analysing the risk of thermal bleaching for the reefs off a particular coast, account would need to be taken of such factors as local bathymetry, tides, and currents that predispose the area to amelioration or exacerbation of regional heating (see Salm et al., this volume). Knowledge of the dominant coral species types would be the basis of an assessment of inherent resistance to, and tolerance of, bleaching (Loya et al. [2001]; McClanahan et al. [2004]). The final assessment of vulnerability would also benefit from an evaluation of the local natural history of the reef, and its degree of connectedness to strong sources of the right types of larvae.

5. Conclusion

The implications of global climate changes for tropical coastal systems have been clearly signalled by IPCC. Figure 6 (from Jones [2003]) provides a synopsis of possibilities and a framework for human response. This framework is consistent with Article 2 of

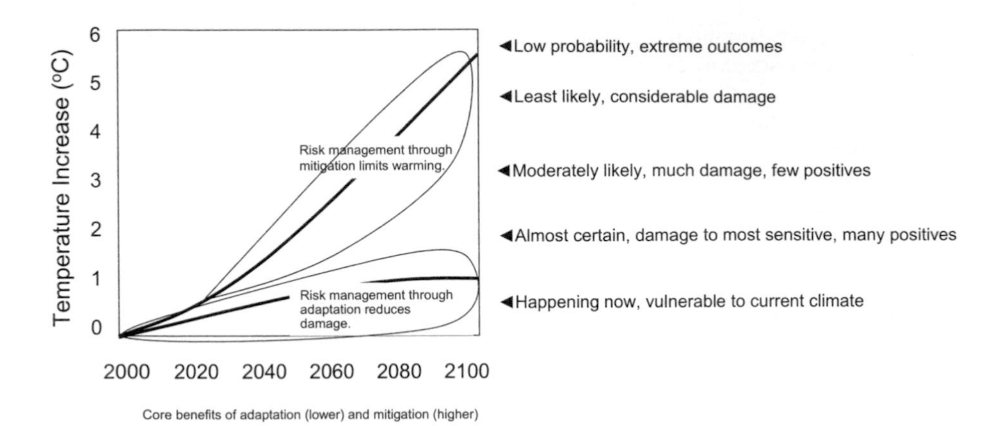

Figure 6. Synthesis of risk assessment approach to global warming. The left part of the figure shows global warming extremes of the six SRES greenhouse gas emission marker scenarios with the zones of maximum benefit for adaptation and mitigation. The right side shows likelihood based on threshold exceedance as a function of global warming and the consequences of global warming reaching that particular level based on the conclusions of IPCC WG II. Risk is a function of probability and consequence. After Jones [2003].

the United Nations Framework Convention on Climate Change, can be used to investigate individual activities and can be aggregated to the global level. It also shows that adaptation and mitigation actions from society should not be delayed for want of greater precision and certainty in climate modelling. They can proceed while continued risk assessments at the local and global scale are pursued. The framework also highlights the need for vulnerability assessment, irrespective of a precise assessment of likelihood of occurrence [Grübler and Nakicenovic, 2001]. We hope this chapter has identified both aspects of vulnerability for the major shallow tropical marine ecosystems, and a useful conceptual framework for putting local assessments into the global context.

Despite all the uncertainties, Figure 6 does show that the sustained mitigation of greenhouse gases is imperative to reduce the likelihood of the highest potential warming. Earliest mitigation efforts will always yield the largest gross benefits in terms of damage reduction, both short- and long-term. Our approach can guide the prioritisation of adaptation and mitigation options according to the greatest need, and it can be refined as new information becomes available.

For the ecological sciences and for management of the coastal zone, perhaps our greatest challenge now is to determine place-specific implications for particular ecosystems and their services to human populations. Such knowledge would underpin the will and capacity of humans to manage the local coastal system to continue to support human populations and sustain production of a reasonable level of goods and services. Variability in climate related impacts within regions reflects patchiness in stresses, patchiness in vulnerability, or both. Here, we have used a risk-based approach to link notions of plausible change coastal systems, through to the global warming targets that would be needed, assuming we define plausible outcomes that are acceptable. The framework we have described works on a number of levels. At one level, it can be used by people in political arenas to articulate vulnerabilities of local ecological and amenity values of coastal habitats in relation to the development of global greenhouse gas policy. At another, it can be used by local community leaders and natural resource managers as a template for assessing risks, making plans and implementing actions, bay by bay, reef by reef, and coast by coast.

References

Adey, W. H., and R. Burke, Holocene bioherms of the Lesser Antilles – geologic of control of development, in *Reefs and Related Carbonates- Ecology and Sedimentology,* edited by S. H. Frost, M. P. Weiss, and J. B. Saunders. *Am. Assoc. Petrol. Geol.*, Studies in Geology, 4, 67-81, 1977.

Allee, R., M. Dethier, D. Brown, L. Deegan, G. Ford, T. Hourigan, J. Maragos, C. Schoch, K. Sealey, R. Twilley, M. Weinstein, and M. Yoklavich, *Marine and Estuarine Ecosystem and Habitat Classification*, NOAA Technical Memorandum NMFS-F/SPO-43, 2000.

Alongi, D. M., *Coastal Ecosystem Processes*, Marine Science Series, CRC Press, Boca Raton. pp. 419, 1998.

Andersson, A. J., F. T. Mackenzie, and L. M. Ver, Solution of shallow-water carbonates, an insignificant buffer against rising atmospheric CO_2, *Geology*, 31, 513-516, 2003.

Archer, D., H. Kheshgi, and E. Maier-Reimer, Multiple timescales for neutralization of fossil fuel CO_2, *Geophy. Res. Lett.*, 24, 405-408, 1997.

Ayukai, T., Introduction, carbon fixation and storage in mangroves and their relevance to the global climate change – a case study in Hinchinbrook Channel in northeastern Australia, *Mangroves and Salt Marshes*, 2, 189-190, 1998.

Bacon, P. R., Template for evaluation of impacts of sea level rise on Caribbean coastal wetlands. *Ecol. Eng.*, 3, 171-186, 1994.

Baker, A. C., C. J. Stewart, T. R. McLanahan, and P. W. Glynn, Corals' adaptive response to climate change, *Nature*, 430, 741, 2004.

Barnes, D. J., and B. E. Chalker, Calcification and photosynthesis in reef-building corals and algae, In *Ecosystems of the World 25, Coral Reefs,* edited by Z. Dubinsky, pp. 109-131, Elsevier, Amsterdam, 1990.

Bellwood, D. R., T. P. Hughes, C. Folke, and M. Nystrom, Confronting the coral reef crisis, *Nature*, 429, 827-833, 2004.

Berkelmans, R., Time-integrated thermal bleaching thresholds of reefs and their variation on the Great Barrier Reef, *Mar. Ecol. Prog. Ser.*, 229, 73-82, 2002.

Berkelmans, R., G. De'ath, S. Kininmonth, and W. J. Skirving, A comparison of the 1998 and 2002 bleaching events on the Great Barrier Reef, spatial correlation, patterns and predictions, *Coral Reefs*, 23, 74-83, 2004.

Borowitzka, M. A., Calcium carbonate deposition by reef algae, morphological and physiological aspects, In *Perspectives on Coral Reefs,* edited by D. J. Barnes, pp. 16-28, Brian Clouston Publisher, Manuka, Australia, 1983.

Brown, B. E., Coral bleaching, Causes and consequences, *Coral Reefs* 16 Suppl., S129-S138, 1997.

Buddemeier, R. W., J. A. Kleypas, and R. B. Aronson, *Coral Reefs and Global Climate Change, Potential Contributions of Climate Change to Stresses on Coral Reef Ecosystems*, The Pew Center on Global Climate Change, Arlington, VA, USA. vi + 44 pp, 2004.

Burke, L., Y. Kura, K. Kassem, C. Revenga, M. Spalding, and D. McAllister, *Pilot analysis of global ecosystems. Coastal ecosystems*, World Resources Institute, Washington, DC. xii + 94 pp, 2000.

Carter, R. W. G., and C. D. Woodroffe, Coastal evolution, an introduction, In *Coastal Evolution, Late Quaternary Shoreline Morphodynamics,* edited by R. W. G. Carter and C. D. Woodroffe, pp. 1-32, Cambridge University Press, Cambridge, UK, 1994.

Ceccarelli, D. M., G. P. Jones, and L. J. McCook, Territorial damselfishes as determinants of the structure of benthic communities on coral reefs, *Oceanog. Mar. Biol. Ann. Rev.*, 39, 355-389, 2001.

Chabanet, P., Coral reef fish communities of Mayotte (Western Indian Ocean) two years after the impact of the 1998 bleaching event, *Mar. Freshw. Res.*, 53, 107-113, 2002.
Chappell, J., and C. D. Woodroffe, Macrotidal estuaries, In *Coastal Evolution, Late Quaternary Shoreline Morphodynamics,* edited by R. W. G. Carter and C. D. Woodroffe, pp. 187-218,. Cambridge University Press, Cambridge, New York, 1994.
Clough, B. F., Primary productivity and growth of mangrove forests, in *Tropical Mangrove Ecosystems*, edited by A. I. Robertson and D. M. Alongi, pp. 225-249. American Geophysical Union, Washington DC, USA, 1992.
Clough, B. F., Mangrove forest productivity and biomass accumulation in Hinchinbrook Channel, Australia, *Mangroves and Salt Marshes* 2, 191-198, 1998.
Coles, S. L., and B. E., Brown, Coral bleaching – capacity for acclimatization and adaptation, *Adv. Mar. Biol.*, 46, 183-224, 2003.
Coley, P. D., M. Massa, C. E. Lovelock, and K. Winter, Effects of elevated CO_2 on foliar chemistry of saplings of nine species of tropical trees, *Oecologia*, 133, 62-69, 2002.
Craig, P., C. Birkeland, and S. Belliveau, High temperatures tolerated by a diverse assemblage of shallow-water corals in American Samoa, *Coral Reefs*, 20, 185-189, 2001.
Davies, P. J., Reef growth, In *Perspectives on Coral Reefs,* edited by D. J. Barnes, pp. 69-106, Brian Clouston Publisher, Manuka, Australia, 1983.
Done T. J., P. K. Dayton, A. E. Dayton, and R. Steger, Regional and local variability in recovery of shallow coral communities, Moorea, French Polynesia and central Great Barrier Reef, *Coral Reefs*, 9, 183-192, 1991.
Done, T. J., Coral community adaptability to environmental changes at scales of regions, reefs and reef zones, *American Zoologist*, 39, 66-79, 1999.
Drexler, J. Z., and K. C. Ewel, Effect of the 1997-1998 ENSO-related drought on hydrology and salinity in a Micronesian wetland complex, *Estuaries*, 24, 347-365, 2001.
Duarte, C., The future of seagrass meadows, *Envir. Cons.*, 29, 192-206, 2002.
Dunne, R., and B. E. Brown, The influence of solar radiation on bleaching of shallow water reef corals in the Andaman Sea, 1993-1998, *Coral Reefs*, 20, 201-210, 2001.
Eakin, C. M., Post-El Nino Panamanian reefs, less accretion, more erosion and damselfish protection. *Proc. Seventh Int. Coral Reef Sym., Guam.*, 1, 387-396, 1992.
Ellison, A. M., and E. J. Farnsworth, Anthropogenic disturbance of Caribbean mangrove ecosystems, Past impacts, present trends and future predictions, *Biotropica*, 28, 549-565, 1996.
Ellison, J. Climate change and sea level rise impacts on mangrove ecosystems, In *Global Climate Change and Biodiversity*, edited by R. E. Green, M. Harley, L. Miles, J. Scharlemann, A. Watkinson and O. Watts, pp. 26-27, University of East Anglia, Norwich, UK., April 2003, Summary of papers and discussion. 2003. Available Online at, http//www.jncc.gov.uk/habitats/agency/resources/MJHGlobalclimatechange_14.08.03.pdf
Elvidge, C. D., J. B. Dietz, R. Berkelmans, S. Andrefouet, W. Skirving, A. E. Strong, and B. T. Tuttle, Satellite observation of Keppel Islands (Great Barrier Reef) 2002 coral bleaching using IKONOS data, *Coral Reefs*, 23, 123-132, 2004.
Farnsworth, E. J., A. M. Ellison, and W. K. Gong, Elevated CO_2 alters anatomy, physiology, growth, and reproduction of red mangrove (*Rhizophora mangle* L.), *Oecologia*, 108, 599-609, 1996.
Fisk D. A., and T. J. Done, Taxonomic and bathymetric patterns of bleaching, *Proc. Seventh Int. Coral Reef Congr., Tahiti*, 6, 149-154, 1985.
Fitt, W. K., B. E. Brown, M. E. Warner, and R. P. Dunne, Coral bleaching, interpretation of thermal tolerance limits and thermal thresholds in tropical corals, *Coral Reefs*, 20, 51-65, 2001.
Gates, R. D., and P. J. Edmunds, The physiological mechanisms of acclimatization in tropical reef corals, *American Zoologist*, 39, 30-43, 1999.
Gattuso, J. P., M. Frankignoulle, I. Bourge, S. Romaine, and R. W. Buddemeier, Effect of calcium carbonate saturation of seawater on coral calcification, *Global and Planetary Change*, 18, 37-46, 1998.

Goreau, T. J., and R. L. Hayes, Coral bleaching and ocean hot spots, *Ambio*, 23,176-180, 1994.

Grübler, A., and N. Nakicenovic, Identifying dangers in an uncertain climate, *Nature*, 415, 15, 2001.

Guinotte, J. M., R. W. Buddemeier, and J. A. Kleypas, Future coral reef marginality, temporal and spatial effects of climate change in the Pacific basin, *Coral Reefs*, 22, 551-558, 2003.

Halley, R. B., and K. K. Yates, Will reef sediments buffer corals from increased global CO_2. *Proc. Ninth Int. Coral Reef Sym*, Bali, Indonesia, Abstract. p. 248, 2000.

Halley, R. B., K. K. Yates, and J. C. Brock, South Florida coral-reef sediment dissolution in response to elevated CO_2. *Proc. Tenth Int. Coral Reef Sym* Okinawa, Japan, Abstract. p. 178, 2004.

Harvell, C. D., C. E. Mitchell, J. R. Ward, S. Altizer, A. P. Dobson, R. S. Ostfield, and M. D. Samuel, Climate warming and disease risks for terrestrial and marine biota, *Science*, 296, 2158-2162, 2002.

Hatcher, B. G., and A. W. D. Larkum, An experimental analysis of factors controlling the standing crop of the epilithic algal community on a coral reef, *J. Exp. Mar. Biol. Ecol.*, 69, 61-84, 1983.

He, J.-S., and F. A. Bazzaz, Density-dependent responses of reproductive allocation to elevated atmospheric CO_2 in *Phytolacca Americana*, *New Phytologist*, 157, 229, 2003.

Hoegh-Guldberg, O, Climate change, coral bleaching and the future of the world's coral reefs, *Mar. Freshw. Res.*, 50, 839-866, 1999.

Holling, C. S., Resilience, and stability of ecological systems, *Ann. Rev. Ecol. Syst.*, 4, 1-23, 1973.

Hughes, T., A. H. Baird, D. R. Bellwood, M. Card, S. R. Connolly, C. Folke, R. Grosberg, O. Hoegh-Guldberg, J. B. C. Jackson, J. Kleypas, J. M. Lough, P. Marshall, M. Nystrom, S. R. Palumbi, J. M. Pandolfi, B. Rosen, B., and J. Roughgarden, Climate change, human impacts and the resilience of coral reefs, *Science*, 301, 929-933, 2003.

Hutchings, P. A., Biological destruction of coral reefs, a review, *Coral Reefs*, 4, 239-252, 1986.

IPCC, Climate Change 2001, Impacts, Adaptation and Vulnerability. A Contribution of Working Group II to the Third Assessment Report of the Intergovernmental Panel on Climate Change, In *Climate Change 2001, Impacts, Adaptation and Vulnerability,* edited by J. J. McCarthy, O. F. Canziani, N. A. Leary, D. J. Dokken, and K. S. White, pp. 1-17, Cambridge University Press, Cambridge, UK., pp. 1032, 2001a.

IPCC, *Climate Change 2001, Synthesis Report,* edited by R. T. Watson, D. L. Albritton, T. Barker, I. A. Bashmakov, O. Canziani, R. Christ, U. Cubasch, O. Davidson, H. Gitay, D. Griggs, J. Houghton, J. House, Z. Kundzewicz, M. Lal, N. Leary, C. Magadza, J. J. McCarthy, J. F. B. Mitchell, J. R. Moreira, M. Munasinghe, I. Noble, R. Pachauri, A. B. Pittock, M. Prather, R. G. Richels, J. B. Robinson, J. Sathaye, S. H. Schneider, R. Scholes, T. Stocker, N. Sundararaman, R. Swart, T. Taniguchi, and D. Zhou. Cambridge University Press, Cambridge, UK, pp. 398, 2001b.

Jokiel, P. L., and E. B. Guinther, Effects of temperature on reproduction in the hermatypic coral *Pocillopora damicornis*, *Bull. Mar. Sci.*, 28, 786-789, pp. 398, 1978.

Jones, R. N., An environmental risk assessment/management framework for climate change impact assessments, *Nat. Haz.*, 23, 197-230, 2001.

Jones, R. N., *Managing climate change risks*. Report of Working Group to the OECD Workshop on the Benefits of Climate Policy, Improving Information for Policy Makers. Organisation for Economic Co-operation and Development. Paris, France. ENV/EPOC/GSP(2003) 22/FINAL. pp. 36, 2003.

Kennedy, V. S., R. R. Twilley, J. A. Kleypas, J. H. Cowan Jr., and S. R. Hares, *Coastal and marine systems and global climate change – potential effects on U.S. resources*, The Pew Center on Global Climate Change, Arlington, VA, USA. vi + 52 pp., 2002.

Kleypas, J., and C. Langdon, Overview of CO_2-induced changes in seawater chemistry, *Proc. Ninth Int. Coral Reef Sym Bali, Indonesia* 2, 1085-1089, 2002.

Kleypas, J. A., R. Buddemeier, D. Archer, J. P. Gattuso, C. Langdon, and B. N. Opdyke, Geochemical consequences of increased atmospheric CO_2 on corals and coral reefs, *Science*, 284, 118-120, 1999.

Kleypas, J., R. Buddemeier, and J. P. Gattuso, The future of coral reefs in an age of global change, *Int. J. Earth Sci.*, 90, 426-437, 2001.
Kleypas, J. A., Buddemeier, R., Archer, D., Gattuso, J. P., Langdon, C., and Opdyke, B. N., Geochemical consequences of increased atmospheric CO_2 on corals and coral reefs, *Science*, 284, 118-120, 1999.
Kleypas, J. A., R. W. Buddemeier, C. M. Eakin, J-P. Gattuso, J. Guinotte, O. Hoegh-Guldberg, R. Iglesias-Prieto, P. L. Jokiel, C. Langdon, W. Skirving, and A. E. Strong, Comment on "Coral reef calcification and climate change, The effect of ocean warming", *Geophys. Res. Let.*, 32, L08601, doi:10.1029/2004GL022329, 2005.
Leclercq, N., J. P. Gattuso, and J. Jaubert, CO_2 partial pressure controls the calcification rate of a coral community, *Global Change Biol.*, 6, 329-334, 2000.
Lindahl, U., M. C. Öhman, and C. K. Schelten, The 1997/1998 mass mortality of corals, effects on fish communities on a Tanzanian coral reef, *Mar. Poll. Bull.*, 42, 127-131, 2001.
Little, A. F., M. J. H. van Oppen, and B. L. Willis, Flexibility in algal endosymbiosis shapes growth in reef corals, *Science*, 304, 1492-1494, 2004.
Littler, M. M., and M. S. Doty, Ecological components structuring the Seaward Edges of tropical Pacific reefs, the distribution, communities and productivity of *Porolithon*, *J. Ecol.*, 63, 117-129, 1975.
Loreau, M., S. Naeem, P. Inchausti, J. Bengtsson, J. P. Grime, A. Hector, D. U. Hooper, M. A. Huston, D. Raffaelli, B. Schmid, D. Tilman, and D. A. Wardle, Biodiversity and ecosystem functioning, current knowledge and future challenges, *Science*, 294, 804-808, 2001.
Lough, J. M., and D. J. Barnes, Environmental controls on growth of the massive coral *Porites, J. Exp. Mar. Biol. Ecol.*, 245, 225-243, 2000.
Loya, Y., K. Sakai, K. Yamazoto,Y. Nakano, H. Sembali, and R. van Woesik, Coral bleaching, the winners and losers, *Eco. Lett.*, 4, 1122-131, 2001.
Lugo, A. E., and S. C. Snedaker, The ecology of mangroves, *Ann. Rev. Ecol. Syst.,* 5, 39-64, 1974.
Macintyre, I. G., P. W. Glynn, and R. S. Steneck, A classic Caribbean algal ridge, Holandés Cays, Panamá, an algal coated storm deposit, *Coral Reefs*, 20, 95-105, 2001.
Marshall, P., and A. Baird. A., Bleaching of corals on the Great Barrier Reef, differential susceptibilities among taxa, *Coral Reefs*, 19, 155-163, 2000.
Matsui, N., Estimated stocks of organic carbon in mangrove roots and sediments in Hinchinbrook Channel, Australia, *Mangroves and Salt Marshes*, 2, 199-204, 1998.
McClanahan, T. R., A. H. Baird, and P. A. Marshall, Comparing bleaching and mortality responses of hard corals between southern Kenya and the Great Barrier Reef, Australia, *Mar. Poll. Bull.*, 48, 327-335, 2004.
McClanahan, T., N. Polunin, and T. Done, Ecological states and the resilience of coral reefs, *Cons. Ecol.,* 6(2), 18, 2002. [online] URL, http//www.consecol.org/vol6/iss2/art18
McClanahan, T. R., Primary succession of coral-reef algae: Differing patterns on fished versus unfished reefs, *J. Exp. Mar. Biol. Ecol.*, 218, 77-102, 1997.
McCook, L. J., Macroalgae, nutrients and phase shifts on coral reefs, scientific issues and management consequences for the Great Barrier Reef, *Coral Reefs*, 18, 357-367, 1999.
McInnes, K. L, K. J. E. Walsh, G. D. Hubbert, and T. Beer, Impact of sea-level rise and storm surges on a coastal community, *Nat. Haz.*, 30, 187-207, 2003.
McManus, J. W., and J. F. Polsenberg, Coral-algal phase shifts on coral reef, ecological and environmental aspects, *Prog. Oceanog.*, 60, 263-279, 2004.
McNeil, B. I., R. J. Matear, and D. J. Barnes, Coral reef calcification and climate change, The effect of ocean warming, *Geophys. Res. Lett.*, 31, L22309, 2004, doi:10.1029/2004GL021541.
McRoy, C. P., and C. McMillan, Production ecology and physiology of seagrasses, in *Seagrass Ecosystems, A Scientific Perspective,* edited by C. P. McRoy and C. Helfferich, pp. 53-85, Marcel Dekker Inc., New York and Basel, 1977.
Milliman, J. D., *Carbonate Sediments*, Springer-Verlag, Berlin, pp. 375, 1974.
Mumby, P. J., A. J. Edwards, E. Arias-Gonzales, K. C. Lindeman, P. G. Blackwell, A. Gall, M. I. Gorczynska, A. R. Harborne, C. L. Pescod, H. Renken, C. C. C. Wabnitz, and G. Llewellyn,

Mangroves enhance the biomass of coral reef fish communities in the Caribbean, *Nature*, 427, 533-536, 2004.

Murphy, J. M., D. M. H. Sexton, D. N. Barnett, J. S. Jones, M. J. Webb, M. Collins, and D. A. Stainforth, Quantification of modelling uncertainties in a large ensemble of climate change simulations, *Nature*, 430, 768-772, 2004.

Nakicenovic, N., and R. Swart (Editors). *Emissions Scenarios, Special Report of the Intergovernmental Panel on Climate Change*, Cambridge University Press, Cambridge, UK., pp. 570, 2000.

Nyström, M., C. Folke, and F. Moberg, Coral reef disturbance and resilience in a human-dominated environment, *Trends Ecol. Evol.*, 15, 413-417, 2000.

Pandolfi, J. M., Response of Pleistocene coral reefs to environmental change over long temporal scales, *Am. Zool.*, 39, 113-130, 1999.

Pandolfi, J. M., R. H. Bradbury, E. Sala, T. P. Hughes, K. A. Bjorndal, R. G. Cooke, D. McArdle, L. McClenachan, M. J. Newman, G. Paredes, R. R. Warner, and J. B. Jackson, Global trajectories of the long-term decline of coral reef ecosystems, *Science*, 301, 955-958, 2003.

Peltier, W. R., On eustatic sea level history, Last Glacial Maximum to Holocene, *Quat. Sci. Rev.*, 21, 377-396, 2002.

Pittock, A. B., Coral reefs and environmental change: Adaptation to what? *Amer. Zool.*, 39, 10-29, 1999.

Pittock, A. B., and R. N. Jones, Adaptation to what and why? *Envir. Mon. Res.*, 61, 9-35, 2000.

Pittock, A. B., R. N. Jones, and C. D. Mitchell, Probabilities will help us plan for climate change, *Nature*, 413, 249, 2001.

Preen, A. R., W. J. Lee Long, and R. G. Coles, Flood and cyclone related loss, and partial recovery of more than 1000 km^2 of seagrass in Effects of cyclone Sadie on seagrasses at Hervey Bay, Queensland, Australia, *Aq. Bot.*, 52, 3-17, 1995.

Rosenberg, E., and Y. Ben-Haim, Microbial diseases of corals and global warming, *Env. Microbiol.*, 4, 318-326, 2002.

Rowan, R., Thermal adaptation in coral reef symbionts, *Nature*, 430, 742, 2004.

Salm, R. V., S. E. Smith, and G. Llewellyn, Mitigating the impact of coral bleaching through marine protected area design, In *Coral Bleaching, Causes, Consequences and Response. Selected papers presented at the 9th International Coral Reef Symposium on "Coral Bleaching, Assessing and Linking Ecological and Socioeconomic Impacts, Future Trends and Mitigation Planning,* edited by H. Z. Schuttenberg, pp. 81-88, Coastal Management Report #2230, Coastal Resources Center, University of Rhode Island, USA. pp. 102, 2001.

Salm, R.V., T. J. Done, and E. McLeod, Marine protected area (MPA) planning in a changing climate, (this volume), 2006.

Sammarco, P. W., J. S. Levinton, and J. C. Ogden, Grazing and control of coral reef community structure by *Diadema antillarum* Philippi (Echinodermata, Echinoidea), a preliminary study, *J. Mar. Res.*, 32, 47-53, 1974.

Sammarco, P. W., *Diadema* and its relationship to coral spat mortality, grazing, competition, and biological disturbance, *J. Exp. Mar. Biol. Ecol.*, 45, 245-272, 1980.

Sarmiento, J. L., R. Slater, R. Barber, L. Bopp, S. C. Doney, A. C. Hirst, J. Kleypas, R. Matear, U. Mikolajewicz, P. Monfray, V. Soldatov, S. A. Spall, and R. Stouffer. Response of ocean ecosystems to climate warming, *Global Biogeo. Cycles*, 18, GB3003, doi:10.1029/2003GB002134, 2004.

Schneider, S. H., What is 'dangerous' climate change? *Nature*, 411, 17-19, 2001.

Semeniuk, V., Predicting the effect of sea-level rise on mangroves in Northwestern Australia, *J. Coastal Res.*, 10, 1050-1076, 1994.

Sheppard, C. R. C., Predicted recurrences of mass coral mortality in the Indian Ocean, *Nature*, 425, 294-297, 2003.

Sheppard, C. R. C., M. Spalding, C. Bradshaw, and S. Wilson, Erosion vs. recovery of coral reefs after 1998 El Niño, Chagos Reefs, Indian Ocean, *Ambio*, 31, 40-48, 2002.

Shick, J. H., M. P. Lesser, and P. L. Jokiel, Effects of ultraviolet radiation on corals and other coral reef organisms, *Global Change Biol.*, 2, 527-545, 1996.

Short, F. T., and H. A. Neckles, The effects of global climate change in seagrasses, *Aquat. Bot.*, 63, 169-196, 1999.

Skirving, W., and C. R., Steinberg, Hydrodynamics of a bleaching event, (this volume), 2006.

Snedaker, S. C., Mangroves and climate change in the Florida and Caribbean region, scenarios and hypotheses, *Hydrobiol.*, 295, 43-49, 1995.

Spalding, M. D., and G. E. Jarvis, The impact of the 1998 coral mortality on reef fish communities in the Seychelles, *Mar. Poll. Bull.*, 44, 309-321, 2002.

Spalding. M. D., F. Blasco, and C. D. Field, *Atlas of Mangroves of the World*, The International Society for Mangrove Ecosystems, Okinawa, Japan, pp. 178, 1997.

Steneck, R. S., Is herbivore loss more damaging to reefs than hurricanes? Case study from two Caribbean reef systems (1978-1988), *Proc. Colloquium on Global Aspects of Coral Reefs, Health, Hazards and History*, compiled by R. N. Ginsburg, pp. 220-226, Rosenstiel School of Marine and Atmospheric Science, University of Miami, Miami, USA, 1994.

Szmant, A. M., and N. J. Gassman, The effect of prolonged "beaching" on the tissue biomass and reproduction of the reef coral *Montastrea annularis*, *Coral Reefs*, 8, 217-224, 1990.

Twilley, R. R., S. C. Snedaker, A. Yáñez-Arancibia, and E. Medina, Biodiversity and ecosystem process in tropical estuaries, perspectives of mangrove ecosystems, *Functional Roles of Biodiversity. A Global Perspective*, edited by H. A. Mooney, J. H. Cushman, E. M. Medina, O. Sala and E. D. Schulze, pp. 327-392, John Wiley and Sons, Chichester, 1996.

Veron, J. E. N., *Corals of the World*, Australian Institute of Marine Science, Townsville, Australia, 2000.

West, J. M., and R. V. Salm, Resistance and resilience to coral bleaching, Implications for coral reef conservation and management, *Cons. Biol.*, 17, 956-967, 2003.

Wilkinson, C. R., *Status of the coral reefs of the world*, Australian Institute of Marine Science, Townsville, Australia, 2002.

Wilkinson, C., O. Linden, H. Cesar, G. Hodgson, J. Rubens, and A. E. Strong, Ecological and socioeconomic impacts of 1998 coral mortality in the Indian Ocean, An ENSO impact and a warning of future change? *Ambio*, 28, 188-196, 1999.

Williams, I. D., N. V. C. Polunin, and V. J. Hendrick, Limits to grazing by herbivorous fishes and the impact of low coral cover on macroalgal abundance on a coral reef in Belize, *Mar. Ecol. Prog. Ser.*, 222, 187-296, 2001.

Woodroffe, C., Mangrove sediments and geomorphology, In *Tropical Mangrove Ecosystems*, edited by A. I. Robertson, and D. M. Alongi, pp. 7-41, American Geophysical Union, Washington DC., USA, 1992.

Woodroffe, C. D., D. R. Stoddart, T. Spencer, T. P. Scoffin, and A. W. Tudhope, Holocene emergence in the Cook Islands, South Pacific, *Coral Reefs*, 9, 31-39, 1990.

Wooldridge, S., and T. J. Done, Learning to predict large-scale coral bleaching from past events, A Bayesian approach using remotely sensed data, in-situ data, and environmental proxies, *Coral Reefs*, 23, 96-108, 2004.

Yates K. K., and L. L. Robbins, Microbial lime-mud production and its relation to climate change, In *Geological perspectives of global climate change*, edited by L. C. Gerhard LC, W. E. Harrison WE and B. M., Hanson, *Am. Assoc. Petrol. Geol.*, Studies in Geology, 247, 267-283, 2001.

Yu, K., J. X. Zhaob,, J. X., T. S. Liue, T. S., G. J. Weic, G. J., P. X. Wangd, P. X., and K. D. Collerson., K. D., High-frequency winter cooling and reef coral mortality during the Holocene climatic optimum, *Earth Plan. Sci. Lett.*, 224, 143-155, 2004.

Zepp, R. G., *UV exposure of coral assemblages in the Florida Keys*. U.S. Environmental Protection Agency, Florida, USA, pp. 46, 2003.

Zepp, R. G., T. V. Callaghan, and D. J. Erickson, Interactive effects of ozone depletion and climatic change on biogeochemical cycles, *Photochem. Photobiol. Sci.*, 2, 51-61, 2003.

3

Coral Reef Records of Past Climatic Change

C. Mark Eakin and Andréa G. Grottoli

Abstract

Coral skeletons serve as excellent natural archives of paleonvironmental conditions in tropical and subtropical waters. The isotopic, trace, and minor elemental composition of coral skeletons can vary with environmental conditions such as temperature, salinity, cloud cover, river discharge, upwelling, and ocean circulation. As such, coral cores offer a suite of proxy records with the potential for reconstructing paleoclimatic and paleoceanographic conditions on interannual-to-centennial timescales. Living colonies can provide several centuries of continuous paleo-recordings and have been combined with fossil corals to reveal conditions over recent millennia and earlier periods. This chapter provides an overview of reconstructing environmental parameters from coral cores and some of the limitations of the various techniques. Corals have proven their worth as reliable recorders of past environmental conditions, but limitations exist due to the way that coral paleoclimatic records are collected and analyzed.

A Brief History of Coral Paleoclimatology

Coral skeletons serve as excellent natural archives of paleonvironmental conditions in tropical and subtropical waters. Living colonies can provide several centuries of continuous paleo-recordings and have been combined with fossil corals to reveal conditions over recent millennia and earlier periods.

The early 1970s, scientists discovered that corals regularly alternated the density of their calcium carbonate skeletons between seasons. With this discovery, sclerochronology (record of coral skeleton growth) was developed as an indicator of environmental conditions [Barnes, 1973; Buddemeier et al., 1974; Knutson et al., 1972; Weber et al., 1975]. In one of the first applications of this new tool, Hudson et al. [1976] made the connection to climate, or at least weather, by identifying stress bands recorded in corals during a cold winter in Florida in 1969-70. At about the same time, Weber and Woodhead [1972] applied geochemical analyses that had been in use in paleoceanographic studies to coral skeletons. Fairbanks and Dodge [1979] combined this with sclerochronology and found regular periodicity in geochemical ratios of $^{18}O/^{16}O$ ($\delta^{18}O$) of corals from Jamaica, Barbados, and Bermuda, with warm waters and high density skeletal bands corresponding to skeletons depleted in ^{18}O. While ratios of $^{13}C/^{12}C$ ($\delta^{13}C$) were positively correlated at two sites, they were inversely correlated at the third – perhaps the first signal that $\delta^{13}C$

Coral Reefs and Climate Change: Science and Management
Coastal and Estuarine Studies 61
Copyright 2006 by the American Geophysical Union.
10.1029/61CE04

variability was not driven by temperature but by some other environmental and/or biological parameter. Although sclerochronology continues to have direct applications as a temperature proxy [Lough and Barnes, 1997; Slowey and Crowley, 1995], most paleoclimatic data from corals now use geochemical analysis of their skeletons. More recently, sclerosponges have been investigated as paleoenvironmental recorders of both the surface and intermediate ocean waters. To date, approximately 100 coral skeletal and several sclerosponge isotopic and/or elemental records of 20 years of longer have been produced, most of which are available from the World Data Center for Paleoclimatology and other sources (Plate 1a).

Introduction, Collection, Preparation, and Chronology Establishment

Corals

In the tropical oceans, the isotopic, trace and minor elemental composition of coral skeletons can vary as a result of environmental conditions such as temperature, salinity, cloud cover, river discharge, upwelling, and ocean circulation. As such, coral cores offer a suite of proxy records with potential for reconstructing tropical paleoclimatic and paleoceanographic conditions on interannual-to-centennial timescales. Massive, symbiotic stony corals are good tropical climate proxy recorders because: 1) they are widely distributed throughout the tropics, 2) their continuous annual skeletal banding pattern offers excellent chronological control, 3) they incorporate a variety of climate tracers from which paleo-ocean temperature, salinity, cloud cover, upwelling, ocean circulation, ocean mixing patterns, and other climatic and oceanic features can be reconstructed, 4) their proxy records can nearly match instrumental records for fidelity, 5) their records can span several centuries, and 6) their high skeletal growth rate (usually ranging from 5-25 mm /year) permits sub-seasonal sampling resolution. Thus proxy records in corals provide the best means of obtaining long seasonal-to-centennial timescale paleoclimatic information in the tropics.

When used in large numbers, like tree-ring site chronologies, growth records can be used as proxies for changing climatic conditions [Lough and Barnes, 1997]. In groups or individually, skeletal growth, or sclerochronology, provides information on environmental stress with corals growing faster in years of favorable conditions and more slowly under stressful conditions [Hudson et al., 1989; Eakin et al., 1994].

Geochemical records preserved in the coral skeleton are most commonly used to reconstruct paleotemperature records, but are also commonly used to reconstruct salinity, winds and upwelling, runoff, pollutants, and ocean mixing (Table 1 and following sections). In addition, the tropical habitat of corals and tropical origin of the El Niño-Southern Oscillation (ENSO) System has resulted in a natural pairing of corals as a source of proxy data on pre-instrumental ENSO variability. This continues to be an important area of study, as ENSO clearly dominates interannual- to decadal-scale climatic variability [Quadrelli and Wallace, 2004].

Not only do corals serve as proxies for modern climate, they have also been used to reconstruct conditions in the Holocene, Last Glacial Maximum, and last Interglacial. However, calibration issues typically prevent the geochemical proxies from being used to directly reconstruct temperatures, so these records are primarily used as relative indicators of temperature or other climatic variability. With living corals, the geochemical proxies can be compared with instrumental and remotely sensed temperatures, allowing calibration with varied levels of precision.

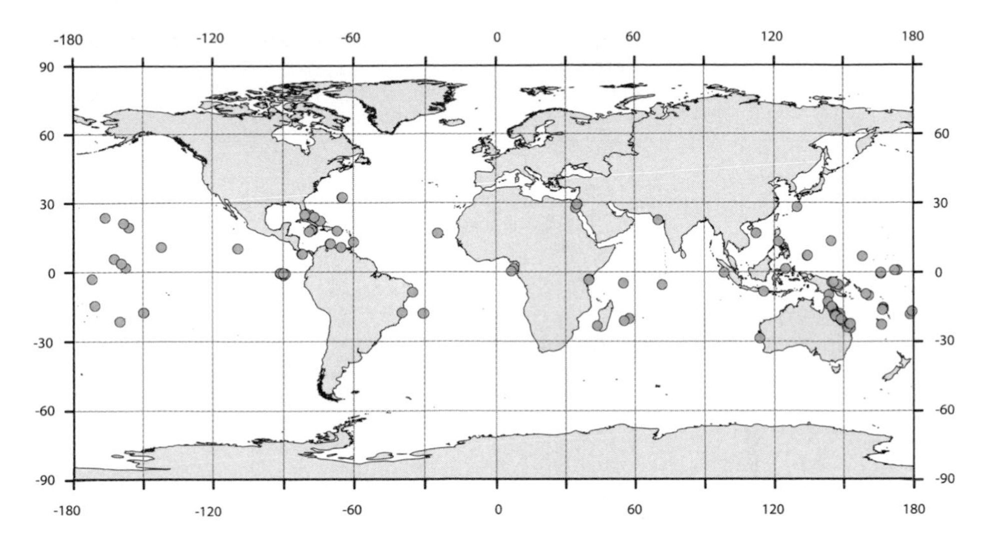

Plate 1a. Locations of coral and sclerosponge records archived in the World Data Center for Paleoclimatology (http://www.ncdc.noaa.gov/paleo/corals.html) as of 1 June 2006.

Plate 1b-d. Illustration of coral cores and their extraction. From left to right: (b) *Porites lutea* being cored by one of the authors (AGG) in Saipan (July 2003), (c) x-radiograph of a *Montastraea* coral core from Florida [Halley et al., 1994], and (d) an image of coral luminescence in a *Porites lutea* core from Kenya (unpublished image of record from Cole et al., 2000).

TABLE 1. Environmental variable(s) that can be reconstructed from coral skeletal isotopes, trace and minor elements, and growth records.

Proxy	Environmental variable
Isotopes	
$\delta^{18}O$	Sea surface temperature, sea surface salinity
$\delta^{13}C$	Light (e.g., seasonal cloud cover), plankton intake
$\Delta^{14}C$	Ocean ventilation, water mass circulation
$\delta^{11}B$	pH
Trace and Minor Elements	
Sr/Ca	Sea surface temperature
U/Ca	Sea surface temperature
Mg/Ca	Sea surface temperature
Mn/Ca	Wind anomalies, upwelling
Cd/Ca	Upwelling, contamination
Ba/Ca	Upwelling, river outflow
Pb/Ca	Gasoline burning
Skeleton	
Skeletal growth bands	Light (seasonal changes), stress, water motion, sedimentation, sea surface temperature
Luminescence	River outflow, ocean productivity

Method: Massive stony corals of the genera *Porites, Pavona*, and *Montastraea* are most commonly targeted for paleoclimatic studies because they form large mounding colonies with distinct annual bands, can grow for several hundred years, and are common. Continuous records of past tropical climate conditions are obtained by extracting a core from an individual massive coral head along its major axis of growth. Typically, this involves drilling a corer through the top center of the coral head to its initial point of growth (Plate 1b-d). The extracted core is cut longitudinally into slabs ranging in thickness from 0.5-1 cm, cleaned with water and dried, then X-rayed. X-ray positive prints reveal the banding pattern of the slab and are used: 1) as a guide for sample drilling and 2) to establish a chronology for the entire coral record when the banding pattern is clear. In some cases, the x-rays serve as environmental proxies in and of themselves (see *Coral Growth as an Indicator of Environmental Stress*). In a few cases, composite records have been made by using multiple cores to extend the record length beyond that available from a single core [Cobb et al., 2003; Dunbar et al., 1994].

For geochemical analysis, carbonate powder samples are extracted along the major axis of growth by grinding the skeletal material with a dental drill bit. For high-resolution paleo-reconstructions, samples are extracted every millimeter or less along the entire length of the core. Since corals grow about 5-15 mm per year, this sampling method can yield sub-seasonal resolution. Much higher resolution sampling is possible by micro-drilling or using laser ablation techniques that yield samples at approximately weekly temporal resolution [Sinclair et al., 1998], but this is not commonly performed. The coral carbonate powder is then analyzed for one or more isotopes or elements to build any one of a number of possible paleo-proxy records (Table 1). In most cases, the stable oxygen ($\delta^{18}O$) and carbon ($\delta^{13}C$) values of each sample are measured. Since the $\delta^{13}C$ and/or $\delta^{18}O$ composition of corals usually have a strong seasonal component, they are often used to establish the chronology in the absence of banding, and/or to confirm or adjust the chronology established from skeletal bands.

Sclerosponges

Sclerosponges are slow growing calcareous sponges that are normally found in protected, shaded microhabitats on reefs and walls, ranging from sub-surface to 1000m depths, and can often be found exposed to open oceanic waters. Their growth rates range from 0.1-1.6 mm / year and can grow for up to 1000 years [Benavides and Druffel, 1986; Böhm et al., 1996, 2002; Fallon et al., 2003; Grottoli unpublished; Hughes and Thayer, 2001; Reitner et al., 1996; Swart et al., 1998, 2002; Willenz and Hartmen, 1985; Wörheide et al., 1997]. Published growth rates for *Acanthocheatetes wellsi, Astroclera willeyana,* and *Ceratoporella nicholsoni* are 0.05-1.6, 0.35-1.6, and 0.1-0.4 mm/year, respectively. Sclerosponges are emerging as good paleoclimatic recorders and complement the information provided by faster growing corals [Swart et al., 1998a]. The slower growth often limits sclerosponge resolution (but see exception to this in Grottoli, 2006) but allows many sclerosponges to record several centuries of information. They are found throughout the tropics across a wide depth range (surface to 1000m) and can live for several centuries. In addition, sclerosponges appear to accrete their calcium carbonate skeleton in isotopic equilibrium with seawater due to the absence of photosynthesis and relatively low metabolic activity, thus more directly recording oceanic chemistry than corals [δ^{18}O: Moore et al., 2000; δ^{13}C: Böhm et al., 1996; Druffel and Benavides, 1986; Wörheide et al., 1997]. As such, sclerosponges appear to be reliable proxy recorders of seawater dissolved inorganic carbon δ^{13}C (DI-δ^{13}C) [i.e., Druffel and Benavides, 1986] and seawater δ^{18}O ($\delta^{18}O_{sw}$) [Swart et al., 2002]. Recent work suggests that the ratio of strontium to calcium (Sr/Ca) in sclerosponges may be a proxy for ocean temperature [Haase-Schramm et al., 2003; Swart et al., 2002; Rosenheim et al., 2005]. Because sclerosponges are not light limited with no detectable metabolic fractionation effects on isotopic or elemental signatures, sclerosponge-derived proxy records can lend great insights into the regional and temporal variability of oceanographic features such as thermocline depth, upwelling, ocean ventilation, and the uptake rate of anthropogenic carbon from the atmosphere into the oceans. This provides an improved knowledge of the spatial and temporal variability in surface and depth integrated DI-δ^{13}C inventory changes that extends the well-used WOCE data set hundreds of years in the past for tropical reef locations.

Method: Sclerosponges of the *Ceratoporella* (Caribbean), *Acanthocheatetes* (Pacific), and *Astroclera* (Pacific) genera are the most commonly used for paleoclimatic reconstruction. The specimens are typically collected across a depth range of 1-300 m using either SCUBA or submersible vehicles. A sclerosponge specimen is then cut in half or into slabs along its major axis of growth. X-rays are typically not very informative in sclerosponges as their banding is not annual. However, by using a combination of x-rays and following skeletal microstructure, the sclerosponge slab is sampled by milling the skeleton along the maximum axis of growth at 0.1 mm increments or less using a high-precision microdrill down the length of the slab [Grottoli in press; Hughes and Thayer, 2001; Swart et al., 2002]. The sclerosponge carbonate powder samples are then analyzed for one or more isotopes or elements to build paleoceanographic records. The chronology is established by using a combination of radiogenic dating, stable isotope records, and estimated growth rates. The bomb-radiocarbon signature (Δ^{14}C) introduced into the skeleton from atmospheric bomb testing (1955-1963) is used to help anchor the chronology. In some cases, the ratio of Uranium to Thorium (U/Th) is also used to date the pre-bomb sections of sclerosponge records. Since exact annual dating is not possible in sclerosponges, paleoreconstructions from these records tend to focus on decadal-to-centennial timescale processes. Similar to corals, other geochemical proxies may also be analyzed.

Coral Paleotemperature and Salinity

The $\delta^{18}O$ and Sr/Ca signatures in corals have proven to be very reliable paleo-temperature proxies [Druffel, 1997; Gagan et al., 2000; Grottoli, 2001]. Since large corals suitable for paleoclimatic reconstructions are typically limited to the top ~20m of the ocean, their proxy records reflect near-surface ocean conditions found on coral reefs. $\delta^{18}O$ has the longest history as a temperature proxy and, up until recently, was more common and less expensive than elemental measurements. Unfortunately, coral skeletal $\delta^{18}O$ is influenced by both temperature and salinity, and is therefore not a pure indicator of temperature. This confounding influence is minimized when the relative contribution from either salinity or temperature variability is low or the two variables combine to increase the change in coral $\delta^{18}O$ [Felis et al., 2000; Le Bec et al., 2000; Swart et al., 1998b]. Fortunately, recent improvements in instrumentation have made the application of strontium to calcium (Sr/Ca) in coral skeletons practical as a coral paleothermometer [Beck et al., 1992]. Sr/Ca ratios are not influenced by salinity, making it a direct paleothermometer with few sources of outside noise. This also provides the possibility of detecting salinity changes through the difference between these two geochemical tracers [Gagan et al., 1998; Linsley et al., 2004; McCulloch et al., 1994; Quinn and Sampson, 2002; Ren, 2002; Swart et al., 1999]. A few individuals have applied two other paleothermometers: uranium to calcium (U/Ca) and magnesium to calcium (Mg/Ca) ratios. U/Ca ratios may provide the same level of accuracy or precision of temperature reconstructions as Sr/Ca, but this is still being confirmed [Corrège, et al., 2000; Hendy et al., 2002; Min et al., 1995; Shen and Dunbar, 1995]. Mg/Ca [Mitsuguchi et al., 1996] ratios are probably not accurate enough to replace other proxies [Meibom, 2004; Schrag, 1999]. While each of these has value alone, the greatest strength may lie in a multiproxy approach. Solow and Huppert [2004] recently proposed a technique that may improve the error by combining several elemental paleothermometers. Finally, while coral skeletal $\delta^{18}O$ has a direct value as a paleothermometer, it also can provide information on upwelling [i.e., Guilderson and Schrag, 1989b].

Stable Isotope Analyses ($\delta^{18}O$ and $\delta^{13}C$)

Skeletal $\delta^{18}O$ in marine organisms is affected by temperature-induced fractionation and $\delta^{18}O$ of the surrounding water [Epstein et al., 1953]. As temperature increases, the ^{18}O fraction of the coral skeleton decreases [Kim and O'Neil, 1997]. Based on empirical studies, a 1°C increase in water temperature corresponds to an average decrease of about 0.22% in coral $\delta^{18}O$ though that slope can vary significantly from 0.15 to 0.24% among species and locations [i.e., Linsley et al., 1999; Wellington and Dunbar, 1995; Wellington et al., 1996]. The $\delta^{18}O$ in seawater and coral skeletons also decreases as salinity decreases, because precipitation is depleted in $\delta^{18}O$ relative to seawater, especially in tropical environments where greater convection of moisture reduces the ^{18}O in cloud-borne moisture. In both corals and sclerosponges, the interpretation of the skeletal $\delta^{18}O$ record depends on the hydrological regime of the collection site. While not a paleothermometer, skeletal $\delta^{13}C$ in corals is typically analyzed along with $\delta^{18}O$ (see discussion in *Cloud Cover and Feeding*).

Method: Carbonate powder samples are analyzed for $\delta^{18}O$ (the per mil deviation of the ratio of $^{18}O/^{16}O$ relative to the Peedee Belemnite (VPDB) Limestone Standard) and $\delta^{13}C$ (ratio of $^{13}C/^{12}C$ relative to PDB) by acidifying the sample in 100% orthophosphoric acid and measuring the resulting CO_2 with a mass spectrometer. Automated common acid bath or carbonate Kiel devices are commonly used, especially for small samples. Typically, 10% of samples are run in duplicate to ensure reproducibility and the accepted precision of replicate analyses is now ≤0.05% for $\delta^{13}C$ and ≤0.09% for $\delta^{18}O$. Today, as little as ~80-150 μg of sample material is now used for stable isotope analyses.

Strontium to Calcium and Uranium to Calcium

Strontium and uranium have long residence times in seawater. Because this provides relatively constant concentrations in surface seawater, the observed fluctuation in coral Sr/Ca has been attributed to changes in SST [Weber, 1973]. The application of U/Ca is more recent, but shows promise [Corrège et al., 2000; Hendy et al., 2002; Min et al., 1995; Shen and Dunbar, 1995]. Sr and U replace Ca in the skeletal aragonite through thermodynamic processes, with less replacement of Ca by the heavier elements at higher temperatures [Beck et al., 1992; Weber, 1973]. Although it is a well-established geothermometer, SST-Sr/Ca calibration curves often vary dramatically among species and/or locations, and may be influenced by growth rate [e.g., Beck et al., 1992; Boiseau et al., 1997; deVilliers et al., 1995; Gagan et al., 2000; Goodkin et al., 2005; Marshall and McCulloch, 2002]. Thus Sr/Ca needs to be calibrated to local or regional temperature records for each genus or species and at every location.

Method: Carbonate powder samples are dissolved in acid and the Sr/Ca or U/Ca ratio in the solution is measured using a variety of techniques. TIMS (thermal ionization mass spectrometry), ICP-MS (inductively coupled plasma mass spectrometry), and ICP-OES (inductively coupled plasma optical emission spectroscopy) are some of the tools used to measure Sr/Ca and U/Ca values in carbonates. In most cases, small samples (100-1000 μg) are used for each analysis. The currently accepted precision for Sr/Ca analyses of duplicate corals or sclerosponge samples is ±0.02% (1σ).

Other Applications of Coral Proxy Paleodata

In addition to providing proxy records of paleo-sea surface temperature, coral records have been used to reconstruct paleo-circulation, -salinity, -pH, -runoff, -cloud cover, and –nutrients from both modern and fossil corals.

Cloud Cover and Feeding ($\delta^{13}C$)

While less commonly reported than $\delta^{18}O$, most researchers collect data on $\delta^{13}C$ at the same time as $\delta^{18}O$. Photosynthesis (light) and feeding (which directly affects the respired $\delta^{13}C$) are the primary influences on coral skeletal $\delta^{13}C$. For corals collected from shallow, non-upwelling sites where coral feeding is relatively constant, changes in $\delta^{13}C$ are predominantly driven by changes in photosynthesis, providing a record of seasonal changes in cloud cover or turbidity (i.e., rainy season vs. dry season) [Cole et al., 1990; Grottoli, 1999, 2002; Grottoli and Wellington, 1999; McConnaughey, 1989a,b]. In addition, because of the loss of algal symbiotes during bleaching, decreases in photosynthesis and/or changes in feeding may be recorded in the skeleton of bleached corals [Grottoli et al., 2004; Rodrigues and Grottoli, 2006]. In upwelling regions where feeding opportunities can vary dramatically, changes in $\delta^{13}C$ have been used as an indicator of vertical mixing through change in coral food source [Felis et al., 1998] and in DI-$\delta^{13}C$ [Abram et al., 2003]. In the latter case, a dramatic change in DI-$\delta^{13}C$ off Indonesia, attributed to iron fertilization from the 1997 forest fires, resulted from unprecedented blooms of red tide algae. These elevated levels of primary production caused a shift in the seawater DI-$\delta^{13}C$ that was recorded in coral skeletons. Unlike corals, sclerosponges appear to deposit their skeleton in isotopic equilibrium with seawater and records the $\delta^{13}C$ of the surrounding DI-$\delta^{13}C$ (see anthropogenic carbon input section below for details).

Ocean Circulation: Radiocarbon Analyses ($\Delta^{14}C$)

^{14}C is produced naturally in the stratosphere and was also produced as a result of thermonuclear weapons explosions in the atmosphere in the 1950s and early 1960s. The base of the $\Delta^{14}C$ -bomb is clearly identifiable as ~1955 in coral and sclerosponge carbonate records [i.e., Druffel, 1981; Druffel and Linick, 1978; Fallon et al., 2003]. During the pre-bomb period, coral $\Delta^{14}C$ records reflect a declining trend in the 20th century termed the Suess Effect: the decrease in ^{14}C in the atmosphere due to dilution of natural ^{14}C by the addition of ^{14}C-free fossil fuel CO_2 [i.e., Druffel and Griffin, 1993, 1999; Druffel et al., 2001]. Post-1955, the bomb-curve signal can be used to help confirm/establish coral and sclerosponge chronologies. As a new proxy, $\Delta^{14}C$ is an excellent tracer for detecting upwelling and changes in seawater circulation since deep water has a lower $\Delta^{14}C$ value than surface water [i.e., Druffel and Griffin, 1993; Fallon et al., 2003; Grottoli et al., 2003; Grumet et al., 2004; Guilderson and Schrag, 1998a]. For example, in the eastern equatorial Pacific Ocean, increased upwelling or increases in the proportion of deep water transported to the surface results in a decrease in the $\Delta^{14}C$ of the skeleton, making corals in these locations excellent recorders of upwelling and of changes in upwelling regimes. $\Delta^{14}C$ evidence from the Galapagos indicates that there was a major shift in the source of upwelling water in 1976 when the decadal-scale Pacific Decadal Oscillation [PDO, Mantua et al., 1997] switched from a negative to a positive phase [Guilderson and Schrag, 1998b]. Work by Grottoli et al. [2003] shows that a PDO switch from positive to negative in the late 1940s also appears to be associated with a major change in the source water upwelling in the central equatorial Pacific. At Rarotonga, a $\Delta^{14}C$ seasonal variation of 10-15‰ indicates that vertical mixing occurs each year [Guilderson et al., 2000]. Combining coral $\Delta^{14}C$ data from Rarotonga, Galapagos, and the Solomon Sea, Guilderson et al. [2004] constructed a mixing model that shows variations in the amount of eastern Pacific water entering the Solomon Sea both on El Niño and decadal timescales.

Method: Approximately 7-10 mg of coral or sclerosponge carbonate powder is acidified under vacuum to produce CO_2 gas. For analysis by accelerator mass spectrometry (AMS), the CO_2 is reduced with hydrogen gas on iron or cobalt metal catalyst to produce a graphite target that is analyzed by AMS [Vogel et al., 1987]. For gas counting analyses, the CO_2 is measured directly. In all cases, results are reported as $\Delta^{14}C$ (the per mil deviation of $^{14}C/^{12}C$ of the sample relative to that of 95% Oxalic Acid-1 standard) [Stuiver and Polach, 1977]. All $\Delta^{14}C$ values are corrected for fractionation to a $\delta^{13}C$ of –25‰.

Anthropogenic Carbon Input Rates Into the Tropical Ocean

Estimates of the uptake of CO_2 by the oceans and of the imbalance between air-sea CO_2 reservoirs contain a large degree of uncertainty that greatly influences our understanding of oceanic CO_2 uptake. This is largely attributed to a lack of instrumental data on seawater DI-$\delta^{13}C$ and isotopic air-sea disequilibrium [Gruber and Keeling, 2001; Heimann and Maier-Reimer, 1996; Kortzinger et al., 2003; Quay et al., 1992; Quay et al., 2003; Tans et al., 1993]. Local carbon inventory changes can be reliably estimated by monitoring the change in DI-$\delta^{13}C$ but not as well by DI-$\Delta^{14}C$ [Heimann and Maier-Reimer, 1996; Kheshgi et al., 1999]. An improved knowledge of DI-$\delta^{13}C$ inventory changes over space and time would decrease the uncertainty in CO_2 uptake models [Gruber and Keeling, 2001; Heimann and Maier-Reimer, 1996; Quay et al., 1992, 2003; Sonnerup et al., 1999; Tans et al., 1993]. In addition, the rate of DI-$\delta^{13}C$ change prior to 1970 is unknown from instrumental data. Sclerosponges, are uniquely suited to filling these data gaps when they are exposed to open ocean conditions.

Though not yet directly calibrated, *Ceratoporella nicholsoni* (Jamaica) and *Acanthochaetetes wellsi* (New Caledonia) sclerosponges appear to precipitate their skeletal $\delta^{13}C$ in isotopic equilibrium with the surrounding seawater [Böhm et al., 1996, 2002; Druffel and Benavides, 1986; Fallon et al., 2003; Lazareth et al., 2000; Wörheide et al., 1997]. These authors have found that the $\delta^{13}C$ of sclerosponges growing within the mixed-layer has decreased with decreases in atmospheric $\delta^{13}C_{CO_2}$ over the past few centuries (Suess Effect). In the Great Barrier Reef, Caribbean and New Caledonia, sclerosponge $\delta^{13}C$ decreased by 0.5-1.0% from 1800-1990 with most of the decrease occurring in the latter part of the past century [Böhm et al., 1996, 2002; Druffel and Benavides, 1986; Joachimski et al., 1995; Lazareth et al., 2000; Wörheide et al., 1997]. Regional variability in the decreases in sclerosponge $\delta^{13}C$ is indicative of differences in the proportional contribution of ocean-atmosphere CO_2 flux, mixed-layer depth, water mass mixing to the DI-$\delta^{13}C$ at each site, as well as chronological accuracy. However, none of these published records are annually resolved nor directly calibrated with seawater DI-$\delta^{13}C$, thus limiting our ability to fully and quantitatively interpret or detect any sub-decadal variability in sclerosponge $\delta^{13}C$. Once calibrated, sclerosponges have the potential to fill at least part of that data gap by providing a significant archive of quantitatively robust proxy DI-$\delta^{13}C$ for the past centuries throughout the tropical Pacific and across depths of several hundred meters.

Upwelling, Nutrients, and pH: Other Elemental Ratios and Analyses

In addition to Sr/Ca and U/Ca ratios, other elemental ratios that are measured for paleoceanographic reconstructions include: cadmium to calcium (Cd/Ca), barium to Ca (Ba/Ca), manganese to Ca (Mn/Ca), and magnesium to Ca (Mg/Ca). Since the focus of this review is on the more commonly reported coral and sclerosponge paleo-records, we will only briefly cover these other elemental records here. A large suite of analytical techniques exists for measuring these elemental ratios including ARS-1 (graphite furnace atomic absorption), ICP-MS (inductively coupled plasma mass spectrometry), and LA-MS (laser ablation mass spectrometry). Ba/Ca, Mn/Ca and Cd/Ca have been used as upwelling and/or nutrient proxies [Fallon et al., 1999; Reuer et al., 2003; Shen et al., 1987, 1991, 1992a,b] and Ba/Ca records have recently been found to be excellent recorders of river discharge in corals growing in the flood plume of rivers. McCulloch et al. [2003] used barium records as a proxy for river sediment discharge in a 250-year long Great Barrier Reef *Porites* and found substantial increases in the sediment content of river floods in the period since European settlement of Australia. Abram et al. [2003] used peaks in Mn, lanthanum, and yttrium as indications of an extreme outbreak of red tide that resulted in coral and fish deaths in Indonesia.

Shen et al. [1987, 1992a] pioneered the idea of using Cd/Ca as a paleo-upwelling recorder. Cd concentrations are higher in the deep ocean than in the surface ocean where it is biodepleted, so Cd appears to be a better recorder of upwelling than Ba because Cd input from land is relatively low compared to Ba. Cd/Ca records from corals not only show the frequency of upwelling events, but their relative duration and relative intensity. Further calibration development of this proxy would greatly enhance our ability to quantify upwelling and compare Cd/Ca records among corals and among locations.

Ocean acidification has recently been identified as a major concern for corals and other calcifying organisms [Kleypas et al., 2006, Kleypas and Langdon, this volume]. A new proxy with exciting possibilities is the use of Boron stable isotopes ($\delta^{11}B$) as a proxy for pH. Gaillardet and Allègre [1995] first tested this proxy in corals, but it has not yet been widely applied. Most recently, Hönisch et al. [2004] performed the necessary laboratory and field experiments to calibrate this new proxy and Pelejero et al. [2005] published the first multicentury pH reconstruction from a coral.

Lead and Other Metal Contaminants

Coral skeletons can trap particulate contaminants, providing a record of contaminant input into reefal waters through airborne or waterborne pathways. Early work by Dodge and Gilbert [1984] led to the application of this approach to reconstructing lead input into the atmosphere since 1850 (primarily from gasoline burning) using contamination of a long coral core from Bermuda and shorter cores from other locations around the globe [Shen and Boyle, 1987]. Detection of lead and other heavy metals has been used to measure changes in local- to regional-scale human activity and gasoline use at various locations, primarily as an indicator of airborne contamination [Desenfant et al. 2003; Medina-Elizalde et al., 2002; Reuer et al., 2003]. Shen et al. 1987 also found that cadmium not only can be used as an indicator of upwelling, but also can be used as an indicator of airborne industrial pollutants. In a different application, Fallon et al. [2002], used lead in coral skeletons to detect lead and other contaminant elements entering the water directly from mining activity in Papua New Guinea. The concentration of lead in the ocean steadily increased from 1880-1979 due to the combustion of leaded-gasoline. This lead concentration peak is unambiguous and has been used to date sclerosponge records in the Caribbean [Swart et al., 2002].

Runoff: Detection through Luminescent Banding

Luminescent banding has been found in corals from many locations. Initially, Isdale [1984] reported that fluorescent bands in coral skeletons were an indicator of flow from the Burdekin River. Many researchers began to apply this elsewhere, finding luminescence from non-riverine sources [Theodorou, 1995; Tudhope et al., 1996] and discovering that both fluorescent and phosphorescent compounds contribute to the signal [Wild et al., 2000]. Isdale et al. [1998] were able to reconstruct 83% of the annual (water year) variability of Burdekin River flow using two long coral records, and Lough et al. [2002] have found high correlations between several rivers in northeastern Australia and luminescence in nearshore corals along the Great Barrier Reef. Coral luminescence has been applied to precipitation and related variables in the Red Sea [Klein et al., 1990], the Arabian Sea [Tudhope et al., 1996], the Caribbean [Nyberg, 2002], Florida [Smith et al., 1989], the South China Sea [Peng et al., 2002], and Papua New Guinea [Scoffin et al., 1989; Tudhope et al., 1995].

While Isdale identified the fluorescent bands as humic and fulvic compounds resulting from decaying land plants, this has been debated by other authors. Barnes and Taylor [2001] concluded that changes in skeletal density caused fluorescent. Humic and fulvic acids from biological sources did not. Additionally, it was quite some time before this proxy was successfully applied to other locations. It appears that coral fluorescence only captures river flow under a narrow range of input rates, making it a very useful proxy, but only at a limited number of sites. Non-riverine sources may also produce luminescent bands in corals. Tudhope et al. [1996] found luminescence in corals from arid Arabian Sea corals, concluding that the bands were related to the breakdown of highly seasonal plankton blooms.

Coral Growth as an Indicator of Environmental Stress

While geochemical records are more frequently used for climatic reconstruction, coral growth continues to be used to detect broadly defined stress events. This usually involves the use of multiple cores from multiple colonies [Lough and Barnes, 1997]. However, single long records can prove useful in particular environments. The latter was the case of a centuries-long record of wind-induced mixing from Bermuda [Pätzold et al., 1999]. More generalized stress has been detected through analysis of coral growth, applying sclerochronology to indicate clear changes in corals' responses to parameters such as sediments, nutrients, temperature, and salinity. Hudson pioneered the use of coral growth bands as an indicator of environmental perturbations in the Florida Keys, finding distinct stress bands and multi-year growth reductions related to natural events such as cold winters and hurricanes and human events such as dredge and fill activities and railroad construction [Hudson, 1981; Hudson et al., 1976, 1989]. Dodge and Lang [1983] applied dendrochronological (tree-ring) methods to corals from the East Flower Gardens Bank and found both anticipated annual signals and also decreased growth in years with high discharge from the nearby Atchafalaya river.

These studies led to further use of stress banding as an indicator of disturbance. Eakin et al. [1994] measured the growth rate of a series of corals collected along the southeastern coast of Aruba both upstream and downstream of a major coastal oil refinery. They were able to relate clear changes in growth of the nearby and downstream corals with major changes in refinery operations, probably mostly an impact from sedimentation. However, increased sedimentation does not always result in reduced coral growth. Barnes and Lough [1999] found a decrease in coral growth in Papua New Guinea corals near a gold mine, but there was no relationship between growth and distance from the mine. They concluded that growth decreases resulted from regional changes and were not related to mine-based sediments. Eakin et al. [1993] looked at stress bands and growth rates in Persian Gulf corals and found evidence of stress during the 1991 Gulf War and the Iraqi burning of Kuwaiti oil fields. Coral bleaching also is suspected of reducing coral growth resulting in stress bands [Abram et al., 2003; Leder et al., 1991; also see Future Applications].

Fossil Corals

Recently, the use of fossil corals has provided windows into periods when past climates were either different from today or when similar conditions occurred in the distant past. Most prominently, these include snapshots of climatic variability in the Holocene and

Younger Dryas [Beck, 1997; Corrège et al., 2004; Felis et al., 2004; Gagan et al., 2004; McCulloch et al., 1996; Moustafa et al., 2000; Tudhope et al., 2001; Woodroffe and Gagan, 2000; Woodroffe et al., 2003; Yu et al., 2004;], the last glacial maximum [Gagan et al., 2004; Tudhope et al., 2001], or the last interglacial [Felis et al., 2004; McCulloch et al., 1999; McCulloch and Esat, 2000; Tudhope et al., 2001]. These have provided important, seasonally resolved insights into how the climate system operated during those periods. Unfortunately, they only provide windows of variability and are neither exactly dated (i.e., chronology is typically anchored within +/– 50 years), nor do they provide records continuous to modern times. In one case, work is progressing on a long, continuous coral proxy record of the last millennium. Cobb et al. [2003] spliced together fossil coral materials to develop a series of overlapping records for Palmyra Island, with the eventual hope to develop a continuous, millennial-scale, annually resolved tropical record. We hope other researchers will seek out additional locations where this approach may be applied.

Climatic Data for Retrospective Monitoring

Despite the value of coral skeletons as recorders of past climate and environmental stress, most funding and research have focused on reconstructing large-scale climate patterns. They typically have not been considered as necessary parts of monitoring programs. As a result, of over 200 coral paleoclimatic data sets in the holdings of the World Data Center for Paleoclimatology (http://www.ncdc.noaa.gov/paleo/corals.html) on 1 June 2006, only 10 were from areas of U.S. management interest. Fortunately, the U.S. National Oceanic and Atmospheric Administration (NOAA) recently developed an integrated program of monitoring coral reefs. The Coral Reef Watch [Liu, 2006; Strong et al., 2002, this volume; http://coralreefwatch.noaa.gov/) includes a wide range of observations on coral reefs including the use of coral paleoclimatic data to provide "retrospective" monitoring of reefs before monitoring was implemented [Eakin et al., in press]. Paleoclimatic records need to be included in monitoring programs in the future so that marine protected area management can benefit from paleo records.

Gaps and Challenges for the Future

While corals have proven their worth as reliable recorders of past environmental conditions, limitations exist due to the way that coral paleoclimatic records are frequently collected and analyzed. In general, laboratory precision is no longer a major issue [Lough, 2004]. The important limiting factors now tend to be the frequent use of single cores for analysis [Lough, 2004] and insufficient temporal resolution [Felis and Pätzold, 2004]. This is due to the potentially large range in isotopic [Grottoli, 1999, 2000, 2001] and Sr/Ca [Stephans et al., 2004] natural variability among coral heads, locations, species, and depth. Among-colony natural variability in other trace elements has not been investigated, but is likely to be important as well. It cannot be stated too strongly: multiple colonies need to be cored to help identify local perturbations that limit the reliability of geochemical proxies as recorders of climate. Ideally, this should involve multiple records from each of multiple (at least 3) colonies [Lough, 2004]. Additionally, much more information is available when records are analyzed at sub-seasonal resolutions. Felis and Pätzold [2004] discuss particular value of bimonthly resolution, which is probably the best balance between resolution and cost. While in some cases multiple old colonies are not available, both under replication and insufficient temporal resolution are frequently the result of budgetary restraints placed on the projects. Funding organizations need to realize that while paleo reconstructions from corals are quite cost-effective, they must be funded at levels that provide reliable records.

We can now unravel temperature and salinity from coral geochemical records with high fidelity. Still unsolved, however, is the ability to identify past bleaching events in coral geochemical or skeletal growth records. While we can determine levels of stress that we believe should be sufficient to cause bleaching, we cannot yet identify when an individual coral bleached. Such an ability would provide a tremendous increase in the information we can glean from corals relative to both their past history of bleaching, and the relationship between bleaching and natural climatic variability.

Some studies have addressed this question and yielded clues to possible markers. Leder et al. [1991] found distinct skeletal and isotopic signatures to bleaching events in *Montastraea annularis* from Florida. Most importantly, this study and a subsequent one using *Porites lutea* in Thailand [Allison et al., 1996] and P. compressa and M. capitata in Hawaii [Rodrigues and Grottoli, 2006] {note: This is a much more conclusive study than the previous two as well.} revealed that reduced calcification at the time of bleaching may limit the ability of bleached corals to record high temperature excursions. Both of these studies, and subsequent work from Australia and Japan [Suzuki et al., 2003] and Hawaii [Grottoli et al., 2004], revealed visible changes in $\delta^{13}C$ that probably were related to metabolic and dietary changes during and subsequent to bleaching. These studies indicate that skeletal stress bands, $\delta^{18}O$ and $\delta^{13}C$ changes may provide clues to past bleaching. However, all of these records are sufficiently variable that the solution to the problem is still unresolved.

More recent work on branching *Porites divericata* has revealed a shift in the ratio of certain trace metals in the skeleton of bleached colonies [Burr, 2002]. Burr found statistically significant differences in amounts of strontium, selenium and silver from bleached and unbleached *P. divericata*. Changes in selenium and silver may provide useful bleaching indicators that are independent of the $\delta^{18}O$ and Sr/Ca paleothermometers, but the work also raises some concern that Sr/Ca ratios may be modified through bleaching. Burr also found that the skeletal microstructure of the bleached corals has a "melted" appearance similar to that seen in bleached foraminifera [Toler and Hallock, 1998]. Such malformations may be related to the reduced calcification seen after bleaching and may provide distinct bleaching clues.

The concern over lost growth bands [per Leder et al., 1991] has prompted Halley and Hudson (pers. comm.) to compare luminescence bands to growth bands to try to identify loss of bands due to bleaching. Their preliminary work indicates no loss of bands between 1878 and 1986 in cores from Biscayne National Park, Florida, suggesting that bleaching is strictly a recent phenomenon at those sites.

Conclusions

The use of coral skeletons as recorders of paleoenvironmental information has come a long way in the thirty years since its inception. They continue to be used as proxies of past climate, especially large scale climate such as ENSO and the North Atlantic Oscillation. The use of corals as environmental monitors both as part of comprehensive programs and for detecting land-based stress is increasing, as are many other uses of coral paleodata. The use of sclerosponges is just gaining popularity, but work so far suggests that it will also become a valuable source of paleoclimatic information. Despite the use of these proxies so far, there are still many new areas that need to be explored and new applications left to be discovered.

Acknowledgments. We thank the many authors who have contributed their data to the World Data Center for Paleoclimatology in Boulder, CO. Without their participation in data sharing, this and many other publications would not be possible. We thank the NOAA/MASC Library for helping us find copies of publications. CME also thanks Wendy Gross for providing figures and other support. AG also thanks the Woodrow Wilson Foundation Early Career Fellowship grant, NSF grant #OCE-0610487, and the Mellon Foundation. The manuscript contents are solely the opinions of the author(s) and do not constitute a statement of policy, decision, or position on behalf of NOAA or the U. S. Government.

References

Abram, N. J., M. K. Gagan, M. T. McCulloch, J. Chappell, and W. S. Hantoro, Coral reef death during the 1997 Indian Ocean Dipole linked to Indonesian wildfires. *Science*, 301, 952-955, 2003.

Allison, N., A. W. Tudhope, and A. E. Fallick, Factors influencing the stable carbon and oxygen isotopic composition of *Porites lutea* coral skeletons from Phuket, South Thailand. *Coral Reefs*, 15, 43-57, 1996.

Barnes, D. J., Growth in colonial scleractinians. *Bulletin of Marine Science*, 23, 280-298, 1973.

Barnes, D. J., and J. M. Lough, *Porites* growth characteristics in a changed environment: Misima Island, Papua New Guinea. *Coral Reefs*, 18, 213-218, 1999.

Beck, J. W., R. L. Edwards, E. Ito, F. W. Taylor, J. Recy, F. Rougerie, P. Joannot, and C. Henin, Sea-surface temperature from coral skeletal strontium/calcium ratios, *Science*, 257, 644-647, 1992.

Benavides, L. M. and E. R. M. Druffel, Sclerosponge growth rate as determined by ^{210}Pd and D^{14}C chronologies. *Coral Reefs*. 4, 221-224, 1986.

Böhm, F., M. M. Joachimski, H. Lehnert, G. Morgenroth, W. Kretschmer, J. Vaceot, and W. C. Dullo, Carbon isotope records from extant Caribbean and South Pacific sponges: Evolution of δ^{13}C in surface water DIC. *Earth Planetary Science Letters*, 139, 291-303, 1996.

Böhm, F., A. Haase-Schramm, A. Eisenhauer, W-C. Dullo, M. M. Joachimski, H. Lehnert, and J. Reitner, Evidence for preindustrial variations in the marine surface water carbonate system from coralline sponges. *Geochemistry, Geophysics, Geosystems*, 3: doi:10.1029/2001GC000264, 2002.

Boiseau, M. A., and et al., Sr/Ca and δ^{18}O ratios measured from *Acropora nobilis* and *Porites lutea*: Is Sr/Ca paleothermometry always reliable?, *Comptes Rendues de L'Academie des Sciences Serie II FAscicule A-Sciences do la Terre et des Planetes*, 325, 747-752, 1997.

Buddemeier, R., J. Maragos, and Knutson, Radiographic studies of reef coral exoskeletons. *J. Exp. Mar. Biol. Ecol*, 14, 179-200, 1974.

Burr, S. A., Skeletal proxies for bleaching-related stress in a scleractinian coral. Geological Society of America Annual Meeting, 2002.

Cobb, K. M., C. D. Charles, H. Cheng, and R. L. Edwards, El Niño-Southern Oscillation and tropical Pacific climate during the last millennium. *Nature*, 424, 271-276, 2003.

Cole, J. E., and R. G. Fairbanks, The Southern Oscillation recorded in the delta ^{18}0 of corals from Tarawa atoll. *Paleoceanography*, 5, 669-683, 1990.

Cole, J. E., R. B. Dunbar, T. R. McClanahan, N. A. Muthiga. Tropical Pacific forcing of decadal SST variability in the western Indian Ocean over the past two centuries. *Science*, 287, 617-619, 2000.

Corrège, T., T. Delcroix, J. Récy, W. Beck, G. Cabioch, and F. Le Cornec, Evidence for stronger El Niño-Southern Oscillation (ENSO) events in a mid-Holocene massive coral. *Paleoceanography*, 15, 465-470, 2000.

Corrège, T., M. K. Gagan, J. W. Beck, G. S. Burr, G. Cabioch, and F. L. Cornec, Interdecadal variation in the extent of South Pacific tropical waters during the Younger Dryas event. *Nature*, 428, 927-929, 2004.
Desenfant, F., G. F. Camoin, and A. Veron, Pollutant lead transport and input to the Caribbean during the 20th century. *Journal de Physique IV*, 107, 369-372, 2003.
deVilliers, S., B. K. Nelson, and A. R. Chivas, Biological controls on coral Sr/Ca and $\delta^{18}O$ reconstructions of sea surface temperatures. *Science*, 269, 1995.
Dodge, R. E., and T. R. Gilbert, Chronology of lead pollution contained in banded coral skeletons. *Marine Biology*, 82, 9-13, 1984.
Dodge, R. E., and J. C. Lang, Environmental correlates of hermatypic coral (*Montastrea annularis*) growth on the East Flower Gardens Bank, northwest Gulf of Mexico. *Limnology and Oceanography*, 28, 228-240, 1983.
Druffel, E. M., Radiocarbon in annual coral rings from the eastern tropical Pacific Ocean. *Geophysical Research Letters*, 8, 59-62, 1981.
Druffel, E. R. M., Geochemistry of corals: Proxies of past ocean chemistry, ocean circulation, and climate. *Proceedings of the National Academy of Sciences of the United States of America*, 94, 8354-8361, 1997.
Druffel, E. R. M., and L. M. Benavides, Input of excess CO_2 to the surface ocean based on $^{13}C/^{12}C$ ratios in a banded Jamaican sclerosponge. *Nature*, 321, 58-61, 1986.
Druffel, E. R. M., and S. Griffin, Large variations of surface ocean radiocarbon: evidence of circulation changes in the southwestern Pacific. *Journal of Geophysical Research-Oceans*, 98, 20,249-20, 20249-20259, 1993.
Druffel, E. R. M., and S. Griffin, Variability of surface ocean radiocarbon and stable isotopes in the southwestern Pacific. *Journal of Geophysical Research-Oceans*, 104, c10, 23607-23613, 1999.
Druffel, E. M., and T. W. Linick, Radiocarbon in annual coral rings of Florida. *Geophysical Research Letters*, 5, 913-916, 1978.
Druffel, E. R. M., S. Griffin, T. P. Guilderson, M. Kashgarian, J. Southon, and D. P. Schrag, Changes of subtropical North Pacific radiocarbon and correlation with climate variability. *Radiocarbon*, 43, 15-25, 2001.
Dunbar, R. B., G. M. Wellington, M. W. Colgan, and P. W. Glynn, Eastern Pacific sea surface temperature since 1600 A.D.: The $\delta^{18}O$ record of climate variability in Galapagos corals. *Paleoceanography*, 9, 291-315, 1994.
Eakin, C. M., J. S. Feingold, and P. W. Glynn, Oil refinery impacts on coral reef communities in Aruba, N.A. in *Proceedings of the Colloquium on Global Aspects of Coral Reefs: Health, Hazards and History, 1993*, edited by Ginsberg, R. N., pp. 139-145, University of Miami, Miami, 1994.
Eakin, C. M., M. L. Reaka-Kudla, and M. A. Chen, Growth and bioerosion of coral in the ROPME Sea Area following the 1991 Gulf Oil Spill. In *Scientific Workshop on Results of the R/V Mt. Mitchell Cruise*, Regional Organization for Protection of the Marine Environment, Kuwait, Kuwait City, 1993.
Eakin, C. M., P. K. Swart, T. M. Quinn, K. P. Helmle, J. M. Smith, and R. E. Dodge. Application of paleoclimatology to coral reef monitoring and management. *Proceedings of the 10th International Coral Reef Symposium, Okinawa*, in press, 2005.
Epstein, S., R. Buchsbaum, H. A. Lowenstam, and H. C. Urey, Revised carbonate-water isotopic temperature scale. *Bulletin of the Geological Society of America*, 64, 1315-1326, 1953.
Fairbanks, R. G., and R. E. Dodge, Annual periodicity of the $^{18}O/^{16}O$ and $^{13}C/^{12}C$ ratios in the coral *Montastrea annularis*, *Geochimica et Cosmochimica Acta*, 43, 1009-1020, 1979.
Fallon, S. J. M., M. T. McCulloch, R. v. Woesik, and D. J. Sinclair, Corals at their latitudinal limits: laser ablation trace element systematics in *Porites* from Shirigai Bay. Japan, *Earth and Planetary Science Letters*, 172, 221-238, 1999.

Fallon, S. J., J. C. White, and M. T. McCulloch, *Porites* corals as recorders of mining and environmental impacts: Misima Island, Papua New Guinea. *Geochimica et Cosmochimica Acta*, 66, 45-62, 2002.

Fallon, S. J., T. P. Guilderson, and K. Caldeira, Carbon isotope constraints on vertical mixing and air-sea CO_2 exchange. *Geophysical Research Letters*, 30, 2289, 2003.

Felis, T., and J. Pätzold, Climate reconstructions from annually banded corals. In *Global Environmental Change in the Ocean and on Land*, edited by Shiyomi, M. et al., pp. 205-227, TERRAPUB, 2004.

Felis, T., J. Pätzold, Y. Loya, and G. Wefer, Vertical water mass mixing and plankton blooms recorded in skeletal stable carbon isotopes of a Red Sea coral. *Journal of Geophysical Research*, 103, 30731-30739, 1998.

Felis, T., J. Pätzold, Y. Loya, M. Fine, A. H. Nawar, and G. Wefer, A coral oxygen isotope record from the northern Red Sea (documenting NAO, ENSO, and North Pacific teleconnections) on Middle East climate variability since the year 1750. *Paleoceanography*, 15(6), 679-694, 2000.

Felis, T., G. Lohmann, H. Kuhnert, S. J. Lorenz, D. Scholz, J. Pätzold, S. A. Al-Rousan, and S. M. Al-Moghrabi, Increased seasonality in Middle East temperatures during the last interglacial period. *Nature*, 429, 164-168, 2004.

Gagan, M. K., L. K. Ayliffe, D. Hopley, J. A. Cali, G. E. Mortimer, J. Chappell, M. T. McCulloch, and M. J. Head, Temperature and surface-ocean water balance of the mid-Holocene tropical western Pacific. *Science,* 279: 1014-1018, 1998.

Gagan, M. K., L. K. Ayliffe, J. W. Beck, J. E. Cole, E. R. M. Druffel, R. B. Dunbar, and D. P. Schrag, New views of tropical paleoclimates from corals. *Quaternary Science Reviews*, 19, 45-64, 2000.

Gagan, M. K., E. J. Hendy, S. G. Haberle, and W. S. Hantoro, Post-glacial evolution of the Indo-Pacific Warm Pool and El Nino-Southern Oscillation. *Quaternary International*, 118-19, 127-143, 2004.

Gaillardet, J. and C. J. Allègre, Boron isotopic compositions of corals: seawater or diagenesis record? *Earth Planetary Science Letters*, 136, 665-676, 1995.

Goodkin, N. F., K. A. Hughen, A. L. Cohen, and S. R. Smith, Record of Little Ice Age sea surface temperatures at Bermuda using a growth-dependent calibration of coral Sr/Ca. *Paleoceanography*, 20, PA4016, doi:10.1029/2005PA001140, 2005.

Grottoli, A. G., Variability of stable isotopes and maximum linear extension in reef-coral skeletons at Kaneohe Bay, Hawaii. *Marine Biology*, 135, 437-449, 1999.

Grottoli, A. G., Stable carbon isotopes ($\delta^{13}C$) in coral skeletons. *Oceanography*, 13, 93-97, 2000.

Grottoli, A. G., Climate: Past climate from corals. in *Encyclopedia of Ocean Sciences*, edited by Steele, J., S. Thorpe and K. Turekian, pp. 2098-2107, Academic Press. London, 2001.

Grottoli, A. G., Effect of light and brine shrimp on skeletal $\delta^{13}C$ in the Hawaiian coral *Porites compressa*: a tank experiment. *Geochimica et Cosmochimica Acta*, 66, 1955-1967, 2002.

Grottoli, A. G., Monthly resolved stable oxygen isotope record in a Palauan sclerosponge *Acanthocheatetes wellsi* for the period 1977-2001. *Proceedings of the 10th International Coral Reef Symposium, Okinawa*, 572-579, 2006.

Grottoli, A. G., and G. M. Wellington, Effect of light and zooplankton on skeletal $\delta^{13}C$ values in the eastern Pacific corals *Pavona clavus* and *Pavona gigantea. Coral Reefs*, 18, 29-41, 1999.

Grottoli, A. G., S. T. Gille, E. R. M. Druffel, and R. D. B. Dunbar, Decadal timescale shift in a central equatorial Pacific coral radiocarbon record. *Radiocarbon*, 45, 91-99, 2003.

Grottoli, A. G., L. J. Rodrigues, and C. Juarez, Lipids and stable carbon isotopes in two species of Hawaiian corals, *Porites compressa* and *Montipora verrucosa*, following a bleaching event., *Marine Biology*, 145, 621-631, 2004.

Gruber, N., and C. D. Keeling, An improved estimate of the isotopic air-sea disequilibrium of CO_2: Implications for the oceanic uptake of anthropogenic CO_2. *Geophysical Research Letters*, 28, 555-558, 2001.

Grumet, N. S., N. J. Abram, J. W. Beck, R. B. Dunbar, M. K. Gagan, T. P. Guilderson, W. S. Hantoro, and B. W. Suwargadi, Coral radiocarbon records of Indian Ocean water mass mixing and wind-induced upwelling along the coast of Sumatra, Indonesia. *Journal of Geophysical Research-Oceans*, 109, 2004.

Guilderson, T. P., and D. P. Schrag, Radiocarbon variability in the western equatorial Pacific inferred from a high-resolution coral record from Nauru Island. *Journal of Geophysical Research*, 103, 24,641-624,650, 1998a.

Guilderson, T. P., and D. P. Schrag, Abrupt shift in subsurface temperatures in the tropical Pacific associated with changes in El Niño. *Science*, 281, 240-243, 1998b.

Guilderson, T. P., D. P. Schrag, E. Goddard, M. Kashgarian, G. M. Wellington, and B. K. Linsley, Southwest subtropical Pacific surface water radiocarbon in a high-resolution coral. *Radiocarbon*, 42, 249-256, 2000.

Guilderson, T. P., D. P. Schrag, and M. A. Cane, Surface water mixing in the Solomon Sea as documented by a high-resolution coral ^{14}C record. *Journal of Climate*, 17, 1147-1156, 2004.

Haase-Schramm, A., F. Böhm, A. Eisenhauer, W. C. Dullo, M. M. Joachimski, B. Hansen, and J. Reitner, Sr/Ca ratios and oxygen isotopes from sclerosponges: Temperature history of the Caribbean mixed layer and thermocline during the Little Ice Age. *Paleoceanography*, 18, 2003.

Halley R. B., P. W. Swart, R. E. Dodge, and J. H. Hudson, Decade-scale trend in seawater salinity revealed through delta ^{18}O of *Montastrea annularis* annual growth bands. *Bulletin of Marine Science*, 54, 670-678, 1994.

Heimann, M., and E. Maier-Reimer, On the relations between the oceanic uptake of CO_2 and its carbon isotopes. *Global Biogeochemical Cycles*, 10, 89-110, 1996.

Hendy, E. J., M. K. Gagan, C. A. Alibert, M. T. McCulloch, J. M. Lough, and P. J. Isdale, Abrupt decrease in tropical Pacific Sea surface salinity at end of Little Ice Age. *Science*, 295, 1511-1514, 2002.

Hönisch, B., N. G. Hemming, A. G. Grottoli, A. Amat, G. N. Hanson, and J. Bijma, Assessing scleractinian corals as recorders for paleo-pH: Empirical calibration and vital effects, *Geochimica et Cosmochimica Acta,* 68, 3675-3685, 2004.

Hudson, J. H., Growth rates in *Montastrea annularis*: a record of environmental change in Key Largo Coral Reef Marine Sanctuary, Florida. *Bulletin of Marine Science*, 31, 444-459, 1981.

Hudson, J. H., E. A. Shinn, R. B. Halley, and B. Lidz, Sclerochronology: a tool for interpreting past environments. *Geology*, 4, 361-364, 1976.

Hudson, J. H., G. V. N. Powell, M. B. Robblee, and T. J. Smith III, A 107-year-old coral from Florida Bay: barometer of natural and man-induced catastrophes?. *Bulletin of Marine Science*, 44, 283-291, 1989.

Hughes, G. B. and C. W. Thayer, Sclerosponges: potential high-resolution recorders of marine paleotemperatures. In *Geological perspectives of global climate change*, edited by Gerhard, L. C., W. E. Harrison and B. M. Hanson, pp. 137-151, AAPG Studies in Geology, Tulsa, OK, 2001.

Isdale, P., Fluorescent bands in massive corals record centuries of coastal rainfall. *Nature*, 310, 578-579, 1984.

Isdale, P. J., B. J. Stewart, and J. M. Lough, Palaeohydrological variation in a tropical river catchment: a reconstruction using fluorescent bands in corals of the Great Barrier Reef, Australia. *The Holocene*, 8, 1-8, 1998.

Joachimski, M. M., F. Böhm, and H. Lehnert, Longterm isotopic trends from Caribbean desmosponges: evidence for isotopic disequilibrium between surface waters and atmosphere. *Proceedings of the 2nd European Regional Meeting ISRS*, 29, 141-147, 1995.

Kheshgi, H. S., A. K. Jain, and D. J. Wuebbles, Model-based estimation of the global carbon budget and its uncertainty from carbon dioxide and carbon isotope records. *Journal of Geophysical Research*, 104, 31127-31143, 1999.

Kim, S. -T., and J. R. O'Neil, Temperature dependence of δ^{18}O. *Geochimica et Cosmochimica Acta*, 61, 3461-3475, 1997.

Klein, R., Y. Loya, G. Gvirtzman, P. Isdale, and M. Susic, Seasonal rainfall in the Sinai Desert during the late Quaternary inferred from fluorescent bands in fossil corals. *Nature*, 345, 145-150, 1990.

Kleypas, J. A., and C. Langdon, Chapter 5: Coral Reefs and Changing Seawater Chemistry, *Corals and Climate Change, Geophysical Monographs,* American Geophysical Union, Washington, DC, this volume.

Kleypas, J. A., R. A. Feely, V .J. Fabry, C. Langdon, C. L. Sabine, and L .L. Robbins. Impacts of Ocean Acidification on Coral Reefs and Other Marine Calcifiers: A Guide for Future Research, report of a workshop held 18-20 April 2005, St. Petersburg, FL, sponsored by NSF, NOAA, and the U.S. Geological Survey, pp. 88, 2006.

Knutson, D. W., R. W. Buddemeier, and S. V. Smith, Coral chronometers: seasonal growth bands in reef corals. *Science*, 177, 270-272, 1972.

Kortzinger, A., P. D. Quay, and R. E. Sonnerup, Relationship between anthropogenic CO_2 and the ^{13}C Suess effect in the North Atlantic Ocean, *Global Biogeochemical Cycles*, 17, 1005, doi:1010.1029/2001GB001427, 002003, 2003.

Lazareth, C., P. Willenz, J. Navez, E. Keppens, F. Dehairs, and L. Andre, Sclerosponges as a new potential recorder of environmental changes: Lead in *Ceratoporella nicholsoni*, *Geology*, 28, 515-518, 2000.

Le Bec, N., A. Juillet-Leclerc, T. Corrège, D. Blamart, and T. Delcroix, A coral δ^{18}O record of ENSO driven sea surface salinity variability in Fiji (south-western tropical Pacific). *Geophysical Research Letters*, 27, 3897-3900, 2000.

Leder, J. J., A. M. Szmant, and P. K. Swart, The effect of prolonged "bleaching" on skeletal banding and stable isotope composition in *Montastrea annularis*. Preliminary observations. *Coral Reefs*, 10, 19-27, 1991.

Linsley, B., R. Messier, and R. Dunbar, Assessing between colony oxygen isotope variability in the coral *Porites lobata* at Clipperton Atoll. *Coral Reefs*, 18, 13-27, 1999.

Linsley, B. K., G. M. Wellington, D. P. Schrag, L. Ren, M. J. Salinger, and A. W. Tudhope, Geochemical evidence from corals for changes in the amplitude and spatial pattern of South Pacific interdecadal climate variability over the last 300 years, *Climate Dynamics*, 22, 1-11, 2004.

Liu, G., A. E. Strong, W. Skirving, L. F. Arzayus. Overview of NOAA Coral Reef Watch Program's near-real-time satellite global coral bleaching monitoring activities. *Proceedings of the 10th International Coral Reef Symposium, Okinawa*, in press, 2005.

Lough, J. M., A strategy to improve the contribution of coral data to high-resolution paleoclimatology, *Palaeogeography, Palaeoclimatology and Palaeoecology*, 204, 115-143, 2004.

Lough, J. M., and D. J. Barnes, Several centuries of variation in skeletal extension, density and calcification in massive *Porites* colonies from the Great Barrier Reef: a proxy for seawater temperature and a background of variability against which to identify unnatural change. *Journal of Experimental Marine Biology and Ecology*, 211, 29-67, 1997.

Lough, J. M., D. J. Barnes, and F. A. McAllister, Luminescent lines in corals from the Great Barrier Reef provide spatial and temporal records of reefs affected by land runoff. *Coral Reefs*, 21, 333-343, 2002.

Mantua, N. J., S. R. Hare, Y. Zhang, J. M. Wallace, and R. C. Francis, A Pacific interdecadal climate oscillation with impacts on salmon production. *Bull. Amer. Meteorol. Soc.*, 78, 1069-1079, 1997.

Marshall, J. F., and M. T. McCulloch, An assessment of the Sr/Ca ratio in shallow water hermatypic corals as a proxy for sea surface temperature. *Geochimica et Cosmochimica Acta*, 66, 3263-3280, 2002.

McConnaughey, T., ^{13}C and ^{18}O isotopic disequilibrium in biological carbonates: I. Patterns. *Geochimica et Cosmochimica Acta*, 53, 151-162, 1989a.

McConnaughey, T., ^{13}C and ^{18}O isotopic disequilibrium in biological carbonates: II. *In vitro* simulation of kinetic isotope effects. *Geochimica et Cosmochimica Acta*, 53, 163-171, 1989b.

McCulloch, M. T., and T. Esat, The coral record of last interglacial sea levels and sea surface temperatures. *Chemical Geology*, 169, 107-129, 2000.

McCulloch, M. T., M. K. Gagan, G. E. Mortimer, A. R. Chivas, and P. J. Isdale, A high-resolution Sr/Ca and $\delta^{18}O$ coral record from the Great Barrier Reef, Australia, and the 1982-1983 El Niño. *Geochimica et Cosmochimica Acta*, 58, 2747-2754, 1994.

McCulloch, M., G. Mortimer, T. Esat, L. Xianhua, B. Pillans, and J. Chappell, High resolution windows into early Holocene climate: Sr/Ca coral records from the Huon Peninsula. *Earth and Planetary Science Letters*, 138, 169-178, 1996.

McCulloch, M. T., A. W. Tudhope, T. M. Esat, G. E. Mortimer, J. Chappell, B. Pillans, A. R. Chivas, and A. Omura, Coral record of equatorial sea-surface temperatures during the penultimate deglaciation at Huon Peninsula. *Science*, 283, 202-204, 1999.

McCulloch, M., S. Fallon, T. Wyndham, E. Hendy, J. Lough, and D. Barnes, Coral record of increased sediment flux to the inner Great Barrier Reef since European settlement. *Nature*, 421, 727-730, 2003.

Medina-Elizalde, M., G. Gold-Bouchot, and V. Ceja-Moreno, Lead contamination in the Mexican Caribbean recorded by the coral *Montastrea annularis* (Ellis and Solander). *Marine Pollution Bulletin*, 44, 2002.

Meibom, A., J. Cuif, F. Hillion, B. R. Constantz, A. Juillet-Leclerc, Y. Dauphin, T. Watanabe, and R. B. Dunbar, Distribution of magnesium in coral skeleton, *Geophysical Research Letters*, 31, L23306, doi:10.1029/2004GL021313, 2004.

Min, G. R., R. L. Edwards, F. W. Taylor, J. Recy, C. D. Gallup, and J. W. Beck, Annual cycles of U/Ca in coral skeletons and U/Ca thermometry. *Geochimica et Cosmochimica Acta*, 59, 2025-2042, 1995.

Mitsuguchi, T., E. Matsumoto, O. Abe, T. Uchida, and P. J. Isdale, Mg/Ca thermometry in coral skeletons. *Science*, 274, 961-963, 1996.

Moore, M. D., C. D. Charles, J. L. Rubenstone, and R. G. Fairbanks, U/Th-dated sclerosponges from the Indonesian Seaway record subsurface adjustments to west Pacific winds. *Paleoceanography*, 15, 404-416, 2000.

Moustafa, Y. A., J. Pätzold, Y. Loya, and G. Wefer, Mid-Holocene stable isotope record of corals from the northern Red Sea. *International Journal of Earth Sciences*, 88, 742-751, 2000.

Nyberg, J., Luminescence intensity in coral skeletons from Mona Island in the Caribbean Sea and its link to precipitation and wind speed. *Philosophical Transactions of the Royal Society of London Series a-Mathematical Physical and Engineering Sciences*, 360, 749-766, 2002.

Pätzold, J., T. Bickert, B. Flemming, H. Grobe, and G. Wefer, Holozänes klima des Nordatlantiks rekonstruiert aus massiven korallen von Bermuda. *Natur und Museum*, 129, 165-177, 1999.

Pelejero, C., E. Calvo, M. T. McCulloch, J. F. Marshall, M. K. Gagan, J. M. Lough, and B. N. Opdyke, Preindustrial to modern iterdecadal variability in coral reef pH, *Science,* 309: 2204-2207, 2005.

Peng, Z. C., X. X. He, Z. F. Zhang, J. Zhou, L. S. Sheng, and H. Gao, Correlation of coral fluorescence with nearshore rainfall and runoff in Hainan Island, South China Sea. *Progress in Natural Science*, 12, 41-44, 2002.

Quadrelli, R., and J. M. Wallace, A simplified linear framework for interpreting patterns of Northern Hemisphere wintertime climate variability. *Journal of Climate*, 17, 3728-3744, 2004.

Quay, P. D., B. Tilbrook, and C. S. Wong, Oceanic uptake of fossil fuel CO_2: Carbon-13 evidence. *Science*, 256, 74-79, 1992.

Quay, P. D., R. Sonnerup, T. Westby, J. Stutsman, and A. McNichol, Changes in the $^{13}C/^{12}C$ of dissolved inorganic carbon in the ocean as a tracer of anthropogenic CO_2 uptake. *Global Biogeochemical Cycles*, 71, doi:10.1029/2001GB001817, 2003.

Quinn, T. M., and D. E. Sampson, A multiproxy approach to reconstructing sea surface conditions using coral skeleton geochemistry. *Paleoceanography*, 17, 1062, PA000528, 2002.

Reitner, J., G. Gautret, and P. Gautret, Skeletal formation in the modern but ultraconservative chaetetid sponge *Spirastrella (Acanthochaetetes) wellsi* (Demospongiae, Porifera). *Facies*, 34, 193-208, 1996.

Ren, L., B. K. Linsley, G. M. Wellington, D. P. Schrag, and O. Hoegh-Guldberg, Deconvolving the $\delta^{18}O$ seawater component from subseasonal coral $\delta^{18}O$ and Sr/Ca at Rarotonga in the southwestern subtropical Pacific for the period 1726 to 1997. *Geochimica et Cosmochimica Acta*, 67, 1609-1621, 2002.

Reuer, M. K., E. A. Boyle, and J. E. Cole, A mid-twentieth century reduction in tropical upwelling inferred from coralline trace element proxies. *Earth and Planetary Science Letters*, 210, 437-452, 2003.

Rodrigues, L. J., and A. G. Grottoli, Calcification rate and the stable carbon, oxygen, and nitrogen isotopes in the skeleton, host tissue, and zooxanthellae of bleached and recovering Hawaiian corals. *Geochimica et Cosmochimica Acta*, 70, 2781-2789, 2006.

Rosenheim, B. E., P. K. Swart, S. R. Thorrold, A. Eisenhauer, and P. Willenz, Salinity change in the subtropical Atlantic: Secular increase and teleconnections to the North Atlantic Oscillation. *Geophysical Research Letters*, 32, L02603, doi:10.1029/2004GL021499, 2005.

Schrag, D. P., Rapid analysis of high-precision Sr/Ca ratios in scleractinian corals and other marine carbonates. *Paleoceanography*, 14, 97-102, 1999.

Scoffin, T. P., A. W. Tudhope, and B. E. Brown, Fluorescent and skeletal density banding in *Porites lutea* from Papua New Guinea and Indonesia. *Coral Reefs*, 7, 169-178, 1989.

Shen, G. T., and E. A. Boyle, Lead in corals: reconstruction of historical industrial fluxes to the surface ocean. *Earth and Planetary Science Letters*, 82, 289-304, 1987.

Shen, G. T., E. A. Boyle, and D. W. Lea, Cadmium in corals as a tracer of historical upwelling and industrial fallout. *Nature*, 328, 794-796, 1987.

Shen, G. T., and R. B. Dunbar, Environmental controls on uranium in reef corals. *Geochimica et Cosmochimica Acta*, 59, 2009-2024, 1995.

Shen, G. T., T. M. Campbell, R. B. Dunbar, G. M. Wellington, M. W. Colgan, and P. W. Glynn, Paleochemistry of manganese in corals from the Galapagos Islands. *Coral Reefs*, 10, 91-100, 1991.

Shen, G. T., J. E. Cole, D. W. Lea, L. J. Linn, T. A. McConnaughey, and R. G. Fairbanks, Surface ocean variability at Galápagos from 1936-1982: Calibration of geochemical tracers in corals. *Paleoceanography*, 563-588, 1992a.

Shen, G. T., L. J. Linn, T. M. Campbell, J. E. Cole, and R. G. Fairbanks, A chemical indicator of trade wind reversal in corals from the western tropical Pacific. *Journal of Geophysical Research, C, Oceans*, 97, 12689-12697, 1992b.

Sinclair D. J., L. P. J. Kinsley, and M. T. McCulloch, High resolution analysis of trace elements in corals by laser-ablation ICP-MS. *Geochimica et Cosmochimica Acta* 62, 1889-1901, 1998.

Slowey, N. C., and T. J. Crowley, Interdecadal variability of Northern Hemisphere circulation recorded by Gulf of Mexico corals. *Geophysical Research Letters*, 22, 2345-2348, 1995.

Smith, T. J., J. H. Hudson, M. B. Robblee, G. V. N. Powell, and P. J. Isdale, Freshwater flow from the Everglades to Florida Bay: A historical reconstruction based on fluorescent banding in the coral *Solonastrea bournoni*. *Bulletin of Marine Science*, 44, 374-282, 1989.

Solow, A. and A. Huppert, Optimal multiproxy reconstruction of sea surface temperature from corals, *Paleoceanography*, 19, PA4004, 2004.

Sonnerup, R. E., P. D. Quay, A. P. McNichol, J. L. Bullister, T. A. Westby, and H. L. Anderson, Reconstructing the oceanic ^{13}C Suess effect. *Global Biogeochemical Cycles*, 13, 857-872, 1999.

Stephans, C. L., T. M. Quinn, F. W. Taylor, and T. Corrège, Assessing the reproducibility of coral-based climate records. *Geophysical Research Letters*, 31, L18210, doi:10.1029/2004GL020343, 2004.

Strong, A. E., G. Liu, J. Meyer, J. C. Hendee, and D. Sasko. Coral Reef Watch 2002. *Bulletin of Marine Science, 75*, 259-268, 2004.

Strong, A. E., F. Arzayus, and W. Skirving, Chapter 8: Identifying coral bleaching remotely via Coral Reef Watch – Improved integration and implications for changing climate, *Corals and Climate Change, Geophysical Monographs,* American Geophysical Union, Washington, DC, this volume.

Stuiver, M., and H. A. Polach, Discussion reporting of ^{14}C data. *Radiocarbon*, 19, 355-363, 1977.

Suzuki, A., M. K. Gagan, K. Fabricius, P. J. Isdale, I. Yukino, and H. Kawahata, Skeletal isotope microprofiles of growth perturbations in *Porites* corals during the 1997-1998 mass bleaching event. *Coral Reefs*, 22, 357-369, 2003.

Swart, P. K., J. L. Rubenstone, C. D. Charles, and J. Reitner, Sclerosponges: A new proxy indicator of climate, *Rep. 12*, pp. 19, NOAA Climate and Global Change Program, 1998a.

Swart, P. K., K. S. White, D. Enfield, R. E. Dodge, and P. Milne, Stable oxygen isotopic composition of corals from the Gulf of Guinea as indicators of periods of extreme precipitation conditions in the sub-Sahara. *Journal of Geophysical Research*, 103, 27885-27891, 1998b.

Swart, P. K., G. Healy, L. Greer, M. Lutz, A. Saied, D. Anderegg, R. E. Dodge, and D. Rudnick, The use of proxy chemical records in coral skeletons to ascertain past environmental conditions in Florida Bay. *Estuaries*, 22, 384-397, 1999.

Swart, P. K., S. Thorrold, B. Rosenheim, A. Eisenhauer, C. G. A. Harrison, M. Grammer, and C. Latkoczy, Intra-annual variation in the stable oxygen and carbon and trace element composition of sclerosponges. *Paleoceanography*, 17, 1045, doi:1010.1029/2000PA000622, 2002.

Tans, P. P., J. A. Berry, and R. F. Keeling, Oceanic 13C/12C observations: A new window on ocean CO2 uptake. *Global Biogeochemical Cycles*, 7, 353-368, 1993.

Theodorou, N. K., The enigmatic properties of fluorescent banding in massive corals of the species *Porites lutea* from Phuket, Thailand, Ph.D., The University of Edinburgh, 1995.

Toler, S. K., and P. Hallock, Shell malformation in stressed *Amphistegina* populations: Relation to biomineralization and paleoenvironmental potential. *Marine Micropaleontology*, 34, 107-115, 1998.

Tudhope, A. W., G. B. Shimmield, C. P. Chilcott, M. Jebb, A. E. Fallick, and A. N. Dalgleish, Recent changes in climate in the far western equatorial Pacific and their relationship to the Southern Oscillation: oxygen isotope records from massive corals, Papua New Guinea. *Earth and Planetary Science Letters*, 136, 575-590, 1995.

Tudhope, A. W., D. W. Lea, G. B. Shimmield, C. P. Chilcott, and S. Head, Monsoon climate and Arabian sea coastal upwelling recorded in massive corals from southern Oman. *Palaios*, 11, 347-361, 1996.

Tudhope, A. W., C. P. Chilcott, M. T. McCulloch, E. R. Cook, J. Chappell, R. M. Ellam, D. W. Lea, J. M. Lough, and G. B. Shimmield, Variability in the El Niño–Southern Oscillation through a glacial-interglacial cycle. *Science*, 291, 1511-1517, 2001.

Vogel, J. S., D. E. Nelson, and J. R. Southon, ^{14}C background levels in an accelerator mass spectrometry system. *Radiocarbon*, 29, 323-333, 1987.

Weber, J. N., Incorporation of strontium into reef coral skeletal carbonate, *Geochimica et Cosmochimica*, 2173-2190, 1973.

Weber, J. N., and P. M. J. Woodhead, Temperature dependence of Oxygen-18 concentration in reef coral carbonates. *Journal of Geophysical Research*, 77, 463-473, 1972.

Weber, J. N., E. W. White, and P. H. Weber, Correlation of density banding in reef coral skeletons with environmental parameters: the basis for interpretations of chronological records preserved in the coralla of corals. *Paleobiology*, 1, 137-149, 1975.

Wellington, G. M., and R. B. Dunbar, Stable isotopic signature of El-Niño Southern Oscillation events in eastern tropical Pacific reef corals. *Coral Reefs*, 14, 5-25, 1995.

Wellington, G. M., G. Merlen, and R. B. Dunbar, Calibration of stable oxygen isotope signatures in Galapagos corals. *Paleoceanography*, 11, 467-480, 1996.

Wild, F. J., A. C. Jones, and A. W. Tudhope, Investigation of luminescent banding in solid coral: the contribution of phosphorescence. *Coral Reefs*, 19, 132-140, 2000.

Willenz, P. and W. D. Hartmen, Calcification rate of *Ceratoporella nicholsoni* (Porifera: sclerospongiae): an *in situ* study with calcein Proceedings of the 5th International Coral Reef Congress, Tahiti, pp. 113-118, 1985.

Woodroffe, C. D., and M. K. Gagan, Coral microatolls from the central Pacific record late Holocene El Nino. *Geophysical Research Letters*, 27, 1511-1514, 2000.

Woodroffe, C. D., M. R. Beech, and M. K. Gagan, Mid-late Holocene El Niño variability in the equatorial Pacific from coral microatolls. *Geophysical Research Letters*, 30, 2003.

Wörheide, G., P. Gautret, J. Reitner, F. Böhm, M. M. Joachimski, V. Thiel, W. Michaelis, and M. Massault, Basal skeletal formation, role and preservation of intracrystalline organic matrices, and isotopic record in the coralline sponge *Astrosclera willeyana* Lister, 1900, *Boletin De La Real Sociedad Espanola De Historia Natural. Seccion Geologica*, 91, 355-374, 1997.

Yu, K. F., J. X. Zhao, T. S. Liu, G. H. Wei, P. X. Wang, and K. D. Collerson, High-frequency winter cooling and reef coral mortality during the Holocene climatic optimum. *Earth and Planetary Science Letters*, 224, 143-155, 2004.

4

The Cell Physiology of Coral Bleaching

Sophie G. Dove and Ove Hoegh-Guldberg

Abstract

Coral bleaching is one of the most studied phenomena associated with coral reefs. In the broadest sense, it is described as a color change that occurs in response to a range of stressors. Coral bleaching has affected coral reefs in recent mass bleaching events which are due to episodes of warmer than normal sea temperature. These changes are linked to climate change and are exacerbated by light and to a lesser extent by flow and other stressors on coral reefs. While some of the details still remain to be filled in, the Photoinhibition Model of coral bleaching goes a long way to explaining why reef-building corals bleach, although the precise point of lesion during the initial steps remains debatable. Some caution is needed in terms of technology affecting our conclusion as to whether the host or the symbionts (or both) are the most critical element in mass bleaching episodes. Evidence for consistent physiological differences between the clades of *Symbiodinium* remains elusive, as does evidence of the ecological flexibility of symbiosis to changing environmental conditions. While "shuffling" [sensu Baker, 2003] of symbionts indeed occurs, major changes that result in evolutionarily novel symbionts (evolutionary "switching") are rare and do not occur at ecological meaningful time scales. Furthermore, it is not a forgone conclusion that space needs to be created within the coral house in order for "shuffling" to occur, and hence the link between bleaching and shuffling is somewhat tenuous.

Introduction

Climate change is projected to drive major changes in ocean temperature and acidity over this century [IPCC, 2001; Raven et al., 2005]. These conditions are likely to affect both the growth and survivorship of corals, a conclusion that is based on matching the conditions that coral reefs enjoy today and future conditions. This is a bleak future for coral reefs, which in turn depend on Scleractinian (reef-building) corals and their symbionts (dinoflagellates of the genus *Symbiodinium*). While many coral reef biologists agree that climate change poses a serious threat to coral reef ecosystems, there is some debate around the extent of the changes that are likely to occur. Key to resolving these debates is a better understanding of the cellular biology of symbiosis and the processes that lead to its dissociation via a process referred to as "bleaching". Proponents of the Adaptive Bleaching Hypothesis [or ABH; Buddemeier and Fautin, 1993; Baker, 2001], for example, proposed that thermally tolerant endosymbiotic dinoflagellates strains (clades, sub-clades,

Coral Reefs and Climate Change: Science and Management
Coastal and Estuarine Studies 61
Copyright 2006 by the American Geophysical Union.
10.1029/61CE05

sub-sub-clades) will replace (and are replacing) less thermally tolerant varieties of dinoflagellate in host cells, thereby imparting greater thermal tolerance to the holosymbiont (host and symbiont). Critical to this idea is whether or not symbiosis is flexible enough to swap one dinoflagellate or host partner for another. Some authors point to the lack of data to support this proposition and to recent high coral mortality rates associated with temperature anomalies [>3°C.month^{-1}, Hoegh-Guldberg, 1999; Hoegh-Guldberg, 2001]. Corals also exhibit a high degree of specificity between host and symbiont such that the flexibility of a host to take up (not simply shuffle pre-existing populations of dinoflagellates) "thermally tolerant" varieties of dinoflagellates is questionable. This specificity inhibits rapid adaptive improvements in the survival rates of symbiotic corals through the exchange of their symbionts, leaving adaptive change dependent on the somewhat slower genetic selection associated with the coral host.

These questions can only be resolved by a greater insight into the cellular processes that occur during the establishment and, under some circumstances, the disintegration of the symbiosis. This knowledge has the potential to assist us in understanding the ecological outcomes of bleaching, as well as assessing the potential for corals and their symbionts to undergo adaptive change. The combination of this information with higher levels of analysis represents a powerful approach to understanding the changes that are almost certain to occur in coral reefs as the temperature and acidity of tropical seas increases. These avenues of information will ultimately address the question: Will coral dominate reef systems be present fifty years time when the global ocean is considerably warmer and more acidic?

In this chapter, we will review the current understanding of the cellular mechanisms underpinning coral bleaching. In doing this, it becomes evident that the pathways are many and depend not only on the type of stress applied, but also the level and combination of stresses experienced. How these data are being interpreted is also critical. Most of the work so far, for example, has focused on the changes that occur within the dinoflagellate symbiont (*Symbiodinium* spp.), partially because technologies such as Pulsed Amplitude Modulated (PAM) Fluorescence (which is able to detect stress rapidly and non-invasively) make investigating this partner relatively easy. We will also reexamine the proposition that "symbiotic dinoflagellates are usually less heat tolerant than their host" [Fitt et al., 2001]. The host is a critical part of the association that has to some extent not received the attention that is due as we try to understand how changes to the conditions surrounding a coral will drive the disintegration of the symbiosis between host and symbiont. In the process, we will ask what we mean by "less heat tolerant" and the relevance of this notion to the various IPCC scenarios regarding future increases in SST [IPCC, 2001].

Coral Bleaching as a Sign of Distress

Coral bleaching occurs as endosymbiotic dinoflagellate pigmentation declines within the tissues of the host organism [Glynn, 1993], which may be due to the loss of cells and or pigmentation from the host. Typically a 50% or more loss in dinoflagellate pigmentation is required for this paling to be observable to the human eye [e.g., Jones, 1995]. Recently, loss of dinoflagellate pigmentation, below threshold values have been correlated with exponential increases in within tissue light intensity [Enriquez et al., 2005]. It maybe that the sudden increase in the internal light field correlates well with the pigment concentration density at which the white skeleton becomes observable by the human eye. Coral bleaching (observable or not) occurs on seasonal and local scales as well as on a global scale [Brown et al., 2000; Fagoonee et al., 1999; Fitt et al., 2000; Warner et al., 2002]. Coral bleaching may or may not lead to coral mortality, depending on the severity and the length of exposure of the coral to stress. Bleaching is also part broad response within corals and their symbionts to stress and may include changes in growth and reproduction. The environment in which the stress is applied may play a big role in determining the initial lesion point and the specific cascade of biological changes that follow. Similarly, the level of stress may dictate the variety of initial lesion points, some of which may be more detrimental to holosymbiont health than others.

Whilst visually observable mass bleaching events are triggered by elevated temperature (and light levels), there are a number of other factors that initiate coral bleaching. Exposure to low temperatures [Hoegh-Guldberg et al., 2005], elevated UV-B [Lesser, 1997], low salinity [Kerswell and Jones, 2003, Van Woesik et al., 1995], bacterial infections [Kushmaro et al., 1996], herbicides [Jones and Kerswell, 2003], exposure to chemical such as copper ions [Jones, 1997], cyanide [Jones and Hoegh-Guldberg, 1999; Cervino et al., 2003], and caffeine [Sawyer and Muscatine, 2001] will cause bleaching. For this reason, it is important to appreciate that bleaching is not a single disease *per se*, but is rather a generalized sign of distress by coral exposed to a range of physical, chemical and biological stressors.

Changing weather conditions at local scales may also be associated with the general changes to sea temperature. These in turn, may bring about changes to cloud conditions and hence the amount of solar irradiance falling on a reef. Light is central to the favored mechanisms for coral bleaching, with some of the first steps in thermal stress being considered to be associated with the ability of photosystems to process captured excitations [Jones et al., 1998]. Clear skies that are often associated with doldrums conditions lead to increased light intensities and as well as sea temperatures, leading to greater impacts of thermal stress on corals [Hoegh-Guldberg, 1999, Mumby et al., 2001]. These conditions may also drive changes to the flow rate of water across corals, which are also considered to contribute an additional stress to coral reefs during these still periods [Nakamura et al., 2003]. These factors (as discussed elsewhere, see chapter by Van Woesik, this volume) contribute variability within data sets being used to predict changes on coral reefs as reef warm.

Corals, Dinoflagellates and Light

Symbiodinium, the principle symbiont of reef-building corals, is photoautotrophic and uses a range of pigments to capture light. These pigments consist mainly of chlorophylls a and c_2, peridininin and the xanthophylls, diadinoxanthin and diatoxanthin [Olaizola et al., 1994]. These pigments are found in the water soluble peridinin-chl a-protein (PCP) and membrane bound chl a-chl c_2-peridinin-protein complexes (acpPC), both of which form the light harvesting complexes of dinoflagellates. Chl a is also associated the core reaction centres that are involved with charge separation [Iglesias-Prieto and Trench, 1997; Ruffle et al., 2001]. In most plants, increases in photon flux density trigger acclimative (*sensu*, photo-adaptative) changes to antennae pigment-protein complexes that reduces the number of photons harvested for photochemistry. The restructuring of the antennae that optimizes photosynthesis in high light fields results in a reduction in chlorophyll pigmentation (bleaching) for all organisms that have flexible antennae structures, and tends to result in death for those that do not. "Bleaching" in this sense is very much a mechanisms that extends the physiological range of light environments inhabitable by many photoautotrophic corals.

Steady state measurements of oxygen flux are used to determine the relationship between photosynthetic rates (P; µmol O_2 cm^{-2} h^{-1} or µmol O_2 Chl a^{-1} h^{-1}) and irradiance

(I; μmol photons m^{-2} s^{-1}) for corals experiencing an increase or decrease in photon flux provides information on the rate of photo-acclimation to changing light fields. The shape of the Photosynthesis versus Irradiance (P – E) curve is generally consistent across photosynthesizing organisms (Figure 1). Photosynthesis increases linearly until saturation light intensity (I_k) is attained at which point the rate of photosynthesis slows at it nears a maximum value (P_{max}). Shade acclimated corals tend to describe P – I curves with lower I_k values that reflect a more efficient use of available photons [Falkowski and Dubinsky, 1981, Hoegh-Guldberg and Smith, 1989, Anthony and Hoegh-Guldberg, 2003]. Shade acclimated corals also tend to have reduced P_{max} partially because they tend to exhibit slow turnover of carbon fixing enzymes and electron transfer components [Zvalinskii et al., 1980; Beardall and Morris, 1976, Falkowski, 1980].

Acclimation to light varies spatially and temporally within the one coral colony [Hill et al., 2004]. Light environments within a coral colony vary from deep shade to exposed sites, leading to some regions within a coral colony (branch bases) being acclimated to shade while others (e.g., branch tips) are acclimated to sun conditions. In some cases,

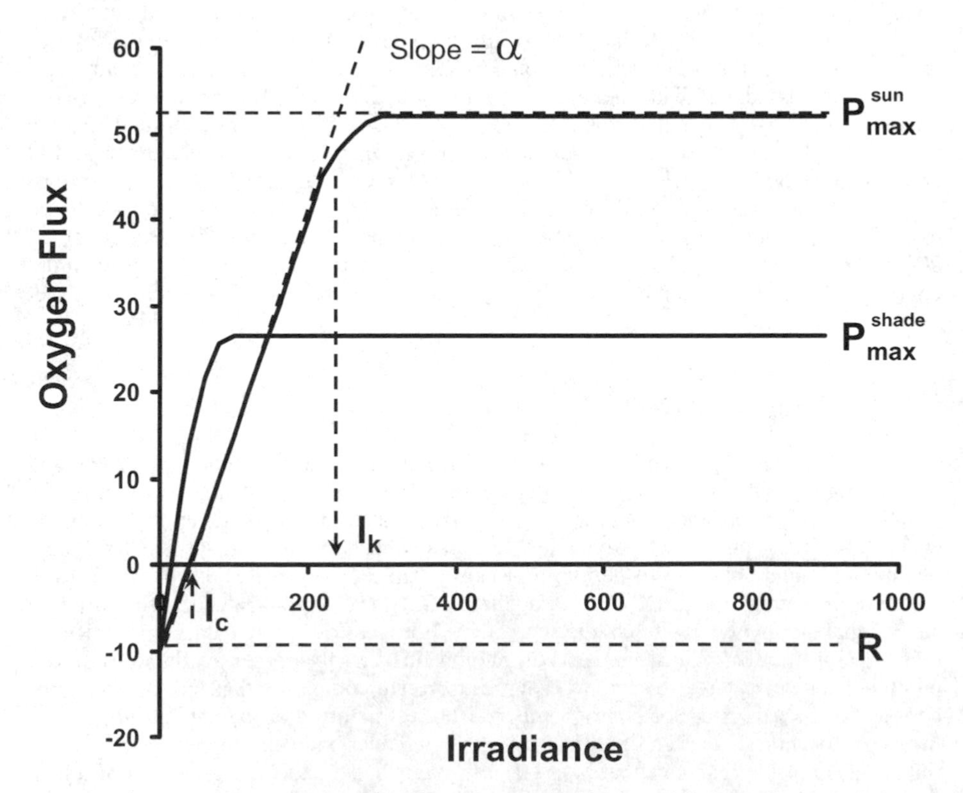

Figure 1. Classic photosynthesis vs irradiance curves for shade and light acclimated photosynthetic organisms. The curve demonstrates the relationship between photosynthetic rates (measured as oxygen flux) and irradiance (I; μmol photons m^{-2} s^{-1}). I_c; the compensation point at which the rate of respiration (R) is equal to the rate of photosynthesis (P). I_k; the saturation irradiance or the irradiance at which the linear part of the curve intersects P_{max}.

reduced relative photosynthetic electron transport based on Chl fluorescence techniques can occur in sites that receive very high light such as branch tips due to differential chlorophyll concentration (Figure 2). In a similar way, corals may adjust their photosynthetic behavior to light conditions over time. The acclimation rate of a coral to the opposing light fields has been explored using transplantation experiments. Falkowski and Dubinsky [1981] revealed that colonies of the coral *Stylophora pistillata* that were transferred from deep water in the Red Sea to the shallows had fully acclimated to increases in photon flux within a month. When corals were transferred from shallow to deep water, however, they took longer to acclimate to the reduced photon flux. In this case, as in the case of Hoegh-Guldberg and Smith [1989], acclimation was achieved by varying chlorophyll *a* per dinoflagellate cell rather than areal dinoflagellate densities. In both cases, visual paling of corals was observed for coral transplanted to higher light regimes. In another study, Anthony and Hoegh-Guldberg [2003] found that acclimation occurred readily within 5-10 days and took the same time in either direction for the coral *Turbinaria mesenterina* from a 2-5 m deep inshore reef that frequently experienced radical changes irradiance due to recurrent wind-driven sediment resuspension equivalent to 90% shading [Anthony and Larcombe, 2002].

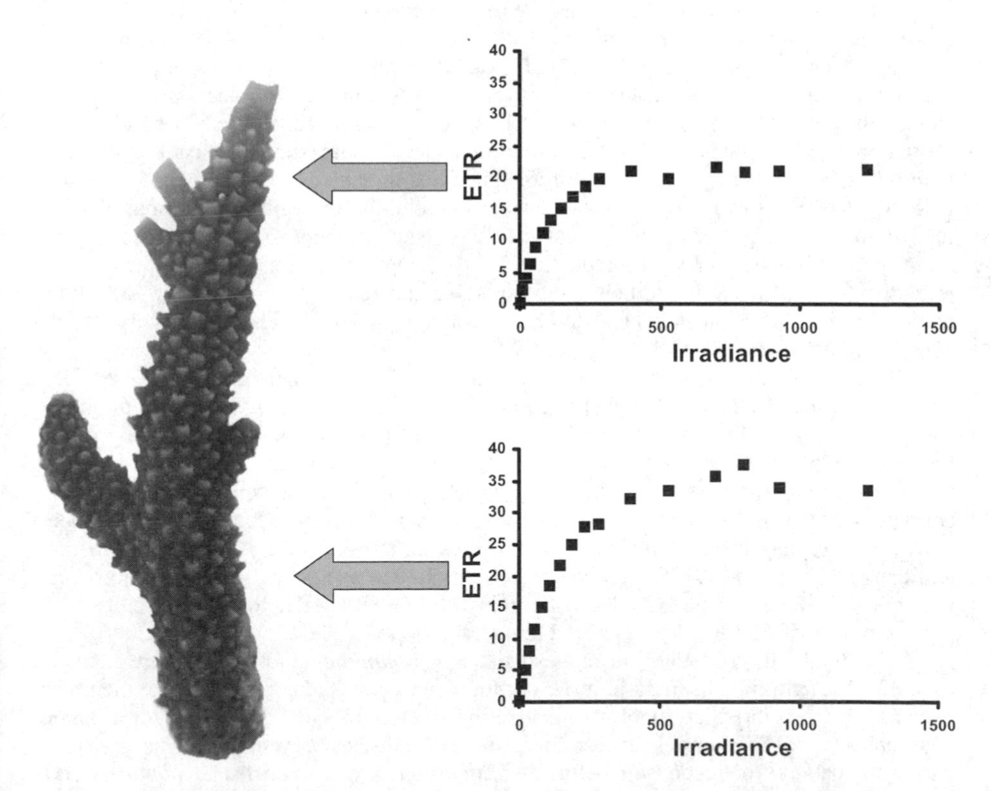

Figure 2. Relative electron transport rates vs irradiance curves determined by Imaging Pulse Amplitude Modulated Fluorometer (WALZ, Germany) comparing branch tip and based for the scleractinian coral *Acropora aspera*.

The compounding effect of light on thermal stress during bleaching was reported several decades ago: stressed colonies typically showed a greater tendency to bleach on their upper more sunlit surfaces when exposed to elevated temperatures [Goenaga and Canals, 1990, Hoegh-Guldberg and Smith, 1989]. Surveys have also reported that shade projected from high islands [e.g., Palau, Bruno et al., 2001] and clouds [e.g., Moorea, Mumby et al., 2001] reduced the extent of bleaching during the 1998 global event. These studies make sense in terms of the photoinhibition model of coral bleaching (Figure 3) in which thermal stress reduces the ability of *Symbiodinium* to process captured excitations, leading to the build-up of oxygen radicals. While many of the reactive oxygen species produced are disarmed by cytoplasmic and membrane bound oxidases, over-production can overwhelm the antioxidative pathways and cause damage within the cells of both symbiont and host. These impacts on cellular physiology probably lead to a series of processes in which symbionts are either disintegrated or dismantled (apoptosis) *in situ* [Dunn et al., 2002; Franklin et al., 2004] or are actively removed from the association [Hoegh-Guldberg and Smith, 1989]. The latter may occur by either expulsion of *Symbiodinium* directly or through the release of host endoderm cells that contain *Symbiodinium* [Gates et al., 1992].

Gates et al. [1992] have offered an alternative mechanism for thermal bleaching that places the sensitivity to heat with the host rather than the dinoflagellates. The mechanism proposes that elevated temperature directly results in the disassociation of host endodermal cell. Schmid et al. [1981] demonstrated this phenomenon in a variety of cnidarians. In this case, the temperature required for disassociation varied from 32-40°C dependant on developmental stage and organism studied. Cell adhesion is controlled by the phosphorylation of proteins involved in adhesion. Sawyer and Muscatine [2001] successfully demonstrated that caffeine which alters protein phosphorylation could be used to emulate the results obtained by Gates et al. [1992] for corals under thermal stress. The cautionary note here is that hosts have an upper thermal tolerance that may be exceeded in the near future if IPCC predictions eventuate. Even under the Photoinhibition Model of thermal bleaching, the primary target of thermal stress can lie outside the photosynthesizing endosymbiont, possibly associated with a reduction in host supplied key photosynthetic cofactors [e.g., Ca^{2+}; Bumann and Oesterhelt, 1995], or substrates [e.g., CO_2; Dunn et al., 2002]; or changes to host physiology that alter internal photon flux densities [PFD; Dove et al., 2006; Enriquez et al., 2005].

The Photoinhibition Model of coral bleaching leads to a number of ecological predictions, as outlined by Hoegh-Guldberg [1999, Table 1]. These include the effect of shade and position on a colony but also include the effect of tissue thickness, colony morphology and polyp behaviour, all of which affect the light environment of *Symbiodinium* and hence the amount of stress during bleaching. The proposed mechanism also suggests that coral species that have mechanisms (e.g., host pigmentation) by which they shade their *Symbiodinium* may also be more resilient to thermal stress [Salih et al., 1998; Hoegh-Guldberg and Jones, 1999]. The photo-protection afforded by host pigmentation, however, may not be sustained above 32-33°C for corals of the Great Barrier Reef as these proteins appear to be unstable *in vivo* at elevated temperature [Dove, 2004; Dove et al., 2006].

The Photoinhibition Model also suggests that *Symbiodinium* genotypes that are adapted to high light environments may show greater thermal tolerance due potentially to the presence of a larger antioxidant pool for inactivating reactive oxygen species and/or a larger xanthophylls pool for dumping excess energy to heat. This has been suggested as an explanation for patterns of response to stress within colonies of *Montastraea annularis* and *M. faveolata* [Rowan et al., 1997]. These two corals host more than one clade of *Symbiodinium* within their tissues. Rowan et al. [1997] found that patterns of bleaching were correlated with the specific elimination of symbiont clades that dominated the low

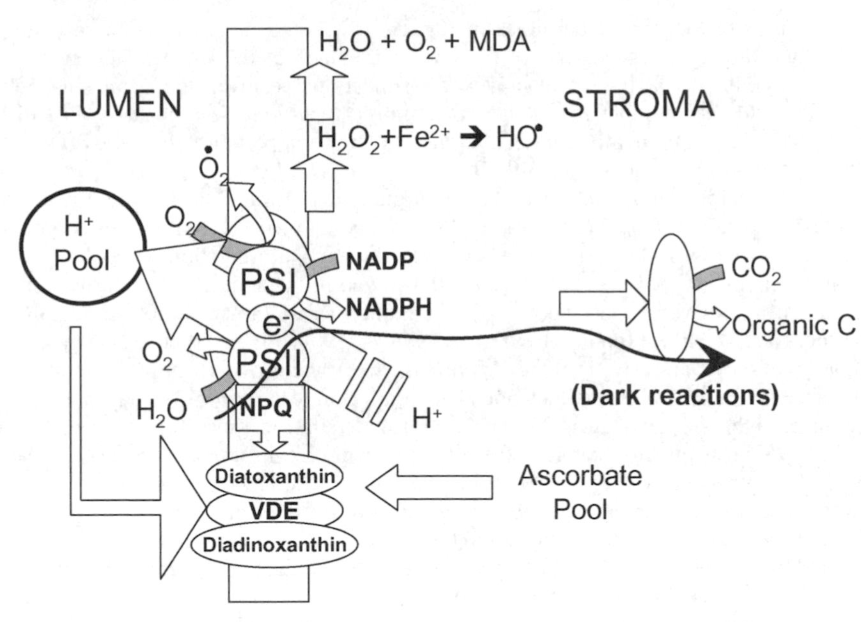

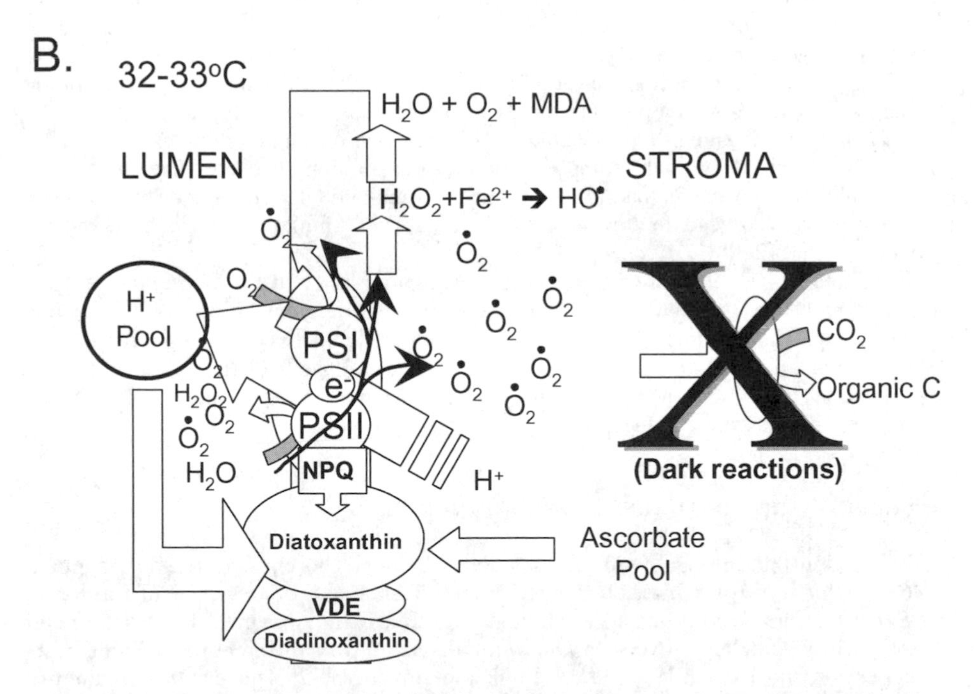

Figure 3. One of several proposed mechanisms describing how thermal stress results in photoinhibition leading to the production of reactive oxygen species (e.g., O_2^-, H_2O_2). A. 27-28°C; excitation pressure at PSII is effectively dissipated by assimilatory photochemical quenching in the dark reaction of photosynthesis. B. 32-33°C; thermal damage inhibits assimilatory photochemical quenching, resulting in the formation of reactive oxygen species at the reaction centres of photosystem I (PSI) and II (PSII).

light regions of the coral from areas of highest irradiance, while a symbiont clade that dominated the high-light regions of the coral was unaffected. The assumption that light tolerance confers heat tolerance, however, is questionable given the increasing number of studies that are demonstrating that for cultured *Symbiodinium sp.*, tolerance to high PFD does not automatically confer tolerance to high temperature [Iglesias-Prieto et al., 1992; Iglesias-Prieto and Trench, 1997].

The photoinhibition model for bleaching argues that elevated heat or light result in an increase in photon induced excitation pressure between the two algal photosystems (PSII and PSI). In photosynthesis, the energy produced from captured photons can be quenched along three distinct pathways (1) disserpated to heat, (2) handed over to molecular oxygen to form reactive oxygen species (3) taken up by assimilatory photochemical quenching inclusive of carbon fixation [review by Nigoyi, 1999; Figure 3]. The energy handling capacity of these pathways is however limited; set by the growth irradiance conditions of the organism; and sometimes but not always adjustable within a 3-5 day period under increased light [higher plants, Walters, 2004] or excitation [unicellular algae, Maxwell et al., 1995] pressure. How can it be therefore that a high-light-tolerant symbiont that ought to be able to dissipate more energy to heat and inactivate ROS with greater efficiency is not necessarily heat tolerant? Two non-exclusive answers have been provided in the literature: Tchernov et al. [2004] offer the option that dinoflagellate heat tolerance is dictated by the poly-unsaturated fatty acid composition of the thylakoid membrane whereby some but not all membranes are able to withstand three or more days of heat stress without disintegrating: leading to a consequential loss of all energy dissipation pathways that depend on a pH gradient across this membrane and leaving O_2 as the only remaining recipient for captured energy (Figure 3). Dove et al. [2006] suggest that symbiotic dinoflagellates follow the example of higher plants rather than that of non-symbiotic algae in so far as excitation pressure increases on their own are not sufficient to drive the restructuring of the antennae that includes an increase in the xanthophylls pool [Walters, 2004]. Xanthophylls not only dump energy via the xanthophylls cycle, they also inhibit lipid peroxidation of the thylakoid membrane [Havaux and Nigoyi, 1999]. Their absence can therefore also lead to downstream consequences similar to those observed by Tchernov et al. [2004].

In summary, the weight of evidence is suggesting that different *Symbiodinium* can have different properties that may to some extent explain the differential susceptibility of corals to bleaching. The questions of whether there is any match between these properties and *Symbiodinium* phylogeny has hardly begun although dinoflagellates from clade D, that as yet, have not been portioned into many sub-clades have frequently been linked in the literature with thermal tolerance.

Genetic Variation Among *Symbiodinium*

The notion that all symbiotic dinoflagellates belonged to one species [Freudenthal, 1962; Taylor, 1969] was first challenged Robert Trench at the University of California at Santa Barbara, who began to accumulate differences in isozyme, ultrastructure and host specificity studies between *Symbiodinium* isolated from different invertebrate hosts [Schoenberg and Trench, 1976, 1980; Blank and Trench, 1985]. The idea that all dinoflagellate symbionts belonged to the one species, *Symbiodinium microadriaticum*, was conclusively rejected by molecular analysis of ribosomal DNA, initially conducted by Rowan and Powers [1991]. Rowan and Knowlton [1995] went on to demonstrate at a molecular level that not only was there a variety of dinoflagellates or clades hosted by corals, but that an individual coral could host multiple different clades, thereby confirming a suggestion made by Buddemeier and Fautin [1993]. Rowan et al. [1997] further demonstrated that clade distribution over the surface of two species of coral (*Montastraea annularis* and *M. faveolata*) correlated with regional light intensities. These studies combined to produce the conclusion that clades A and B symbionts are high light specialists, with clade C taking up the role of the low light specialist. These ideas strongly reinforced the idea that *Symbiodinium* was essentially determining the environmental distribution of the holobiont [Baker, 2001, 2003; Rowan et al., 1997]. Coral species that are able to express both high and low light specializing symbionts, may acclimate (photoadaption, in the non-evolutionary sense) to changing photon flux via the differential regulation of these symbiont types within their cells instead of relying on the flexibility of one symbiont type to restructure its light harvesting antennae [Baker, 2001; Hoegh-Guldberg et al., 2002].

The physiological capability of *Symbiodinium* was explored in cultures created from host organisms. *Symbiodinium* has a free-living stage in addition to the coccoid symbiotic stage. The abilities of dinoflagellate cultures (each isolated from different invertebrate hosts) to acclimate to different light levels began in the early eighties [Chang and Trench, 1982; Iglesias Prieto et al., 1992 and Iglesias-Prieto and Trench, 1997], the data could not however be compared to the results of Rowan et al. [1997], until LaJeunesse [2001] molecularly typed each dinoflagellate culture. In culture, different sub-clades of dinoflagellates show different abilities to acclimate to changes in light conditions. For instance, subclade A1 from *Cassiopeia sp.* readily acclimated to both high and low PAR, whilst subclade A2 from *Zoanthus sp.* was exclusively a high light specialist [Iglesias-Prieto et al., 1992; Iglesias-Prieto and Trench, 1997; LaJeunesse, 2001]. Closer inspection of A1 and A2 therefore lends some inductive support to the claim that clade A phylotypes cope with a high light regime better than, for instance, subclade F1 from the coral, *Montipora verrucosa* that proved to be a low light specialist [Iglesias-Prieto and Trench, 1997; LaJeunesse, 2001]. It is doubtful however, that all clade Cs will conform to the description of "low light specialists" given there dominance in the Pacific in both high and low light environments [LaJeunesse et al., 2003]. On the other hand, some characteristics such as the ability of *Symbiodinium* type A to produce the UV blocking microsporine-like amino acids are consistent with type A being a high light specialist [Banaszak et al., 2000]. This last point brings in the important point that genetic differences at the level of clades A-F among the *Symbiodinium* that form symbioses with coral hosts is likely to be too coarse to map the potential physiological differences among *Symbiodinium* from different hosts [Tchernov et al., 2004]. Despite this observation, clade D has been toted by many has the thermally tolerant *Symbiodinium* clade [Baker et al., 2004]. Clade D, however, also appears to be exceptional in so far as it is not associated with subclade diversity [LaJeunesse et al., 2004, 2005, van Oppen et al., 2005]. Interestingly, whilst clade D *Symbiodinium* appear to be heat tolerant, the trade off for corals hosting this genotype appears to be slower growth rates [van Oppen et al., 2005]. The difference however between bleaching tolerant or sensitive corals irrespective of whether they contain clade D or other *Symbiodinium* is only a matter of 1-2°C and therefore even a moderate increase in temperature as a result of global climate change is likely to impact coral-dinoflagellate assemblages "otherwise known as bleaching resistant" [Ulstrup et al., 2006]. The assumption behind this statement is that some corals can physiologically acclimate to increasing temperature by "shuffling" to a more thermally tolerant *Symbiodinium* genotype that is already present (perhaps at low levels) within their endodermal cells. The assumption merely recognizes that the bleaching threshold provided of +1°C above long term summer maximum [e.g., Hoegh-Guldberg, 1999, 2001] was never meant to be applicable to all corals, with some host-dinoflagellate assemblances potentially having thresholds in excess of +3°C.

Is Bleaching an Adaptive Mechanism?

The discovery of a range of different genetic varieties of *Symbiodinium* has accompanied the idea that reef-building corals may be able to vary their genotypes of *Symbiodinium* to acquire new tolerances with respect to external conditions. This idea was initially encapsulated in the Adaptive Bleaching Hypothesis [Buddemeir and Fautin, 1993] which proposed that bleaching was an adaptive strategy to promote changes in the *Symbiodinium* genotypes present in a host. This idea has evolved somewhat to include a series of processes by which one or several genotypes of *Symbiodinium* become dominant within a host; processes that do not require an actual change to the *Symbiodinium* genotypes hosted in the animal cell. In this respect, host corals that enter stressed conditions could potentially use 'shuffling' or 'switching' to re-sort their symbionts [Baker, 2003]. 'Shuffling' is defined as a quantitative change in the relative abundance of *Symbiodinium* genotypes within a colony, while 'switching' is qualitative change involving *Symbiodinium* genotypes acquired from the environment. These exogenous symbionts may represent types that are new to the colony but not the species. Alternatively, some authors have proposed that corals may be able to take up truly novel *Symbiodinium* genotypes, a process which is referred to as 'evolutionary switching.' While some evidence exists of the first strategy [shuffling, Baker, 2001; Toller et al., 2001], there is less evidence for the second strategy [switching, Lewis and Coffroth, 2004], and there is very little evidence of evolutionary switching operating on ecological timescales. It has however, been clearly important in evolutionary time given that the phylogeny of *Symbiodinium* genotypes rarely if ever tracks host phylogeny [LaJeunesse, 2002; LaJeunesse et al., 2003].

The ability of adult corals to "switch" (uptake exogenous) *Symbiodinium* is questionable: Lewis and Coffroth [2004] experiment was innovative and elegant in so far as adult gorgonians were exposed to a naturally occurring mutant *Symbiodinium*. The study however suffered from the fact that the negative control was contaminated, and the post-bleaching repopulation of the adult coral by mutant *Symbiodinium* tainted by the observation that the error bars on the unexpanded scale are 4-fold larger than the maximum post-bleaching mean population density attained, suggesting that these means are not significantly different from zero.

Evolutionary switching is one of the core ideas promoted by Buddemeir and Fautin [1993] who initially argued that the ABH required that free-living stress-resistant dinoflagellates be able to replace stress-prone resident dinoflagellates. They also argue in this paper that the more severe the bleaching (loss of stress-prone dinoflagellates) then the greater the likelihood of establishing the stress-resistant re-mix. The reasons for severe bleaching were two-fold, firstly severe bleaching would increase the ratio of stress-resistant to stress-prone dinoflagellates in the water column, thereby decreasing the likelihood of re-infection by stress-prone dinoflagellates; secondly, it was deemed necessary to wipe out the resident populations in order to prevent rapid reversion as the stress is removed. Slow reversion or shifts to dinoflagellate populations with faster growth rates was viewed as likely to occur during a prolonged absence of stress, due to an assumed uninterrupted sampling of the water column by the host for new symbionts.

Baker [2001] interpreted his transplantation data as a conclusive demonstration that bleaching (defined as loss of dinoflagellates) was in fact required to increase the survivorship of corals under light stress. It is unfortunate that Baker [2001] settled only for a visual assessment of bleaching. His argument is essentially that shade-acclimated corals moved to high light bleached, then switched symbiont clades and survived, whilst light adapted corals moved to low light did not bleach, did not swap symbiont clades and consequently experienced relatively high mortality. It is clear from Falkowski and Dubinsky [1981] and Hoegh-Guldberg and Smith [1989] that Baker [2001] failed to establish his central premise that the move from low light to high light resulted in a (severe) decrease in dinoflagellate population density, rather than a reduction in chlorophyll per dinoflagellate cell. Retrospectively, it is also easy to criticize the article on the grounds that a failure to swap clades is not equivalent to a failure to shuffle physiological significant algal types. For instance, in the downwards transplant, subclade A1 (high light only specialist) could have been switched for subclade A2 (high and low light specialist). Similar to Ware et al. [1996], Baker [2001] and Buddemeir et al. [2004] allow for the stress-resident dinoflagellate to already be present within the host, that is, for the host-*Symbiodinium* assemblage to undergo no overall genotypic change. Buddemeir and Fautin [1993]'s original argument for severe bleaching (extinction or near extinction of existing dinoflagellate populations) was based on the need to provide conditions that were suitable for the uptake and relatively prolonged stand of a new (not pre-existing) genotype within the host. Remove this requirement and it is not altogether obvious why there is any need to reduce dinoflagellate densities. The stress itself would appear sufficient to confer a physiological advantage to the stress-resistant, over the stress-prone dinoflagellate, with the potential for a decrease in stress-prone density to be matched by an increase in stress-resistant dinoflagellate densities. That is, a switch achieved without the need for drastic loses to the overall dinoflagellate population density and hence to primary productivity.

The explicit model for the ABH provided by Ware et al. [1996] implicitly assumes a fixed carrying capacity within the host for symbionts, with the idea that as stress-prone dinoflagellates die out, they make space available for stress-resident dinoflagellates to occupy. Loss of dinoflagellates from "remaining accommodation" is included as a prerequisite to allow the stress-resident symbiont to remain the dominant strain after the removal of the stress. Any reversion to remaining pockets of stress-prone dinoflagellates would require a similar bleaching event. Fautin and Buddenmeir [2004] clearly state this in assumption "bleaching provides the opportunity for repopulation of a host with different dominant photosymbiont" and again Baker [2003] pushes this notion that bleaching allows the opening of space within potential hosts for colonization by the symbiont that is most able to resist the thermal bleaching stress applied.

Loss of total protein is infrequently measured in thermal bleaching studies, but has been found to occur in many cases were significant reductions in dinoflagellates per unit surface area also occurred [e.g., Dove, 2004]. Seasonal loss of dinoflagellates has also been shown to be accompanied by significant losses of host protein suggesting that dinoflagellate loss can be due to reduction in the size of the house [Fitt et al., 2000; Warner et al., 2002]. There is also some evidence for seasonal shuffling of resident symbionts; as well as evidence for changes in resident dinoflagellate phylotype post-thermal stress [Little et al., 2004, Baker, 2003]; it is not however clear that this shuffling occurred without a reduction in host accommodation, or that "stress-resistant" phylotypes increased their population densities to fill vacated accommodation space. The host is not an empty bag, but is compartmentalized into endodermal cells that hold roughly 1-3 *Symbiodinium* each [Gates and Muscatine, 1992]. Whilst it is possible that *Symbiodinium* may be transferable between cells, the most likely mechanisms by which *Symbiodinium* come to occupy new cells and hence increase in density is via the division of *Symbiodinium*-containing-host-cells which is equivalent to an increase in host cell carrying capacity. Unfortunately, our physiological understanding of the processes involved in the establishment, maintenance and cellular proliferation (or growth) of this coral-dinoflagellate symbiosis is very much in its infancy and mostly based on studies on analogous bacterial-invertebrate or -plant symbioses.

Bleaching and the Mortality of Corals

Why does mortality sometimes follow coral bleaching? There are three non-exclusive possible factors influencing whether corals may die after bleaching. Firstly, the physical stresses may be more intense from one event to another. The second is that proximal factors such as water quality and competition with other benthic organisms such as macrophytes play a role in the mortality of coral reefs following an event. The last brings into focus the possibility that thermal stress throws the symbiosis out of balance and, given enough imbalance from physical stress, leads to a downward trajectory of the coral-algal symbiosis.

Both field and experimental evidence reveal that a combination of the degree to which sea temperature is elevated multiplied by its exposure time predicts mass coral bleaching with great accuracy [Strong et al., 1995, 2000; Hoegh-Guldberg, 1999, 2001]. Anomaly sizes and exposure times can be multiplied together to generate a parameter called degree heating weeks or months (DHW or DHM; Strong et al., 2000; Hoegh-Guldberg, 2001]. DHW values correlated strongly with the degree of impact on reefs during the 1998 global mass bleaching event, such that reefs that had low DHM values (1.1 ± 0.49; Moorea, Cook Islands and the Great Barrier Reef) experienced low mortality, while reefs that experienced high DHM values [3.2 ± 0.47; Palau, Seychelles, Scott Reef, Okinawa; Hoegh-Guldberg, 2004] experienced high mortalities of greater than 70%. These examples demonstrate that higher and/or longer anomalies drive the differences between whether a bleaching event results in a high or low mortality.

The role of proximal factors in affecting the outcome of bleaching for a coral colony is less clear. Among the factors cited for why corals may die is due to being overgrown by microalgae as well as macrophytes in fished areas [e.g., McCook, 1999; Diaz-Pulido and McCook, 2002] or in areas where water quality favors algal growth over corals. Some authors have considered whether protecting corals during and after a thermal event would affect coral cover [McClanahan et al., 2001], with results showing no real effect due to a large degree of variation between sites within protection categories. In this case, fishing has been explored as potentially affecting corals through the elimination of herbivores that would otherwise have kept the macroalgae from out-competing the corals as they recover from a bleaching event. McClanahan et al. [2001] indicated, however, that predation of coral spat and juveniles may be higher in protected areas due to higher levels of scraping herbivores and predators. Water quality in coastal areas has also been cited as an impediment to recovery from bleaching events. Like fishing, there are few experimental studies that are able to show that coral reefs recover faster if bathed in water that is low in nutrients and sediment. As with the protection of grazers, it stands to reason that factors that promote rather than inhibit coral growth will be those that will be important secondary factors in the primary impact and then recovery of reefs following a bleaching event. As with fishing, the demonstration of the influence of these factors in the death of corals following a mass bleaching event remains to be demonstrated conclusively.

The last possibility to consider is that stress can lead to mortality by inducing an imbalance in which symbiosis may have a hard time recovering from. In this case, dysfunction in the symbiosis leads to much reduced performance against other benthic competitors, invading pathogens and fouling organisms. One of the possibilities is that the photic environment within the skeleton becomes intolerable for the remaining algae as symbionts and/or their pigments are lost. Enríquez et al. [2005] demonstrated that corals could undergo substantial seasonal changes in dinoflagellate pigmentation (30-100 mg Chl *a*) without increasing the within tissue light environment (measured as the specific absorption m^2 per mg chl *a*). Further reductions in chlorophyll, however, resulted in an exponential amplification of light within tissue. The extended lifetime of photons traveling within host tissue is believed to results in elevated excitation pressure on remaining dinoflagellates irrespective of their tolerance to the initial heat stress, and substantially increased UV levels with resultant ROS and DNA damage to both host and symbionts alike. These factors may indicate that corals can survive bleaching, so long as they don't exceed these "thresholds of no return". Similar issues may exist with respect to the host where the loss of cells is enough to critically damage the host such that it cannot recover. These intriguing possibilities, however, remain to be explored and verified.

Conclusion

Coral bleaching has been one of the most studied ecological phenomena associated with coral reefs. In this respect, the understanding of the key ecological drivers is fairly well developed. Coral bleaching occurs due to episodes of warmer than normal summer sea temperatures and is exacerbated by high light levels and still, low flow conditions. Less well understood are the relationships between cellular events and the advent of mortality by coral colonies affected by coral bleaching. In this respect, the cellular events hold the key to some of the most vexing questions facing coral biologists and reef managers. If we further our understanding of these events, we stand a chance of understanding why some coral species (e.g., *Porites*) are more resilient and less prone to mortality than others (e.g., *Acropora*). We also stand to learn about the characteristics of the symbiosis that underpins all reef-building corals and whether or not if symbiosis is flexible enough to allow changes to the genotype of *Symbiodinium* to change rapidly to create new host-symbiont combinations that are capable of adapting rapidly to climate change. The answer to these questions is vitally important if we are to project how coral reefs will fare in a century of rapid environmental change.

Acknowledgments. The authors are grateful to the US Coral Reef Task Force and the Bleaching working group of the Coral Reef Targeted Research Program (GEF, World Bank) coordinated and supported by the University of Queensland for support during the preparation of this article.

References

Anthony, K. R. N, and O. Hoegh-Guldberg, Variation in coral photosynthesis, respiration and growth characteristics in contrasting light microhabitats: an analogue to plants in forest gaps and understoreys? *Functional Ecology*, 17, 895-899, 2003.

Anthony, K. R. N., and P. Larcombe, Sediment and coral stress: some mechanisms of adaptation to life on turbid reefs. In: 9th *Int Coral Reef Symp.*, Bali, 2002.

Baker, A. C., Reef corals bleach to survive change. *Nature*, 411: 765-766, 2001.

Baker, A. C., Flexibility and specificity in coral-algal symbiosis: diversity, ecology and biogeography of Symbiodinium. *Annu. Rev. Ecol. Evol. Syst.*, 34, 661-689, 2003.

Baker, A. C., C. J. Starger, T. R. McClanahan, P. W. Glynn, Corals' adaptive response to climate change. *Nature*, 430(7001): 741-741, 2004.

Banaszak, A. T., T. C. LaJeunesse, and R. K. Trench, The synthesis of mycosporine-like amino acids (MAAs) by cultured, symbiotic dinoflagellates. *J. Exp. Mar. Biol. Ecol.*, 249: 219-233, 2000.

Beardall, J. and I. Morris, The concept of light intensity adaptation in marine phytoplankton: some experiments with Phaeodactylum tricornutum. *Mar. Biol.*, 37: 377±387, 1976.

Blank, R. J., and R. K. Trench, Speciation and symbiotic dinoflagellates. *Science*, 229: 656-658, 1985.

Brown, B. E., Coral bleaching: causes and consequences. *Coral Reefs*, 16 Suppl.: S129-S138, 1997.

Brown, B. E., R. P. Dunne, M. S. Goodson, and A. E. Douglas, Bleaching patterns in reef corals. *Nature*, 404, 142-143, 2000.

Bruno, J., C. Siddon, J. Witman, P. Colin, and M. Toscano, El Niño related coral bleaching in Palau, Western Caroline Islands. *Coral Reefs*, 20: 127-136, 2001.

Buddemeier, R. W. and D. G. Fautin, Coral bleaching as an adaptive mechanism—a testable hypothesis. *BioScience*, 43: 320-326, 1993.

Buddemeier, R. W., A. C. Baker, D. G. Fautin, and J. R. Jacobs, The adaptive hypothesis of bleaching. In: *Coral Health and Disease* (E. Rosenberg, ed.). Springer-Verlag, *Berlin*. pp. 427-444, 2004

Bumann, D., and D. Oesterhelt, Destruction of a single chlorophyll is correlated with the photoinhibition of photosystem II with a transiently inactive donor site. *Proc. Natl. Acad. Sci. USA*, 92: 12195-12199, 1995.

Cervino, J. M., R. L. Hayes, M. Honovich, T. J. Goreau, S. Jones, and P. J. Rubec, Changes in zooxanthellae density, morphology, and mitotic index in hermatypic corals and anemones exposed to cyanide. *Marine Pollution Bulletin*, 46: 573-586, 2003.

Chang, S. S., B. B. Prezelin, and R. K. Trench, Mechanisms of photoadaptation in three strains of the symbiotic dinoflagellate *Symbiodinium microadriaticum*. Marine biology. Berlin, *Heidelberg*, 76: 219-229, 1983.

Diaz-Pulido, G., and L. J. McCook, The fate of bleached corals: patterns and dynamics of algal recruitment. *Marine Ecology Progress Series*, 232: 115-128, 2002.

Dove, S. G., Scleractinian corals with photoprotective host pigments are hypersensitive to thermal bleaching. *Mar. Ecol. Progr. Ser*. 272: 99-116, 2004.

Dove, S., J. C. Ortiz, S. Enríquez, M. Fine, P. Fisher, R. Iglesias-Prieto, D. Thornhill, and O. Hoegh-Guldberg, Response of holosymbiont pigments from the scleractinian coral *Montipora monasteriata* to short term heat stress Limnol. *Oceanogr*. 51(2): 1149-1158, 2006.

Dunn, S. R., J. C. Bythell, M. D. A. LeTissier, W. J. Burnett, and J. C. Thomason, Programmed cell death and cell necrosis activity during hyperthermic stress induced bleaching of the symbiotic sea anemone *Aiptasia sp. J. Exp. Mar. Biol. Ecol.*, 272: 29-53, 2002.

Enríquez, S., E. R. Méndez, and R. Iglesias-Prieto, Multiple scattering on coral skeletons enhances light absorption by symbiotic algae. Limnol. *Oceanogr*., 50: 1025-1032, 2005.

Fagoonee, I., H. B. Wilson, M. P. Hassell, and J. R. Turner, The Dynamics of zooxanthellae Populations: A Long-Term Study in the Field Science, 283: 843-845, 1999.

Falkowski, P. G, Light-shade adaptation in marine phytoplankton. *In:* Primary productivity in the sea, pp. 99-120. Ed. By P.G. Falkowski. New York: Plenum Press, 1980

Falkowski, P., and Z. Dubinsky, Light-shade adaptation of Stylophora pistillata, a hermatypic coral from the Gulf of Eilat. *Nature*, 289: 172-174, 1981.

Fitt, W., F. McFarland, M. Warner, and G. Chilcoat, Seasonal patterns of tissue biomass and density of symbiotic dinoflagellates in reef c o rals and relation to coral bleaching. Limnol. *Oceanogr*. 45: 667-687, 2000.

Fitt, W., B. Brown, M. Warner, and R. Dunne, Coral bleaching: interpretation of thermal tolerance limits and thermal thresholds in tropical corals. *Coral Reefs*, 20: 51-65, 2001.

Franklin, D. J., O. Hoegh-Guldberg, R. J. Jones, and J. A. Berges, Cell death and degeneration in the symbiotic dinoflagellates of the coral Stylophora pistillata (Esper) in response to the combined effects of elevated temperature and light. Marine Ecology Progress Series, 272: 117-130, 2004.

Freudenthal, H. D., Symbiodinium gen. nov. Symbiodinium microadriaticum sp. nov., a zooxanthella: taxonomy, life cycle, and morphology, *I. Protozoology*, 9, 45-52, 1962.

Gates, R. D. and L. Muscatine, 3 methods for isolating viable anthozoan endoderm cells with their intracellular sybiotic dinoflagellates. *Coral Reefs*, 11(3): 143-145, 1992

Gates, R. D., G. Baghdasarian, and L. Muscatine, Temperature stress causes host cell detachment in symbiotic cnidarians: implications for coral bleaching. *Biol Bull*, 182: 324-332, 1992.

Glynn, P. W., Coral reef bleaching ecological perspectives. *Coral Reefs*, 12, 1-17, 1993.

Goreau, T. F., Mass expulsion of zooxanthellae from Jamaican reef communities after hurricane Flora. *Science*, 145: 383-6, 1964.

Goenaga, C., and M. Canals, Island-wide coral bleaching in Puerto Rico. *Caribbean Journal of Science,* 26, 171-175, 1990.

Havaux, M., and K. Nigoyi, The violaxanthin cycle protects plants from photooxidative damage by more than one mechanism. *Proc. Natl. Acad. Sci.* USA, 96: 8762-8767, 1999.

Hill, R., A. W. D. Larkum, C. Frankart, M. Kuhl, and P. J. Ralph, Spatial heterogeneity of photosynthesis and the effect of temperature-induced bleaching conditions in three species of corals. *Photosynthesis Research*, 82: 59-72

Hoegh-Guldberg, O., Coral bleaching, Climate Change and the future of the world's Coral Reefs. Review, *Marine and Freshwater Research*, 50: 839-866, 1999.

Hoegh-Guldberg, O., The future of coral reefs: Integrating climate model projections and the recent behaviour of corals and their dinoflagellates. Proceedings of the Ninth International coral reef symposium. October 23-27, 2000, Bali, Indonesia, 2001.

Hoegh-Guldberg, O., Coral reefs in a century of rapid environmental change. *Symbiosis*, 37: 1-31, 2004.

Hoegh-Guldberg, O., and R. Jones, Diurnal patterns of photoinhibition and photoprotection in reef-building corals. *Marine Ecology Progress Series*, 183: 73-86, 1999.

Hoegh Guldberg, O., and G. J. Smith, "The effect of sudden changes in temperature, irradiance and salinity on the population density and export of zooxanthellae from the reef corals Stylophora pistillata (Esper 1797) and Seriatopora hystrix (Dana 1846)." *Exp. Mar. Biol. Ecol.*, 129: 279-303, 1989.

Hoegh-Guldberg, O., R. J. Jones, S. Ward, and W. K. Loh, Is coral bleaching really adaptive? *Nature*, 415: 601-602, 2002.

Hoegh-Guldberg, Ove, Maoz Fine, William Skirving, Ron Johnstone, Sophie Dove, and Alan Strong, Coral bleaching following wintry weather. Limnol. *Oceanogr*. 50: 265-271, 2005.

Iglesias-Prieto, R., Matta, J. L., Robins, W. A. and Trench, R. K., Photosynthetic response to elevated-temperature in the symbiotic dinoflagellate Symbiodinium microadriaticum in culture. *Proceedings of the National Academy of Science*, 89: 302-305, 1992.

Iglesias-Prieto, R., and R. K. Trench, Acclimation and adaptation to irradiance in symbiotic dinoflagellates. II. Response of chlorophyll-protein complexes to different photon-flux densities. *Mar. Biol.* 130: 23-33, 1997.

Intergovernmental Panel on Climate Change (IPCC 2001), Climate Change 2001: Synthesis Report.

Jones, R. J., Hoegh-Guldberg, O., Effects of cyanide on coral photosynthesis: implications for identifying the cause of coral bleaching and for assessing the environmental effects of cyanide fishing. *Marine Ecology Progress Series*, 177: 83-91, 1999.

Jones, R. J., Sublethal stress assessment in scleractinia and the regulatory biology of the coral-algal symbiosis. PhD Thesis, James Cook University, Australia, 1995.

Jones, R. J., Zooxanthellae loss as a bioassay for assessing stress in corals. *Mar Ecol. Progr Ser.*, 158: 51-59, 1997

Jones, R. J., O. Hoegh-Guldberg, A. W. D. Larkum, U. Schreiber, Temperature-induced bleaching of corals begins with impairment of the CO_2 fixation mechanism in zooxanthellae. *Plant Cell Environ*, 21: 1219-1230, 1998.

Jones, R. J., and A. P. Kerswell, Phytotoxicity of Photosystem II (PSII) herbicides to coral. *Mar Ecol Prog Ser*, 261: 149-159, 2003.

Kerswell, A. P., and R. J. Jones, Effects of hypo-osmosis on the coral Stylophora pistillata: nature and cause of 'low-salinity bleaching. *Marine Ecology Progress Series*, 253: 145-154, 2003.

Kushmaro, A., Y. Loya, M. Fine, and E. Rosenberg, Bacterial infection and coral bleaching. *Nature*, 380: 396, 1996

LaJeunesse, T. C. Investigating the biodiversity, ecology, and phylogeny of endosymbiotic dinoflagellates in the genus *Symbiodinium* using the ITS region: in search of a "species" level marker. *J. Phycol*. 37: 866-880, 2001.

LaJeunesse, T. C. Diversity and community structure of symbiotic dinoflagellates from Caribbean coral reefs. *Mar Biol.*, 141: 387-400, 2002.

LaJeunesse, T. C., W. K. Loh, R. van Woesik, O. Hoegh-Guldberg, G. W. Schmidt, and W. K. Fitt. Low symbiont diversity in southern Great Barrier Reef corals relative to those of the Caribbean. *Limnol. Oceanogr*. 48: 2046-2054, 2003.

LaJeunesse, T. C., D. J. Thornhill, E. F. Cox, et al., High diversity and host specificity observed among symbiotic dinoflagellates in reef coral communities from Hawaii Coral Reefs, 23(4): 596-603, 2004.

LaJeunesse, T. C., S. Lee, and S. Bush, et al., Persistence of non-Caribbean algal symbionts in Indo-Pacific mushroom corals released to Jamaica 35 years ago Coral Reefs, 24(1): 157-159, 2005.

Lesser, M. P. Oxidative stress causes coral bleaching during exposure to elevated temperatures. *Coral Reefs*, 16: 187-192, 1997.

Lewis, C. L., and M. A. Coffroth, The acquisition of exogenous algal symbionts by an ctocoral after bleaching, *Science*, 304, 1490-1492, 2004.

Little, A. F., M. J. H. van Oppen, and B. L. Willis, Flexibility in algal endosymbiosis shapes growth in reef corals. *Science*, 304: 1492-1494, 2004.

McClanahan, T. R., N. A. Muthiga, and Mangi, S., Coral and algal changes after the 1998 coral bleaching: interaction with reef management and herbivores on Kenyan reefs. Coral Reefs, 2001.

McCook, L. J., Macroalgae, nutrients and phase shifts on coral reefs: scientific issues and management consequences for the Great Barrier Reef. *Coral Reefs*, 18(4): 357-367, 1999.

Maxwell, D. P., S. Falk, and N. P. A. Huner, Photosystem II Excitation Pressure and Development of Resistance to Photoinhibition (I. Light-Harvesting Complex II Abundance and Zeaxanthin Content in Chlorella vulgaris) *Plant Physiol*. 107: 687-694, 1995.

Mumby, P. J., J. R. M. Chisholm, A. J. Edwards, C. D. Clark, E. B Roark, G. Passeron-Seitre, S. Andréfouët, and J. Jaubert, Mass bleaching-induced coral mortality at Rangiroa Atoll, French Polynesia. *Marine Biology*, 139: 183-189, 2001.

Nakamura, T., H. Yamasaki, R. Van Woesik, Water-flow treatment facilitates recovery from bleaching in the coral Stylophora pistillata. *Marine Ecology Progress Series*, 256: 287-291, 2003.

Niyogi, K. K. Photoprotection revisited. Genetic and molecular approaches. Ann. Rev. Plant Physiol. *Plant Mol. Biol*. 50: 333-359, 1999.

Olaizola, M., J. LaRoche, Z. Kolber, and P. G. Falkowski, Non photochemical fluorescence quenching and the diadinoxanthin cycle in a marine diatom. *Photosyn. Res.*, 41: 357-370, 1994.

Raven, J., K. Caldeira, H. Elderfield, O. Hoegh-Guldberg, P. Liss, U. Riebesell, J. Shepherd, C. Turley, and A. Watson, Ocean acidification due to increasing atmospheric carbon dioxide. Report by The Royal Society, pp. 60, 2005.

Rowan, R., N. Knowlton, Intraspecific diversity and ecological zonation in coral-algal symbiosis. Proceedings of the National Academy of Sciences of the United States of America, 92: 2850-2853, 1995.

Rowan, R., N. Knowlton, A. Baker, J. Jara, Landscape ecology of algal symbionts creates variation in episodes of coral bleaching. *Nature*, 388: 265-269, 1997.

Rowan, R., and D. A. Powers, A molecular genetic classification of zooxanthellae and the evolution of animal-algal symbiosis. *Science*, 251: 1348-1351, 1991.

Ruffle, S. V., J. Wang, H. G. Johnston, T. L. Gustafson, R. S. Hutchison, J. Minagawa, A. Crofts, and R. T. Sayre, Photosystem II Peripheral Accessory Chlorophyll Mutants in *Chlamydomonas reinhardtii*. Biochemical Characterization and Sensitivity to Photo-Inhibition. *Plant Physiol*., vol. 127: 633-644, 2001.

Salih, A., O. Hoegh-Guldberg, and G. Cox, Photoprotection of Symbiotic Dinoflagellates by Fluorescent Pigments in Reef Corals. In: Greenwood, J. G. & Hall, N. J., eds *Proceedings of the Australian Coral Reef Society 75th Anniversary Conference, Heron Island October 1997*. pp. 217-230, 1998.

Sawyer, S., and L. Muscatine, Cellular mechanisms underlying temperature-induced bleaching in the tropical sea anemone Aiptasia pulchella. *The Journal of Experimental Biology*, 204, 3443-3456, 2001.

Strong, A. E., J. A. Preyer, and C. S. Barrientos, Assessing CZCS time-series data globally, regionally, and zonally: 1979-1985, *Adv. Space. Res*., 16(10), 147-150, 1995.

Strong, A. E., E. Kearns and K. K. Gjovig, Sea Surface Temperature Signals from Satellites - An Update. *Geophys. Res. Lett*, 27(11): 1667-1670, 2000.

Taylor, D. L. The nutritional relationship of *Anemonia sulcata* (Pennant) and its dinoflagellate symbiont. *J. Cell Sci.* 4, 751-762, 1969.

Toller, W. W., R. Rowan, and N. Knowlton, Repopulation of zooxanthellae in the Caribbean corals *Montastraea annularis* and *M. faveolata* following experimental and disease-associated bleaching. *Biol Bull*, 201: 360-373, 2001.

Schmid, V., R. Stidwill, A. Bally, B. Marcum, and P. Tardent, Heat dissociation and maceration of marine cnidaria. Roux's Arch. *Dev. Biol*., 190: 143-149, 1981.

Schoenberg, D. A., and R. K. Trench, Specificity of symbioses between marine cnidarians and zooxanthellae. Coelenterate Ecology and Behavior (Mackie, ed.) University of Victoria, British Columbia. pp. 423-32, 1976.

Schoenberg, D. A., and R. K. Trench, Genetic variation in *Symbiodinium (=Gymnodinium) microadriaticum* Freudenthal, and specificity in its symbiosis with marine invertebrates. I. Isozyme and soluble protein patterns of axenic cultures of *Symbiodinium microadriaticum. Proc. R. Soc. Lond.* B 207: 405-427, 1980.

Tchernov, D., M. Y. Gorbunov, C. D Vargas, S. N. Yadav, A. J. Milligan, M. Haggblom, and P. G. Falkowski, Membrane lipids of symbiotic algae are diagnostic of sensitivity to thermal bleaching in corals. *Proc. Natl. Acad. Sci.* USA 101: 13531-13535, 2004.

Ulstrup, K. E., R. Berkelmans, P. J. Ralph, and M. J. H. van Oppen, Variation in bleaching sensitivity of two coral species across a latitudinal gradient on the Great Barrier Reef: the role of zooxanthellae Marine Ecology-Progress Series, 314: 135-148, 2006.

van Oppen, M. J. H., A. J. Mahiny, and T. J. Done, Geographic distribution of zooxanthella types in three coral species on the Great Barrier Reef sampled after the 2002 bleaching event. *Coral Reefs*, 24(3): 482-487, 2005.

Van Woesik, R., L. M. De Vantier, and J. S. Glazebrook, Effects of cyclone `Joy' on nearshore coral communities of the Great Barrier Reef. Marine Ecology Progress Series, 128: 261-270, 1995.

Walters, R. G. Towards an understanding of photosynthetic acclimation. *J. Exp. Botany*, 56: 435-447, 2004.

Ware, J. R., D. G. Fautin, and R. Buddemeier, Patterns of coral bleaching: modeling the adaptive bleaching hypothesis. *Ecological Modelling*, 84, 199-214, 1996.

Warner, M. E., G. C. Chilcoat, F. K. McFarland, and W. K. Fitt, Seasonal fluctuations in the photosynthetic capacity of photosystem II in symbiotic dinoflagellates in the Caribbean reef-building coral *Montastraea. Mar. Biol.* 141: 31-38, 2002.

Zvalinskii, V. I., V. A. Leletkin, E. A. Titlyanov, and M. G. Shaposhnikova, Photosynthesis and adaptation of corals to irradiance. 2. Oxygen exchange. Photosynthetica (CSR) 14: 422-430, 1980.

5

Coral Reefs and Changing Seawater Carbonate Chemistry

Joan A. Kleypas and Chris Langdon

Seawater carbonate chemistry of the mixed layer of the oceans is changing rapidly in response to increases in atmospheric CO_2. The formation and dissolution of calcium carbonate is now known to be strongly affected by these changes, but many questions remain about other controls on biocalcification and inorganic cementation that confound our attempts to make accurate predictions about the effects on both coral reef organisms and reefs themselves. This chapter overviews the current knowledge of the relationship between seawater carbonate chemistry and coral reef calcification, identifies the hurdles in our understanding of the two, and presents a strategy for overcoming those hurdles.

1. Introduction

1.1. History and Background

In the most simplistic context, Earth's carbon cycle can be configured as four different reservoirs: atmosphere, biosphere, hydrosphere and geosphere (Figure 1). Over geologic time, the fluxes of carbon between these reservoirs tend to maintain relative stability in the sizes of the reservoirs. Occasionally, however, a sudden flux of carbon from one reservoir to another upsets the balance, with consequences for Earth's climate and biosphere. The carbon cycle responds with a series of negative feedbacks that cause increased rates of CO_2 uptake in other reservoirs. For example, an increase in atmospheric CO_2 can increase biospheric uptake of carbon, most of which is stored only temporarily, but some of which may be stored for much longer periods (e.g., plant growth which is buried by sediments and eventually becomes converted to fossilized carbon). Over time, these feedbacks bring the cycle back into balance, although not necessarily to the same stasis as before the perturbation.

Fossil fuels were formed through biological processes, but became stored as part of the geologic reservoir over millions of years (most of the oil and coal reserves are more than 50 million years old). Burning the fossilized carbon accelerates the rate of carbon flux from the geosphere to the atmosphere, which greatly outpaces the rates at which other fluxes draw the CO_2 back out of the atmosphere. Thus, the atmospheric reservoir grows, and in turn drives increased fluxes to the biosphere and the ocean (Figure 1). In the biosphere, the increased CO_2 can increase carbon uptake by "fertilizing" plant growth, at least

Coral Reefs and Climate Change: Science and Management
Coastal and Estuarine Studies 61
Copyright 2006 by the American Geophysical Union.
10.1029/61CE06

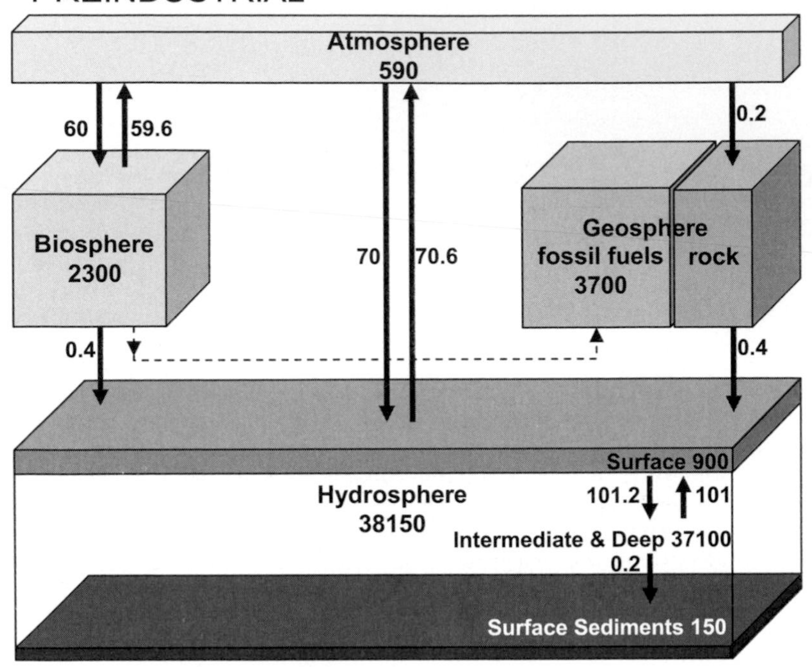

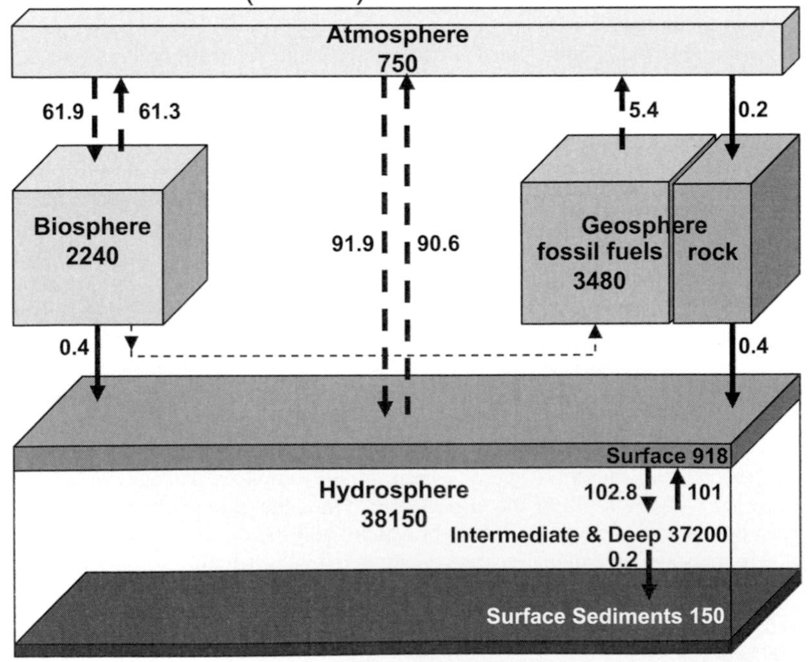

Figure 1. Simplified version of the major reservoirs and fluxes of carbon within the Earth System. Top panel illustrates the preindustrial, quasi-steady state prior to human activities. Lower panel shows present-day state. New and changed fluxes are indicated with dashed arrows.

under conditions where other nutrients are not co-limiting; and by increasing temperature, which under certain conditions can extend the growing season. The bulk of the fluxes from the atmospheric reservoir, however, are to the ocean. Oceanic uptake sequesters more than a third of all fossil fuel emissions, and most of this uptake is geochemical rather than biological. Without oceanic uptake, current day atmospheric CO_2 levels would be about 435 ppmv (parts per million volume), rather than 380 ppmv [Sabine et al., 2004].

The oceans thus have an enormous capacity to buffer changes in atmospheric CO_2. Oceanic uptake of atmospheric CO_2, however, alters the carbonate chemistry of seawater. Fluctuations in atmospheric CO_2 in the past have been accompanied by changes in both seawater carbonate chemistry and in production and preservation of calcium carbonate secreting organisms (hereafter called "calcifiers"). This chapter addresses this process with an emphasis on how future changes in carbonate chemistry will affect calcification rates of coral reef organisms as well as reef-building itself.

1.2. Definitions

Different terms are often used to describe the changing carbonate system in seawater, and some of these are used interchangeably. This is because when one component of the system changes, the other components change proportionally. For example, increasing atmospheric CO_2 concentration drives more CO_2 into seawater, and this causes the carbonate system to adjust so that both pH and carbonate ion concentration decrease. So while *ocean acidification* refers to the lowering of pH, it also infers a reduction in carbonate ion concentration. Another term, *calcium carbonate saturation state* (Ω), is a measure of the ion activity product (IAP) of Ca^{2+} and CO_3^{2-} relative to the apparent solubility product (K′) for a particular calcium carbonate mineral phase (calcite, Mg-calcite, or aragonite):

$$\Omega = [Ca^{2+}]\,[CO_3^{2-}] \,/\, K' \qquad (1)$$

Because Ca^{2+} concentrations are typically 20-30 times that of CO_3^{2-} and do not vary considerably, reduction in $[CO_3^{2-}]$ is essentially synonymous with a reduction in saturation state. The term *carbonate ion concentration* is probably the best for describing atmospheric-induced changes in carbonate chemistry, because experimental evidence shows that under increased atmospheric CO_2 this appears to be the component of the carbonate system that controls coral calcification [Langdon, 2002]. However, the term *saturation state* provides a useful measure of the level of saturation of a particular carbonate mineral (where values <1 indicate undersaturated conditions, and >1 indicate supersaturated). We use both of these terms to describe the carbonate system in seawater.

1.3. Overall Importance of $CaCO_3$ in Marine Carbon Cycle

Most $CaCO_3$ production in the oceans is biogenic, or at least associated with biological activity. Biogenic calcification rates are greatest in the photic zone, while dissolution rates are greatest in the deep ocean where increasing pressure, decreasing temperature, and increased organic matter respiration shifts the saturation state to less than 1. In the open ocean, the marine carbon cycle is affected by the ratio of the "rain rates" of inorganic carbon ($CaCO_3$) and organic carbon (reviewed by Archer [2003]). Shell formation draws down the total alkalinity in the surface ocean (leaving it more acidic), and subsequent sinking and dissolution of the shells increases the total alkalinity at depth. This "calcium carbonate pump" decreases the capacity for the surface ocean to absorb CO_2 from the atmosphere. However, marine organic matter production is also largely restricted to shallow waters, and sinking of organic matter to the deep ocean – the "organic carbon pump" – increases the capacity of the ocean to take up atmospheric CO_2, by removing CO_2 from the surface ocean and transporting it to the deep ocean as organic matter. These two processes may not act independently, however; recent evidence suggests that most of the organic carbon that sinks to the deep ocean is "ballasted" by $CaCO_3$ particles ($CaCO_3$ provides the higher-density weighting to increase sinking rates [Armstrong et al., 2002]).

The rise and fall of calcification on continental shelves, in concert with glacial-interglacial sea level fluctuations, is also considered an important process in the global carbon cycle. The "coral reef hypothesis" [Berger, 1982; Opdyke and Walker, 1992] states that episodic flooding of continental shelves during postglacial sea level rise led to dramatic increases in $CaCO_3$ production, which released significant amounts of CO_2 to the atmosphere (by a process explained in section 2.1). Recent modeling efforts attest to the importance of this mechanism [Ridgwell et al., 2003].

2. Seawater Carbonate Chemistry Changes

Ocean sequestration of fossil-fuel CO_2 emissions is already progressing at a very large scale [Sabine et al., 2004], and carbonate chemistry of the surface mixed layer of the oceans is responding rapidly to this uptake [Feely et al., 2004]. As the debate continues over whether to pump liquefied CO_2 into the deep ocean for temporary storage, the surface ocean has already naturally taken up about 30% of anthropogenic CO_2 from the atmosphere [Sabine et al., 2004]. This has altered the carbonate system in seawater. Evidence for ocean uptake of CO_2 and its effects on seawater carbonate chemistry comes from a variety of methods, including 1) measured changes in CO_2 chemistry in seawater over time; 2) isotopic signals; and 3) modeling studies.

2.1. The Carbonate System in Seawater

Understanding the carbonate system in the ocean requires an understanding of both the equilibrium chemistry of the various inorganic forms of carbon dioxide in seawater, and alkalinity. A complete explanation of the carbonate system in seawater cannot be provided here, and the reader is referred to the thorough descriptions in Zeebe and Wolf-Gladrow [2001]. Instead, this section presents the basics of how increasing carbon dioxide concentrations in the atmosphere are causing changes in the carbonate system that ultimately affect calcification rates on coral reefs.

The carbonate equilibrium in seawater is a complicated system that influences the exchange of CO_2 across the air-sea interface, the interconversion of various inorganic phases of dissolved CO_2, and the formation and dissolution of calcium carbonate (Figure 2). The basic equations describing the carbonate system are:

$$CO_2(g) \leftrightarrow CO_2\,(aq) \tag{2}$$

$$CO_2\,(aq) + H_2O \leftrightarrow H_2CO_3 \tag{3}$$

$$H_2CO_3 \leftrightarrow H^+ + HCO_3^- \tag{4}$$

$$HCO_3^- \leftrightarrow H^+ + CO_3^{2-} \tag{5}$$

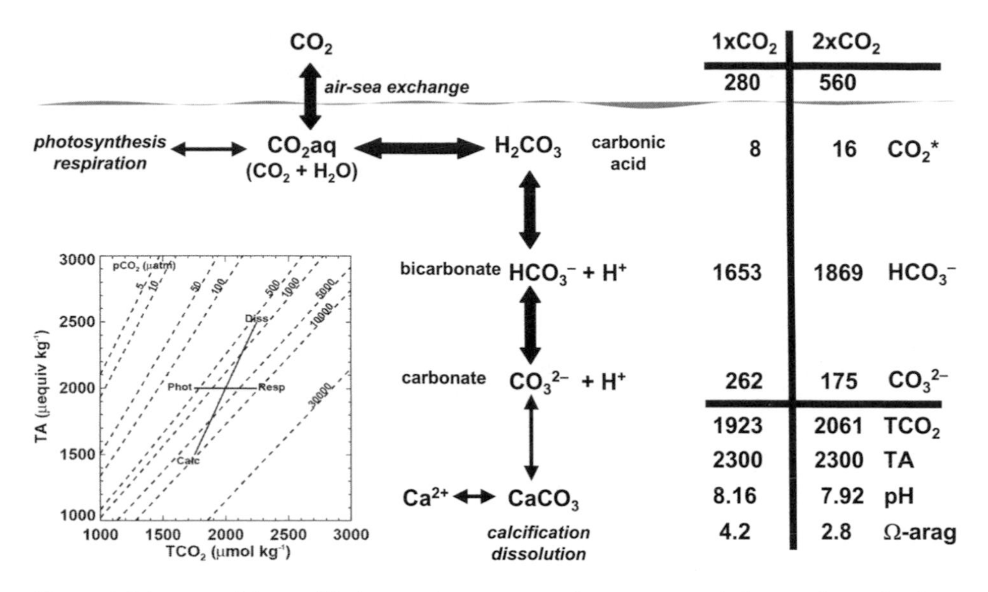

Figure 2. Diagram of the equilibrium carbonate system in seawater, and changes due to forcing by atmospheric CO_2. Table at right shows calculated changes in carbonate chemistry parameters under preindustrial concentrations of atmospheric CO_2 (280 ppmv) and a doubling (560 ppmv), assuming temperatures of 25°C and 26°C, respectively, salinity = 35, and TA = 2300 μequiv kg^{-1}. CO_2* is the total concentration of CO_2(aq) and H_2CO_3. TCO_2 is the sum of CO_2*, HCO_3^- and CO_3^{2-}. Note that a change in atmospheric CO_2 does not affect TA. Inset illustrates the effects of photosynthesis/respiration versus calcification/dissolution on total alkalinity (TA) and total dissolved inorganic carbon (TCO_2). Dashed lines indicate the associated partial pressure of CO_2 in seawater (pCO_2) in μatm.

where:

CO_2 = carbon dioxide, as gas (g) or aqueous (aq)
H_2CO_3 = carbonic acid
HCO_3^- = bicarbonate ion
CO_3^{2-} = carbonate ion

Note that the total concentration of all the dissolved inorganic carbon species is termed "dissolved inorganic carbon" (DIC), or "total CO_2" (TCO_2).

A common inference from Equations (2-5) is that the addition of CO_2 to the water column will ultimately lead to an increase in the carbonate ion content, and given the following equation, one intuitively would consider that this would also lead to an increase in the formation of calcium carbonate ($CaCO_3$).

$$Ca^{2+} + CO_3^{2-} \leftrightarrow CaCO_3 \tag{6}$$

This inference is incorrect, however, because carbonic acid is a weak acid, and all of the various carbon species exist simultaneously (note that all of the above reactions are reversible). The relative proportion of each species, particularly the proportion of

HCO_3^- to CO_3^{2-}, is governed by the pH and the need to maintain the ionic charge balance in seawater. This in turn is closely related to the concept of total alkalinity (TA). TA is defined as "the number of moles of hydrogen ion equivalent to the excess of proton acceptors over proton donors in a kilogram of seawater" [DOE, 1994]; but essentially TA equals the charge difference between conservative cations of strong bases and conservative anions of strong acids:

$$\begin{aligned} TA &= [Na^+] + [K^+] + 2[Mg^{2+}] + 2[Ca^{2+}] - [Cl^-] - 2[SO_4^{2-}] \\ &= [HCO_3^-] + 2[CO_3^{2-}] + [B(OH)_4^-] - [H^+] \end{aligned} \tag{7}$$

Most of the alkalinity (95%) of the right side of Equation (7) is contributed by the first two terms which are also called carbonate alkalinity (CA):

$$CA = [HCO_3^-] + 2[CO_3^{2-}] \tag{8}$$

while the 3rd term in the right side of Equation (7) is called borate alkalinity.
Since

$$CA \approx TA \tag{9}$$

then

$$[HCO_3^-] + 2[CO_3^{2-}] \approx [Na^+] + [K^+] + 2[Mg^{2+}] + 2[Ca^{2+}] - [Cl^-] - 2[SO_4^{2-}] \tag{10}$$

As TA decreases, say by the removal of conservative positive ions (e.g., removal of Ca^{2+} by $CaCO_3$ precipitation), then CA decreases as well, through the conversion of some of the carbonate to bicarbonate. The higher the TA, the more the system moves toward the right in Equation (5), and vice-versa.

TA of seawater changes conservatively with salinity and values are typically around 2300 μmoles kg^{-1} (note that TA is often presented as μequiv kg^{-1} because it is actually the number of moles of H^+ equivalents needed for its titration). High alkalinity values provide seawater with its strong buffering capacity (the resistance to changes in pH as acids or bases are added). By comparison, freshwater systems have low alkalinity values and experience much more dramatic swings in pH.

A direct consequence of the increasing partial pressure of CO_2 in the atmosphere is an increase in seawater TCO_2. Adding CO_2 to seawater does not affect the TA, but will lower the pH because CO_2 reacts with water to produce carbonic acid (Equation 3). The carbonic acid will release protons to produce HCO_3^- (Equation 4); however, in order to maintain charge balance some carbonate ions will be converted to bicarbonate, and this will lower the CO_3^{2-} concentration (Table in Figure 2). As a first-order approximation, the carbonate ion concentration can be estimated as the difference TA – TCO_2. This is because CO_2 and H_2CO_3 are very small compared to HCO_3^- and CO_3^{2-} and CA ≈ TA. An increase in CO_2 thus leads to a decrease in carbonate ion concentration. Figure 2 illustrates how changes in either TA or TCO_2 affect the carbonate ion concentration.

Equation (11) captures the relationship between $CaCO_3$ and CO_2. Addition of CO_2 to the system leads to the dissolution of $CaCO_3$. The reverse reaction, the formation of calcium carbonate, produces CO_2.

$$CaCO_3 + CO_2 + H_2O \leftrightarrow Ca^{2+} + 2HCO_3^- \tag{11}$$

The carbonate system in seawater is influenced by both physical and biological processes – mainly those that affect concentrations of TCO_2 and/or TA. For example, temperature strongly affects Equation (2), as CO_2 solubility increases with decreasing temperature. Biological processes of primary production and respiration affect the system by removing and adding CO_2, respectively; while calcium carbonate formation and dissolution alter both TA (Ca^{2+} removal) and TCO_2 concentration (Figure 2 inset).

2.2. Mechanisms and Time-scales of Future Seawater Carbonate Chemistry Changes

Oceanic uptake of atmospheric CO_2 is restricted to the ocean-atmosphere interface; about 50% of the anthropogenic CO_2 in the oceans is still confined to waters shallower than 400 m and generally to waters above the thermocline [Sabine et al., 2004]. Transport of CO_2 to the deep ocean increases the capacity of the surface ocean to take up more CO_2 from the atmosphere, at least over the time scale of ocean turnover (~1000 years). Transport of CO_2 to the deeper ocean occurs via physical oceanographic (advection, diffusion) or biological processes. For example, the North Atlantic takes up CO_2 disproportionately to its surface area because downward transport removes CO_2-rich surface waters to the ocean interior more rapidly than in other ocean regions [Sabine et al., 2004]. Downward transport is generally higher in mid-latitude regions of bottom and intermediate water formation. The formation of organic matter through photosynthesis ($6H_2O + 6CO_2 \rightarrow C_6H_{12}O_6 + 6O_2$) also removes CO_2 from shallow ocean waters, and downward flux of this material can effectively store CO_2 in the deep ocean for hundreds of years.

The carbonate chemistry of the oceans has constantly evolved over geologic time in response to changes in other carbon reservoirs. Because the surface ocean is in near thermodynamic equilibrium with the atmosphere, the chemical response of the surface ocean to increased atmospheric CO_2 can be estimated from thermodynamic principles (Figure 2). This is the main driver of surface ocean carbonate chemistry, but the added effects of biological activity and ocean mixing complicate these estimates. Biological activity includes both organic matter production and oxidation, which takes up and releases CO_2, respectively, and calcium carbonate production and dissolution, which releases and takes up CO_2, respectively. Predictions in areas of ocean advection, particularly of upwelling and subduction, require estimates of mixing using physical oceanographic techniques.

The time scales of ocean response to increased atmospheric CO_2 vary depending on the processes that transport CO_2 across the air-sea interface, physical processes that transport that CO_2 into the ocean interior, the formation and remineralization of organic matter, and the formation and dissolution of $CaCO_3$. CO_2 equilibration between the atmosphere and surface ocean is generally achieved within about a year, while the deep ocean concentrations reflect the atmospheric concentration when they were last in contact with the atmosphere plus the additional CO_2 released via oxidation of organic matter. Even though the oceans have a high capacity to store CO_2, that capacity is limited by the rate of ocean overturning and the rate of $CaCO_3$ dissolution, so that thousands of years will be required for oceanic processes to restore atmospheric concentrations to near preindustrial levels [Archer et al., 1998].

In the meantime, the carbonate chemistry of shallow oceanic waters will reflect atmospheric CO_2 concentrations. Because coral reef ecosystems are confined to the upper mixed layer of the ocean, they will be readily exposed to these shifts in the ocean carbonate equilibrium.

2.3. Evidence for Seawater Carbonate Chemistry Changes

2.3.1. Direct measurements

Changes in seawater carbonate chemistry over time have been best observed at the oceanographic stations in Hawaii and Bermuda, where monthly measurements have been taken for over ten years. Data from both stations illustrate that changes in seawater carbonate chemistry coincide with changes in atmospheric CO_2 (Figure 3) [Bates, 2001; Keeling et al., 2004].

A large number of observations of open seawater carbonate chemistry have been obtained in the last decade or so. During the 1990's, inorganic carbon measurements from some 10,000 stations were obtained from nearly 100 oceanographic cruises as part of the World Ocean Circulation Experiment (WOCE) and the Joint Global Ocean Flux Survey (JGOFS). These measurements were used to map the distribution of carbon in the oceans, to determine what proportion of the carbon was anthropogenic [Sabine et al., 2004], and to determine temporal changes in the ocean $CaCO_3$ system [Feely et al., 2004] (Plate 1). These changes agree well with predictions. The increased oceanic burden of CO_2 is causing a shift in the $CaCO_3$ system in the oceans, and a shallowing of the $CaCO_3$ saturation horizons [Feely et al., 2004].

2.3.2. Isotopic data

Carbon derived from fossil fuel burning is isotopically lighter than that of the atmosphere, and consequently the $^{13}C/^{12}C$ ratio ($\delta^{13}C$) of the atmosphere is decreasing as more fossil fuel is added to the system. This leads to a parallel decrease in $\delta^{13}C$ of carbon in the ocean. Quantitative estimates of oceanic uptake of carbon from the atmosphere obtained by measuring the decrease in $\delta^{13}C$ of oceanic carbon over time have independently verified rates of oceanic carbon uptake [Quay et al., 1992] estimated by direct measurements and modeling.

2.3.3. Modeled data

Ocean circulation models that include carbon system chemistry have increasingly been used to predict changes in oceanic carbon chemistry over time. The Ocean Carbon Model Intercomparison Project (OCMIP) involved some 14 models that included not only carbonate chemistry, but also a simple biological component [Orr et al., 2001]. Among those models deemed to produce the most realistic simulations (based on comparisons with radiocarbon and CFC-11 ocean tracers), modeled oceanic carbon uptake during the 1980s and 1990s agree well with observations [Matsumoto et al., 2004].

The only model to take mineralogical differences into account was that by Anderssen et al. [2003]. This model was designed to examine the capacity of high-Mg calcite dissolution in shelf sediments to buffer the aragonite and calcite saturation states of the oceans, and illustrated that the extent of buffering was small.

2.4. Mineralogy, Taxonomy, and Biocalcification Mechanisms

The effects of changing seawater carbonate chemistry on calcifying organisms are likely to vary across taxa. This is due to differences in the physiological mechanisms of shell formation, the site of calcification, as well as mineralogical composition and crystal structure.

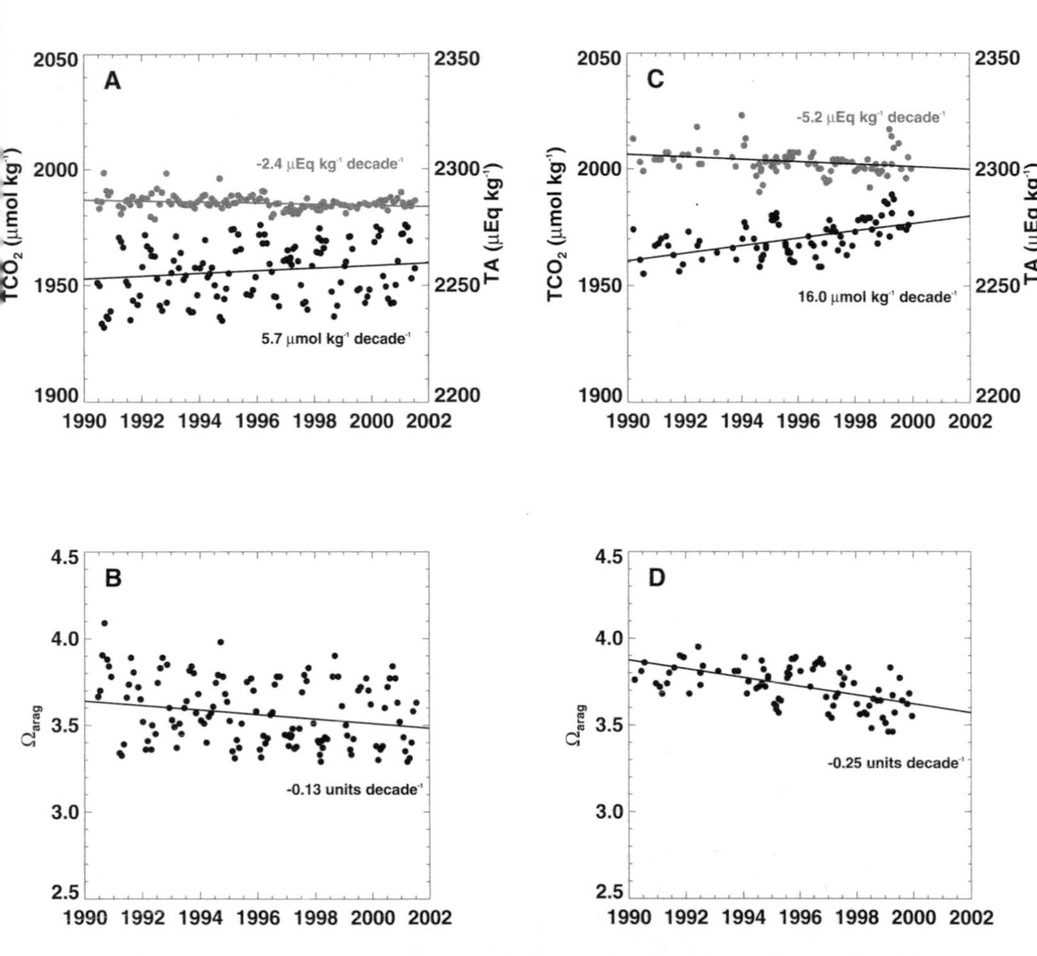

Figure 3. Changes in carbonate chemistry at the Bermuda Atlantic Time Series (BATS) Station and the Hawaiian Ocean Time-series (HOT) Station, for the period 1990 through 2002. A. TCO_2 (black) and TA (gray) at BATS (both normalized to salinity). B. Aragonite saturation state at BATS. C. TCO_2 (black) and TA (gray) at HOT (both normalized to salinity). D. Aragonite saturation state at HOT.

2.4.1. Mineralogy and taxonomy

Marine carbonates on coral reefs exist almost entirely as calcite (also called low-Mg calcite), high-Mg calcite (calcite with ≥5% $MgCO_3$), aragonite, or a combination of calcite and aragonite. Of these, calcite is the most stable in the oceans, while high-Mg calcite with greater than about 12% $MgCO_3$ is the least stable. The two main reef-builders secrete the more soluble forms of $CaCO_3$. Corals secrete aragonite, while calcifying algae secrete either aragonite (green algae such as *Halimeda*) or high-Mg calcite (coralline red algae). Tropical benthic foraminifera, also a major contributor to reef sediments, predominantly secrete high-Mg calcite.

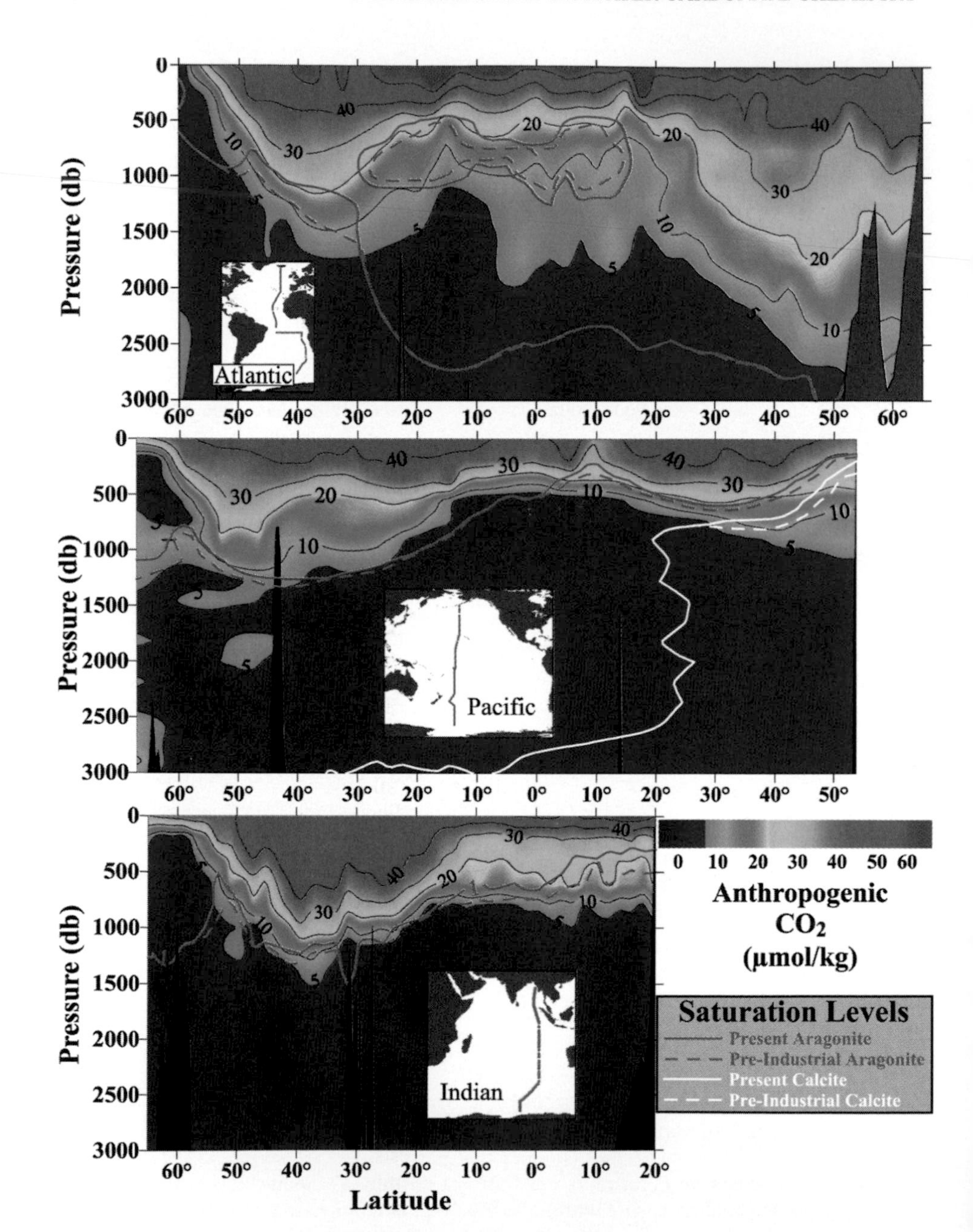

Plate 1. Present-day distribution of anthropogenic CO_2 in the open ocean, showing changes in the depth of the aragonite and calcite saturation horizons (reprinted with permission from Feely et al. [2004], copyright AAAS).

Organisms within major taxonomic groups tend to have the same mineralogy, and this appears to reflect the ocean chemistry conditions, particularly the Mg/Ca ratio, at the time the group evolved [Stanley and Hardie, 1998]. In some groups, skeletal mineralogy reflects changes in ambient conditions; the mineralogy of several species of coralline algae shifted from high-Mg calcite to low-Mg calcite when grown in seawater with a low Mg/Ca ratio [Stanley et al., 2002]. The skeletal bulk of some groups also seems to wax and wane with shifting environmental conditions. Termed "hypercalcifiers" by Stanley and Hardie [1998], these groups include those that tend to secrete massive skeletons and that have "poor control" over calcification (e.g., corals, coralline algae, coccolithophores, and bryozoans; see below).

2.4.2. Biocalcification mechanisms

Mechanisms for biomineralization are sometimes categorized as either biologically-induced or biologically-controlled [Weiner and Dove, 2003]. Biologically-induced calcification occurs as a consequence of biological activities with no direct control by the organism on the calcification process, and the $CaCO_3$ is usually precipitated external to the living cells. Calcification associated with biofilms is a common biologically-induced process. Calcification in the green alga *Halimeda discoidea* has also been described as biologically-induced by pH changes during photosynthesis [De Beer and Larkum, 2001], although the consistent morphology of *Halimeda* segments indicates that there is at least some biological control [Jonathan Erez, pers. comm.].

Biologically-controlled calcification implies that the organism has some control on the calcification process, and is further categorized according to where calcification occurs: extracellular, intercellular, or intracellular [Weiner and Dove, 2003]. Most calcifiers on coral reefs exert some level of control on calcification. Many coralline algae secrete $CaCO_3$ intercellularly (outside or within their cell walls), while corals secrete $CaCO_3$ extracellularly, at the base of the polyps beneath the calicoblastic layer. It seems logical to assume that organisms with less biological control over the calcification process are likely to be more sensitive to environmental change than those in which it is more strongly regulated, but this assumption has not been rigorously tested.

Once formed, skeletal carbonates experience differential rates of dissolution. Dissolution rates are not only affected by mineralogy, but also by thickness, texture, and morphology of the skeletal material, whether it is protected by organic matter, by the crystal structure itself, and by the trace element composition and presence of adhered small particles (see, for example, Bischoff et al. [1993]).

2.5. Natural Variability of the Carbonate System on Coral Reefs

Metabolic processes on coral reefs (photosynthesis, respiration and calcification) impose a powerful control on seawater carbonate chemistry in many reef environments. The greater the surface to volume ratio of the system (reef surface to overlying water volume) – that is, the shallower the water depth – the greater the impact these processes have on the carbonate chemistry (Table 1). The diurnal fluctuations in carbonate system parameters can be dramatic (Figure 4). The greatest reported range in aragonite saturation state (1.83-6.36) was from a shallow (<1m depth) reef flat in Okinawa [Ohde and van Woesik, 1999]. Most of this variation was due to high rates of photosynthetic drawdown of CO_2.

TABLE 1. Reported changes in seawater carbonate chemistry in coral reef waters. TA = total alkalinity.

Study	Location	pCO_2 or fCO_2 µatm	TA reef µeq kg^{-1}	TA offshore µeq kg^{-1}	Comments
Frankignoulle et al., 1996; Gattuso et al., 1996	Moorea lagoon	240-400	–	2378[a]	maximum range of hourly measurements taken over 3 days
"	Yonge Reef	250-700	–	2334[a]	"
Suzuki and Kawahata, 1999	Palau barrier reef	366-414	–	–	difference between lagoon and offshore values
"	Majuro Atoll	345-370	2264-2266[b]	2287[b]	"
"	South Male Atoll	362-368	2291-2317[b]	2317[b]	"
Ohde and Van Woesik, 1999	Rukan sho reef flat	37-813	1714-2348[a]	2294-2336[a]	max range of hourly measurements taken over 5 separate days
"	Rukan sho lagoon	119-378	1974-2319[a]	2294-2336[a]	"
Kawahata et al., 2000	Great Barrier Reef Lagoon				transects of measurements inside and outside GBR
Bates et al., 2001	Bermuda reef flat	340-470	2329-2357[c]	2378[c]	pCO_2: max range of hourly measurements taken over 24 days; TA: range between 0800 and 1600 h of one day of sampling
Kayanne et al., 2003	Shiraho Reef (Ryukyu Is.)	200-600	2100-2270[a]	–	range of nearly continuous measurements taken over 3 days; note that TA remained nearly constant at 2270 at night.

[a] TA normalized to salinity not reported
[b] normalized to S = 35
[c] normalized to S = 36.41

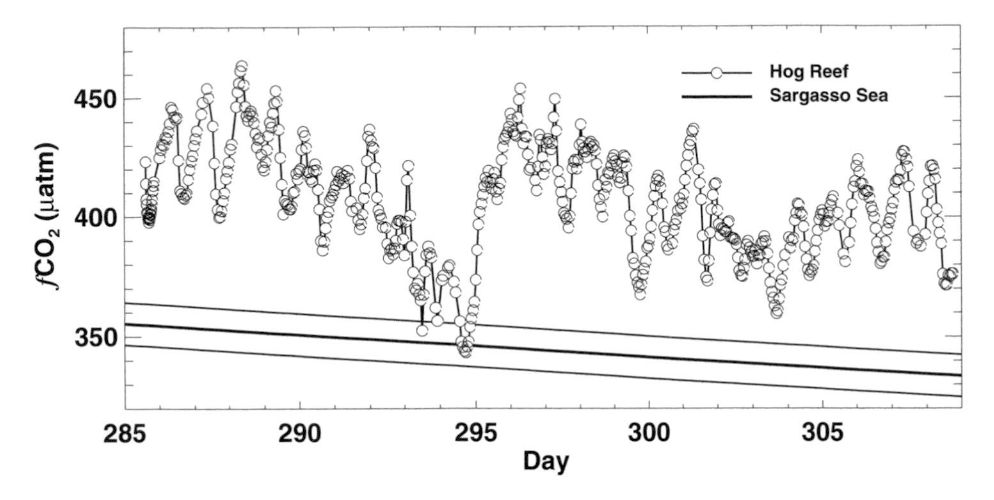

Figure 4. Diurnal fluctuations in fCO_2 on Hog Reef Flat, Bermuda (circles) for a 24-day period in 1998 (fCO_2, the fugacity of CO_2, is approximately equal to pCO_2). The mean seawater fCO_2 for the Sargasso Sea for a 4-year period (1994-1998) is indicated by the straight line (from Bates et al. [2001]; copyright by the American Society of Limnology and Oceanography, Inc.).

2.6. *Misperceptions*

2.6.1. The role of mineral buffers

Various biogeochemical feedbacks within the Earth system act to stabilize atmospheric CO_2 and the carbonate system in seawater. The most effective of these is dissolution of marine carbonates. Increasing ocean acidity leads to increased rates of dissolution of carbonate sediments (Equation 5) that nudge the system back toward equilibrium (this is practically applied in marine aquaria, by circulating waters through a calcium carbonate substrate). Under normal conditions, calcite is the most stable of the carbonate minerals, and high-Mg calcite with ≥15% $MgCO_3$ is the least stable (high-Mg calcite solubility increases with Mg content). That is, as saturation state decreases, the first mineral to dissolve is high-Mg calcite, and it is thus the "first responder" to decreasing saturation state (but as already noted, other factors also affect $CaCO_3$ solubility).

High-Mg calcite is an abundant mineral on coral reefs, mainly due to the predominance of coralline algae, which commonly comprise about 10-45% of reef sediments [Milliman, 1974], and higher percentages on reefs where they are the dominant reef builder. In general, the warmer the water, the greater the Mg content in this mineral, and the higher the Mg content, the greater its solubility [Morse and Mackenzie, 1990]. Measurements of the stoichiometric ion activity product of natural high-Mg calcites ($-\log IAP_{Mg\text{-}calcite}$) vary between –8.50 and –7.95 depending on the source and the mole % $MgCO_3$ [Mucci and Morse, 1984; Walter and Morse, 1984; Bischoff et al., 1987]. These measurements indicate that the most soluble high-Mg calcite, produced by the coralline algae *Amphiroa rigida* (22 mole % $MgCO_3$) is about 1.7-times more soluble than aragonite. Two recent studies support this observation. Morse et al. [2003] determined the solubility of carbonate produced during whitings (the sudden appearance of suspended $CaCO_3$ particles in the water column; see section 3.3.2) on the Great Bahamas Bank to be 1.9-times more soluble than aragonite; and Langdon et al. [2000] found that calcification in the Biosphere 2 coral reef mesocosm, a community dominated by *Amphiroa fragillisma*, dropped to zero when $\Omega_{arag} = 1.7 \pm 0.2$ (Figure 5). Based on these estimates high-Mg calcite begins to dissolve once the partial pressure of CO_2 (pCO_2, that is, the partial pressure of CO_2(g) that is in equilibrium with the seawater) exceeds 1000 μatm.

While the pCO_2 of open ocean seawater should remain below 1000 μatm for the remainder of this century, respiration of organic matter and other processes in sediments elevates pore water pCO_2 to high levels that cause high-Mg calcites in the sediments dissolve. Measurements of pore waters of reef sediments indicate that the carbonate chemistry is generally maintained near equilibrium with high-Mg calcite. High rates of dissolution of high-Mg calcite have also been measured on natural reefs (Great Barrier Reef algal ridges, Chisholm [2000]; Hawaiian reef sediments, Halley and Yates [2000]), and in reef flat sediments [Boucher et al., 1998]. Leclercq et al. [2002] found that sediment dissolution in their mesocosm experiments did not vary with pCO_2 variations between 411 and 918 μatm, although it is not clear whether the mesocosm sediments included high-Mg calcite.

While dissolution thus plays an important role in marine carbonate systems, one misperception is that buffering by carbonate dissolution will prevent a decrease in saturation state. Under normal circumstances, dissolution of a particular carbonate mineral proceeds as long as the seawater remains undersaturated with that mineral; dissolution ceases once saturation ($\Omega = 1$) is achieved. A simple example is a sediment mixture of *Amphiroa* (high-Mg calcite with 22% $MgCO_3$) and coral (aragonite) fragments that is exposed to increasing pCO_2 levels, and maintained at 25°C. Based on the Biosphere 2 experiments described above [Langdon et al., 2000], the high-Mg calcite would begin to dissolve once $\Omega_{hmc} = 1$, which would occur when $pCO_2 \approx 1000$ μatm. At that point, $\Omega_{arag} = 1.7$, so the aragonite

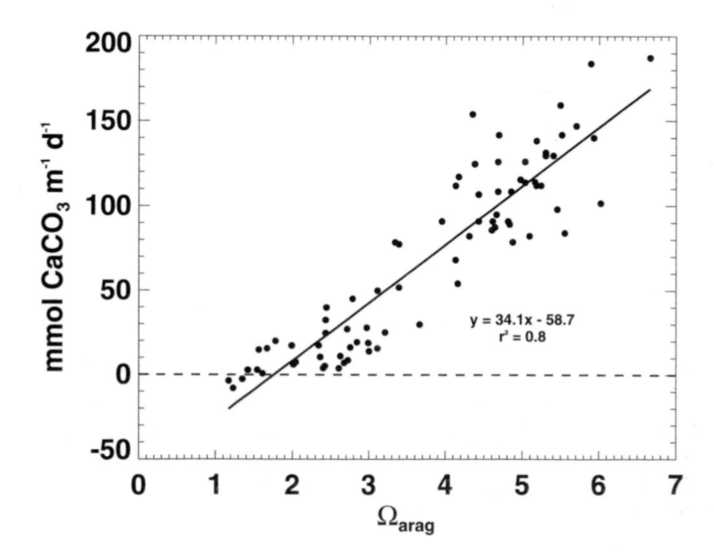

Figure 5. Calcification data from the Biosphere 2 mesocosm. Community was dominated by *Amphiroa fragillisma*, a coralline algae that secretes high-Mg calcite (22% $MgCO_3$). Calcification goes to 0.0 at $\Omega_{arag} = 1.7 \pm 0.2$ (data from Langdon et al. [2000] and other Biosphere 2 results).

sediments would not readily dissolve. As long as the high-Mg calcite dissolved rapidly enough, the saturation states would be maintained at $\Omega_{hmc} = 1$ and $\Omega_{arag} = 1.7$. Once all of the high-Mg calcite was dissolved, the saturation state would decline until $\Omega_{arag} = 1.0$, at which point aragonite sediments would readily dissolve. At no point in this process does dissolution bring the saturation state to levels consistent with preindustrial levels of 280 ppmv ($\Omega_{arag} \sim 4.2$ at 25°C). One can imagine cases where processes of dissolution are separated from the water column, so that the concentration of solutes delivered to the water column are high enough to raise alkalinity to above that of normal seawater (e.g., delivery of high-alkalinity groundwater), but such regions have not been identified.

The effectiveness of dissolution in maintaining the saturation state of a body of water ultimately depends on the residence time of the water. Coral reefs are restricted to full salinity seawater, and depend on regular flushing with open ocean waters. This needed exchange with open ocean waters dilutes the solutes released by dissolution. Some restricted lagoons, for example, may experience significant dissolution between tidal flushings. Observations within the Great Barrier Reef lagoon revealed that lagoon waters reflected a strong calcification signal relative to the open ocean waters adjacent to the lagoon; that is, higher levels of pCO_2, lower alkalinity, and thus a lower saturation state [Kawahata et al., 2000]. However, any dissolution acting to balance the carbonate system was not enough to maintain alkalinity above that of the open ocean waters. Similarly, alkalinity measurements on reef flats and in atoll lagoons are consistently lower than those of adjacent open ocean waters (Table 1). As mentioned earlier, a numerical simulation to investigate the effectiveness of high-Mg calcite dissolution in buffering future ocean pH [Anderssen et al., 2003] found that, although high-Mg calcite dissolution will increase in shelf sediments, the accumulation of alkalinity in surface waters would be small.

Silicate minerals also buffer the carbon system. As atmospheric CO_2 concentrations increase, so does the acidity of rainfall, which increases the rate of chemical weathering of both carbonate and silicate rocks on land, and thus increases the delivery of positive ions (alkalinity) to the ocean. Weathering of silicate rocks consumes atmospheric CO_2. The chemical reaction is:

$$CaSiO_3 + 2CO_2 + 2H_2O \rightarrow Ca^{2+} + 2HCO_3^- + H_2O \quad (6)$$

Geochemists estimate the rate of silicate weathering at 0.06 gigatons of C y^{-1} [Kump et al., 1999]. This process has immense capacity for taking up atmospheric CO_2 but the flux is too small to have any impact on the present build up of atmospheric CO_2. Continental weathering is thought to have drawn down the 100 ppmv increase in pCO_2 since the last glacial maximum (18,000 years ago) by about 6-12 ppmv [Munhoven, 2002].

2.6.2. CO_2 fertilization of zooxanthellae

A common misperception is that, similar to CO_2-fertilization of land plants, increases in CO_2 will enhance growth of the algal symbionts in corals, which in turn will lead to higher calcification rates. This misperception is based on two assumptions about the coral/algal symbiotic relationship: a) that increased photosynthesis increases calcification rates, and b) that zooxanthellar photosynthesis will increase with rising CO_2. The first assumption is based on the fact that zooxanthellate corals calcify faster in the light than in the dark (estimated three times as fast by Gattuso et al. [1999]), and also that calcification in bleached corals slows or stops [Leder et al., 1991]. The inference is that photosynthesis enhances calcification. It should be noted, however, that the mechanisms for light-enhanced calcification have not been adequately resolved, and that the opposite – that calcification enhances photosynthesis – has also been suggested [McConnaughey and Whelan, 1997; Cohen and McConnaughey, 2003] (see Section 4.1).

The importance of the second assumption is somewhat tempered by the fact that zooxanthellae primarily use HCO_3^- as a substrate for photosynthesis, and secondarily use CO_2 (some of which is respired CO_2). While aqueous CO_2 concentrations will increase about 100% in concert with a doubling of atmospheric CO_2, HCO_3^- concentrations will increase only about 14% (Figure 2). This increase in CO_2 and HCO_3^- may in fact stimulate photosynthesis by zooxanthellae, but it is accompanied by a decrease rather than an increase in calcification [Langdon and Atkinson, 2005]. This may be due to competition between zooxanthellae and the host for the same internal pool of dissolved inorganic carbon [Marubini and Davies, 1996; Cruz-Pinon et al., 2003].

2.6.3. Seawater carbonate chemistry is the only variable that affects calcification

Although carbonate saturation state is a likely control on marine biocalcification, at least in some taxa, it is not the only variable that affects calcification rates in corals and other reef builders. In particular, two other variables that are known to affect calcification rates are temperature [Coles and Jokiel, 1978; Houck et al., 1987; Reynaud et al., 2003; Marshall and Clode, 2004], and light [Barnes, 1982; Chalker and Taylor, 1978; Marubini et al., 2001]. These variables help define the "optimum niche" for coral calcification, and their combined effects are addressed in Section 3.3.

3. Calcification Response

3.1. Field Evidence

The first field study showing a correlation between calcification and seawater carbonate chemistry was based on precise observations of the carbonate chemistry of seawater (TA, TCO_2, and pCO_2) in waters overlying the Great Bahamas Bank [Broecker and Takahashi, 1966]. This study illustrated a relationship between the age of a parcel of water based on its salinity or ^{14}C concentration and the draw-down in total alkalinity relative to the oceanic end member (alkalinity drawdown is almost entirely caused by calcification). When Broecker and Takahashi [1966] plotted calcification rate (ΔTA/age of the water parcel) against the IAP of Ca^{2+} and CO_3^{2-} (essentially the numerator of Equation (1)), they were surprised to find a strong linear relationship. At the time they thought such a clear chemical control on the calcification rate could only mean that the calcification was abiotic. We now know that many calcifying organisms also respond linearly to changing IAP or saturation state [Langdon et al., 2000].

Coral "growth rates" have been measured in the field using several methods, but only some of these can be used as a measure of calcification rate. Calcification rates are a measure of the total $CaCO_3$ precipitated per unit area per unit time. Most early works tracked changes in coral growth by measuring extension rate (length per unit time), but this value neglects variation in skeletal density. Other methods that are used to track growth rates are weighing the specimen at regular intervals (e.g., buoyant weight; Jokiel et al. [1978]), or keeping track of the chemistry changes in seawater to account for the CO_3^{2-} uptake by the organism (e.g., alkalinity anomaly; [Smith and Kinsey, 1978; Chisholm and Gattuso, 1991]). Unfortunately, these methods have rarely been applied to corals in the field.

Retrospective records of calcification have, however, been documented from corals in the field. These are obtained from massive corals, by combining skeletal density and skeletal extension rate along the growth axis of a specimen. This technique has been used by Lough and Barnes [1997, 2000] to document historical changes in *Porites* calcification rates along the Great Barrier Reef; and similar techniques have been used by others (Table 2). All of these studies found that calcification rates in these massive corals correlated well with temperature records, and somewhat with irradiance. The original analysis of Lough and Barnes [1997] was from 35 coral cores spanning latitudes 9.4°-23.3°S, and which shared a common period of 1934-1982. Across this latitudinal range, average annual calcification in these corals increased by about 0.16 g cm^{-2} y^{-1} per 1°C. This correlation was reinforced by an additional analysis of an eight-year period on some 245 *Porites* coral heads from 29 Great Barrier Reef sites and extended to 44 sites in the Indo-Pacific [Lough and Barnes, 2000]. The spatial analysis illustrated a strong correlation between calcification rate and temperature (0.33 g cm^{-2} y^{-1} per 1°C; Figure 6; Table 2).

These same studies provided no clear evidence that calcification rates had declined over time. Average calcification rates did decrease over the last few decades, but given that 1) the 1930-1979 average calcification rate was 4% higher than that of 1880-1929, and 2) longer records extending over more than two centuries illustrate considerable variability in calcification, the authors considered the decrease as natural variability rather than an indication of changing seawater carbonate chemistry [Lough and Barnes, 1997, 2000]. The single *Porites* core analyzed by Bessat and Buigues [2003] showed an increase in calcification over time in concert with increasing air temperature records, rather than a decrease as predicted by Kleypas et al. [1999].

These temperature-calcification correlations, particularly the spatial correlations, also include the effects of other variables that co-vary with latitude and temperature, such as aragonite saturation state and light. However, records for these are few, and can only be estimated for most reef locations. What these records highlight is that coral calcification responds to a combination of variables acting both on the seawater carbonate chemistry and on the biological processes of the organisms. Some investigators have predicted that coral calcification will increase rather than decrease as a consequence of fossil fuel burning, because the effects of a temperature increase will outweigh the effects of a saturation state decrease [McNeil et al., 2004]. However, if one considers the complete picture of coral response to increasing temperature – that coral growth and calcification increase to a coral's 'optimum' temperature (close to average summertime maximum) and then decreases rapidly after (Section 3.3.1) – it is likely that the positive response to increased temperature will be short-lived [Kleypas et al., 2005]. In addition, the ability of coral adaptation to keep pace with rising temperature has not been demonstrated, particularly in light of increases in the frequency of coral bleaching over the last few decades [Hoegh-Guldberg, 1999].

3.2. Experimental Evidence

Most experimental work on coral and coral reef ecosystem responses to rising CO_2 has occurred under controlled laboratory conditions, in which seawater carbonate chemistry was manipulated while holding other variables constant (reviewed by Langdon [2002]). Although seawater carbonate chemistry can be manipulated in multiple ways (e.g., adding acid or base; bubbling CO_2 through the system; allowing community metabolism to alter the chemistry) experimental results have consistently shown that coral and coral mesocosm calcification rates are positively correlated with carbonate ion content (Table 3). The calcification response of a coral or a mesocosm to doubled CO_2 varies from about –15%

TABLE 2. Studies of measured calcification rates from massive coral skeletons. G = calcification rate.

Location/ Latitude Range	Species	Time Period	Findings	Reference
Great Barrier Reef 9.4°S-23.3°S	*Porites* spp. cores from 35 large colonies	common period: 1934-1982 total range: 1746-1982	Strong correlation with SST G = 0.16 g cm^{-2} y^{-1} per 1°C weaker correlation with irradiance avg G = 1.72 g cm^{-2} y^{-1} G most strongly correlated with extension	Lough and Barnes, 1997
Great Barrier Reef 29 reefs between 12.4°S-21.7°S	*Porites lobata* *P. lutea* *P. australiensis* *P. solida* 245 colonies	common period: 1979-1986	G = 0.39 g cm^{-2} y^{-1} per 1°C (G = 0.33 g cm^{-2} y^{-1} per 1°C when combined with data from Hawaii and Thailand) avg G = 1.63 g cm^{-2} y^{-1} G most strongly correlated with extension	Lough and Barnes, 2000
Moorea 17.5°S	*Porites lutea* 1 colony	1800-1990	strong correlation with local air temperature records: G = 0.45 g cm^{-2} y^{-1} per 1°C avg G = 1.25 g cm^{-2} y^{-1} G most strongly correlated with extension	Bessat and Buigues, 2001
Caribbean and Gulf of Mexico (12 locations)	*Montastraea annularis*	total range: 1968-1997 (not all specimens share common growth period)	Gulf specimen growth: 0.55 g cm^{-2} y^{-1} per 1°C Caribbean growth: 0.58 g cm^{-2} y^{-1} per 1°C G most strongly correlated with density	Carricart-Ganivet, 2004

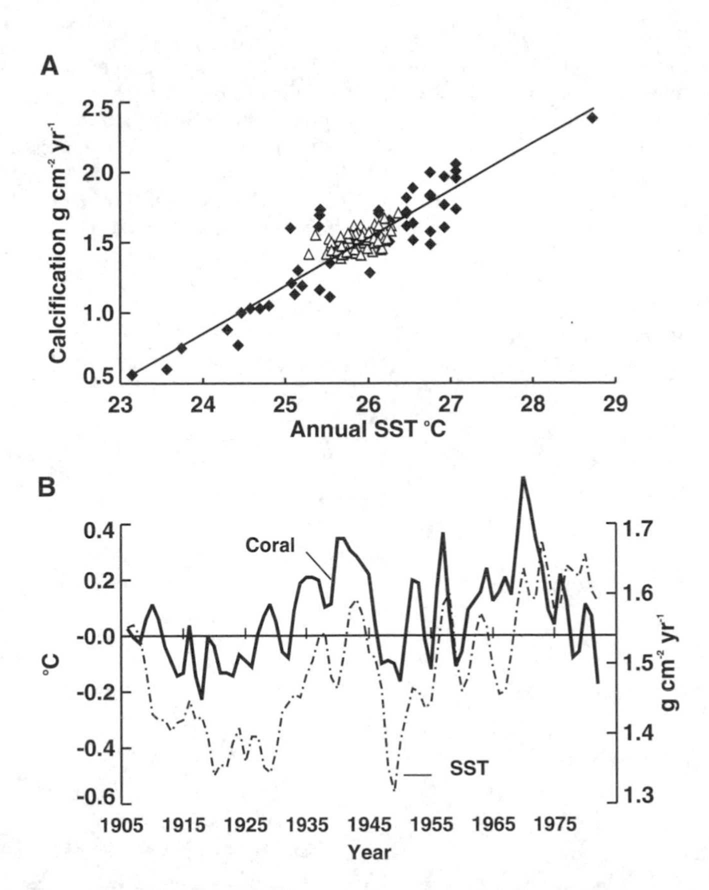

Figure 6. A. Average calcification rate from *Porites* cores versus average sea surface temperature; solid diamonds are the 1979-1986 averages from 44 *Porites* colonies; open triangles are the 1903-1982 averages from ten large *Porites* cores from the Great Barrier Reef (from Lough and Barnes [2000]). B. time-series of the 1906-1982 average calcification rates from ten large *Porites* cores from the Great Barrier Reef, shown with average Great Barrier Reef sea surface temperature for the same period (from Lough and Barnes [1997]). Both figures reprinted with permission from Elsevier.

to –54%. This range almost certainly reflects variation between species (calcification mechanisms, mineralogy, etc.) and environmental conditions (temperature, light, nutrients, etc.). Adaptation to changing seawater carbonate chemistry conditions has not been observed. Although most of these experiments have been conducted over hours to weeks, the Biosphere 2 experiments were conducted over many months, without any indication of adaptation [Langdon, 2000].

Surprisingly few realistic experiments testing calcification rates versus carbonate chemistry have been conducted on calcifying red algae and green algae (e.g., *Halimeda*) of coral reefs. Some of the studies originally cited as illustrating an effect of saturation state on algal calcification [Borowitzka, 1981; Gao, 1993] were conducted over unrealistic chemistry ranges. Borowitzka and Larkum [1976] measured both photosynthesis and calcification

TABLE 3. Measured calcification response in coral reef organisms and mesocosms (based on an earlier version by Langdon, 2002). Percent change in calcification rate (ΔG) was calculated by normalizing calcification rate to percent of the preindustrial value (assuming a preindustrial Ω_{arag} = 4.6). Manipulations of carbonate chemistry are indicated by methods (1) TA held constant, TCO_2 adjusted by bubbling CO_2 gas; (2) TCO_2 held constant, TA adjusted with acid or base; (3) pH held constant, TA and TCO_2 adjusted; (4) CO_2 (aq) held constant, pH, TCO_2, TA allowed to vary; (5) natural community alters seawater carbonate chemistry; and (6) Ca^{2+} concentration adjusted. For reference, in 26°C seawater with pCO_2 = 280 ppmv, average Ω_{arag} is about 4.6; a shift to Ω_{arag} = 3.6 (one unit) is equivalent to an increase in pCO_2 of about 140 ppmv (when pCO_2 = 420 ppmv). HMC = high-Mg calcite.

Organism/ System	Mineralogy	Manip.	Temp °C	Duration	% ΔG per unit decrease in Ω_{arag}	Reference
Calcifying Algae						
Porolithongardineri	HMC	2	19.5-30.0	20-27 d	−10.0[a]	Agegian, 1985
Corals						
Stylophora pistillata	aragonite	6	27	2.5 h	−9.0[b]	Gattuso et al., 1998
S. pistillata	aragonite	1	25.2	5 weeks	−4.5	Reynaud et al., 2003
S. pistillata	aragonite	1	28.2	5 weeks	−38.1	Reynaud et al., 2003
Madracis mirabilis	aragonite	2	26-30	12 d	−22.0	Horst and Edmunds, 2006
Acropora cervicornis	aragonite	1	25	2 months	−27.0	Renegar and Riegl, 2005
Acropora eurystoma	aragonite	2,3,4	24	1-2h	−24.0[c]	Schneider and Erez, 2006
Acropora verweyi	aragonite	2	26.5	8 d	−8.3	Marubini et al., 2003
Porites compressa	aragonite	1	26.5	10 weeks	−10.6	Marubini et al., 2001
P. compressa + Montipora capitata	aragonite	2	23.4	1.5 h	−29.0	Langdon and Atkinson, 2005
P. compressa + M. capitata	aragonite	2	27.3	1.5 h	−24.2	Langdon and Atkinson, 2005
Porites lutea	aragonite	2	25.0	3-6 d	−25.0	Ohde and Hossain, 2004
Porites lutea	aragonite	2	25.0	2-3 d	−22.0	Hossain and Ohde, 2006
Porites rus	aragonite	2	25-28	8 d	−40.0	Horst and Edmunds, 2006
Pavona cactus	aragonite	2	26.5	8 d	−9.0	Marubini et al., 2003

TABLE 3. (*Continued*)

Fungia sp.	aragonite	2	25	2-3 d	–31.0	Hossain and Ohde, 2006
Galaxea fascicularis	aragonite	2	26.5	8 d	–7.6	Marubini et al., 2003
G. fascicularis	aragonite	6	23.0	2-4 h	–39.0	Marshall and Clode, 2002
G. fascicularis	aragonite	6	19.0	2-4 h	–35.0	Marshall and Clode, 2002
Turbinaria reniformis	aragonite	2	26.5	8 d	–5.9	Marubini et al., 2003
Carbonate systems						
Gr. Bahamas Banks*	mixed	5	27.0-28.9	n/a	–37.9	Broecker and Takahashi, 1966 Broecker et al., 2001
Okinawa reef flat	mixed	5	n/a	8 h	–30.1	Ohde and van Woesik, 1999
B2 mesocosm*	mixed, mainly HMC	1,3,5	26.5	days-months	–37.2	Langdon et al., 2000
Monaco mesocosm	mixed	1	26.0	24 h	–7.8	Leclercq et al., 2000
Monaco mesocosm	mixed	1	26.0	9-30 d	–6.5	Leclercq et al., 2002

[a] Based on linear extension
[b] Based on Ω_{arag} in the range 1.9-4.0
[c] Based on light-calcification only; dark calcification value was –27%

in *Halimeda tuna* over various manipulations of seawater carbonate chemistry. Their results indicate that while photosynthesis responds to changes in CO_2 and HCO_3^- concentrations, calcification responds to changes in CO_3^{2-} concentration. Probably the best experimental data on coralline algae are those of Agegian [1985] on *Porolithon gardineri*; her results, which examined both temperature and saturation state, indicate that extension rate (used as a proxy for calcification rate, which may or may not be an appropriate measure in this species) will decline by 40% under doubled CO_2 conditions. This agrees well with the results obtained the in Biosphere 2, which was dominated by coralline red algae.

3.3. Biological versus Inorganic Components of Calcification

On coral reefs, nearly all of the calcium carbonate precipitated is biogenic. Corals, coralline algae, and benthic foraminifera secrete the bulk of the biogenic $CaCO_3$, with echinoderms, mollusks and a few other groups secreting the remainder. Inorganic precipitation of marine cements occurs at much slower rates, but can become a significant component over time.

3.3.1. Biological calcification

As described earlier, biological calcification implies that the organism exerts some level of control on its calcification. Environmental factors that affect the physiology of calcifying organisms should, therefore, also be considered as factors affecting calcification; three of these being light, temperature, and nutrients.

Relatively little is known about the combined effects of carbonate ion concentration and light levels. Only one study to date has looked at the interaction, in a two-factor experiment where *Porites compressa* was grown at four light levels and three saturation states [Marubini et al., 2001]. The results indicated no significant change in the slope of the calcification-Ω_{arag} relationship across the range of light levels tested (Figure 7); i.e., no evidence for interaction between light and saturation state.

Temperature is also known to affect calcification rates in corals. Most species calcify faster with increasing temperature until an optimal temperature is reached; beyond which calcification rates decline [Clausen, 1971; Clausen and Roth, 1975; Jokiel and Coles, 1977]. The field data from *Porites* corals [Lough and Barnes, 2000; Bessat and Buigues, 2001] and from *Montastrea* [Carricart-Ganivet, 2004] illustrate a strong correlation between temperature and calcification that has essentially overridden the predicted decrease in calcification over the Industrial Period. However, only two experimental studies have looked at the interaction between temperature and saturation state, and these produced different results. Reynaud et al. [2003] varied temperature between 25°C and 28°C and varied pCO_2 between 460 and 760 µatm. At 25°C, calcification of *Stylophora pistillata* increased slightly with increased pCO_2, while at 28°C calcification decreased by 50%. Langdon and Atkinson [2005] measured the calcification of an assemblage of *Porites compressa* and *Montipora capitata* under summer (27.3°C) and winter (23.4°C) conditions. They found no significant change in the slope or intercept of the calcification-Ω_{arag}

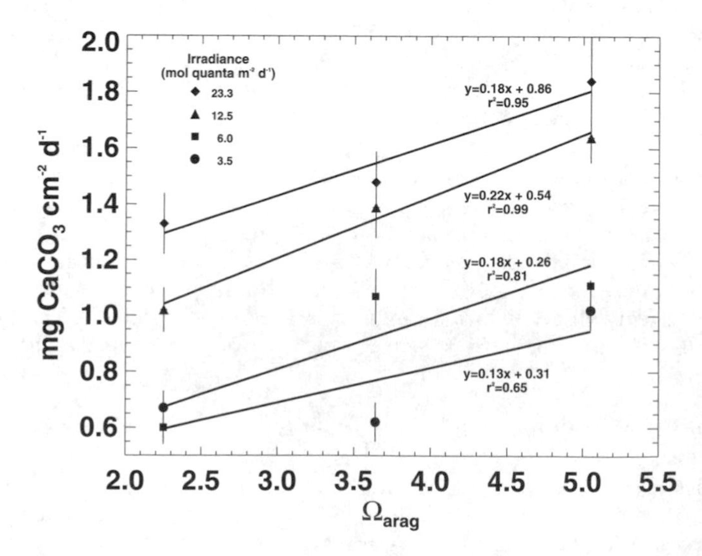

Figure 7. Interaction between altered carbonate chemistry and irradiance on the calcification of *Porites compressa* (data from Marubini et al. [2001]).

relationship between summer and winter temperatures. At present it is not known whether the CO_2–temperature interaction of typical corals is minimal or strong.

Only two studies have considered the interacting effects of nutrients and saturation state on coral calcification. Marubini and Thake [1999] measured the skeletal weight increase of *Porites porites* at three nutrient levels (ambient, +20 μM NO_3 and +20 μM NH_4) and two saturation states (2.6 and 7.4) (Figure 8A). The change in saturation state was achieved by doubling the dissolved inorganic carbon concentration with the addition of bicarbonate. Calcification decreased with nutrient enrichment in the low saturation state treatments (–25% with NH_4 enrichment, and –52% with NO_3 enrichment), but increased in the high saturation state treatments (+12% with NH_4 enrichment, and +33% with NO_3 enrichment). In essence, calcification in *P. porites* was more sensitive to changes in Ω_{arag} under increased nutrient

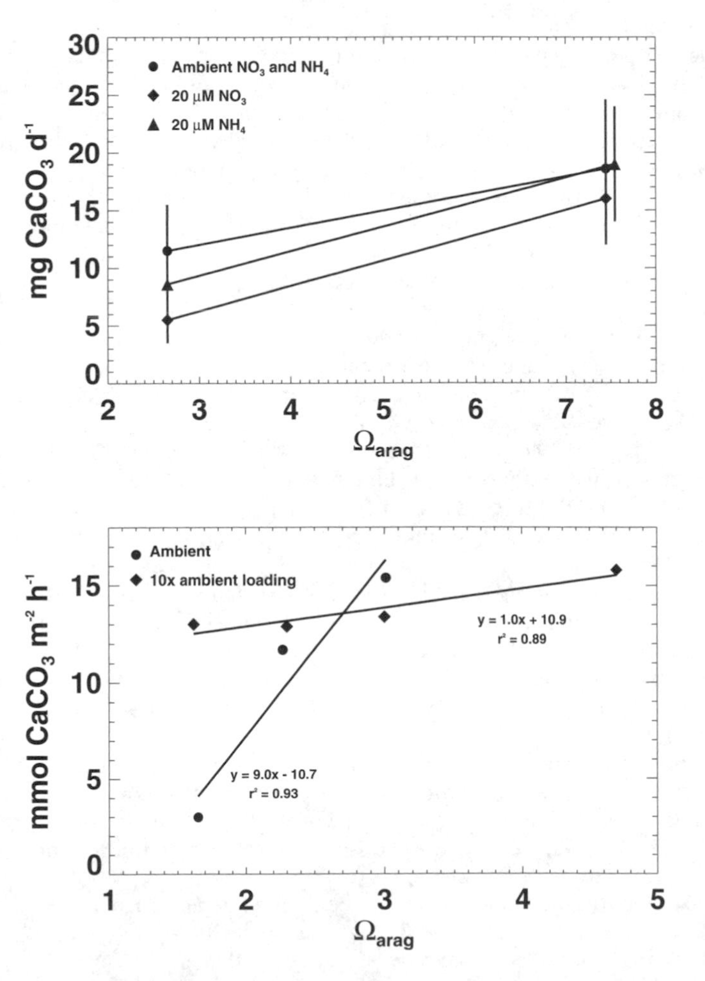

Figure 8. Interactions between altered carbonate chemistry and nutrient enrichment. A. *Porites porites* (data from Marubini and Thake [1999]). B. Mixed assemblages of *Porites compressa* and *Montipora capitata* (data from Langdon and Atkinson [2005]).

conditions. Langdon and Atkinson [2005] observed the opposite interaction (Figure 8B). They added PO_4 and NH_4 in daily pulses and observed the effect on calcification at ambient saturation state and a reduced saturation state produced by doubling the pCO_2 after only four days of nutrient enrichment. They found that nutrient enrichment made the corals *P. compressa* and *M. capitata* much less sensitive to a change in Ω_{arag}.

The experimental details were quite different in these two studies, which likely contributed to the different responses. The different results in these combined Ω_{arag}-nutrient studies, as well as in the combined Ω_{arag}-temperature studies, illustrate the complexity of these interactions, and the need for more controlled studies.

3.3.2. Inorganic calcification

Inorganic calcification within coral reef habitats or carbonate banks includes the formation of marine oöids, the precipitation of marine $CaCO_3$ cements within reef sediments and skeletal framework, and possibly the spontaneous precipitation of crystals within the water column (whitings). Although the origin of whitings remains equivocal [Morse et al., 2003], all three of these inorganic forms of $CaCO_3$ are suspected to be mediated by biological activity, or at least by the presence of organic matter [Folk and Lynch, 2001; Robbins and Blackwelder, 1992]. However, one might assume that precipitation of these "inorganic" forms of calcium carbonate should more closely reflect environmental controls. Indeed, oöids and whitings do seem to be restricted to regions where carbonate saturation state is high; e.g., the Bahamas Bank and the Persian Gulf.

The presence of marine cements within the coral reef environment exhibits spatial variation across reef systems, within individual reefs, and within corals themselves. This is partly controlled by sediment chemistry, and partly by exchange with overlying waters. Pore-water chemistry of reefs tends to be maintained close to equilibrium with aragonite and high-Mg calcite, by processes such as microbially-mediated respiration and sulfate reduction (reviewed by Morse and Mackenzie [1990]; and Morse [2003]). Cementation tends to be greatest where abundant flushing with overlying seawater supplies an adequate supply of ions. Within coral reef systems cementation is thus greatest near the continental shelf edge and lowest within inner reefs (e.g., Belize Barrier Reef: James et al., [1976]; Great Barrier Reef: Marshall, [1985]; Kleypas [1991]). On individual reefs, it tends to be greater on windward margins [Marshall, 1985]. Within individual coral heads, cementation can occur nearly contemporaneously with coral growth, with greater accumulation of inorganic cements at the base of a coral than near the living surface which, incidentally, can contaminate the geochemical records (e.g., isotopes, Ca/Sr ratios, etc. [Müller et al., 2001]). The mineralogy of submarine reef cements is overwhelmingly aragonite and high-Mg calcite.

Patterns of reef cementation across oceanic regions have been noted but have not been explicitly studied. High cementation rates have been reported in regions such as the Red Sea [Friedman et al., 1974], while reefs in the eastern Pacific have low cementation rates [Cortés, 1997]. Whether these regional differences in cementation are related to differences in seawater carbonate chemistry has not been investigated.

The important role of cementation in reef building has often been suggested, but quantitative studies have been few. Submarine cementation rates within reefs are low compared to biological rates [Oberdorfer and Buddemeier, 1986], yet cementation continues long after the biocalcifiers have died, and increases a reef's resistance to erosive hydrologic forces [Lighty, 1985]. It is not clear if cementation protects a reef from bioerosion. Highly exposed and highly indurated coralline algal reefs attract a different suite of borers and bioeroders than poorly cemented reefs in lower-energy environments [Boscence, 1985]. Bioerosion in the high-energy reefs was more extensive than in lower energy environments, but did not destroy the overall reef framework. It is unclear whether these bioeroders were more attracted by the high-indurated reef surface, or by the high-energy environment, but is clear that bioerosion can have a profound effect on reef-building. Following the 1982-83 mortality on many Eastern Pacific reef communities, bioerosion removed thousands of years of coral reef accumulation in a matter of a few years [Glynn, 1994; Eakin, 1996]. Here, bioeroders preferred softer substrates as well as the coralline algae pavements [Reaka-Kudla et al., 1996]. Did lack of cementation play a role in this high rate of erosion, or was this simply a biological process?

Whitings are not considered a major contributor to carbonate production on the Bahamas Bank [Morse et al., 2003]; but carbonate production rates by whitings and oöid formation are poorly quantified.

4. Consequences For Reef Organisms And Communities

4.1. Calcification in Various Reef-Builders

How decreased calcification rates affect biological functioning or organism survival on coral reefs remains essentially unstudied. Predictions about how corals or other calcifying organisms will be affected by decreased calcification rates are conjectural based on the fact that secretion of calcium carbonate by organisms serves some function (or multiple functions) that arguably benefits the organism. Biogenic calcification (shell formation) evolved sometime during the Cambrian period, coincident with a sudden rise in Ca^{2+}, which because high Ca^{2+} is toxic to cellular processes, suggests that it arose as a detoxification mechanism [Brennan et al., 2004]. Organisms have since evolved to put these $CaCO_3$ secretions to good use as skeletal support, protection, and many other functions.

These functions vary from species to species, and indeed within some species over the course of their life cycle. In corals and coralline algae, for example, calcification first appears when the planulae or larvae secrete calcium carbonate to cement the organism to a hard substrate. In experiments with *Porolithon*, recruitment declined under elevated pCO_2 conditions [Agegian, 1985].

Once an organism is established, continued calcification is partitioned into skeletal extension and skeletal strengthening (density). Extension provides support for colony expansion through both individual polyp growth, and colonial expansion of the organism. Skeletal extension elevates the coral above the substrate and into the hydrodynamic regime, and thus increases access to food, nutrients and well-oxygenated waters. Calcification in branching forms also brings them closer to the surface and higher light intensities. Faster growth is one strategy by which corals compete for space on a reef (reviewed by Lang and Chornesky [1990]); with some growing upward and "overtopping" lower-growing species [Glynn and Wellington, 1983]. There is also evidence that in some species, the reflectance by the aragonitic skeleton increases light gathering [Enriquez et al., 2005].

Growth rates may also influence reproduction by affecting either extension rate or density (which increases skeletal resistance to breakage). Some corals do not reproduce until they reach a certain size [Sakai, 1998; Fautin, 2002], while some take advantage of skeletal fragmentation to propagate. Skeletal fragmentation in *Acropora palmata* can actually increase asexual propagation of the species, but it has also been correlated with a lowered potential for sexual reproduction [Lirman, 2000].

Finally, calcification may enhance photosynthesis in some species by providing protons for conversion of HCO_3^- to CO_2(aq) [McConnaughey and Whelan, 1997]. This model was not supported in experiments with *Stylophora pistillata*, where suppression of calcification did not affect its photosynthetic rate [Gattuso et al., 2000], but Cohen and McConnaughey [2003] argue that the photosynthetic need for CO_2 and calcification supply of CO_2 increases the likelihood of a coupling between the two processes. Schneider and Erez [2006] support this argument with experimental data from *Acropora eurystoma*. They propose that calcification enhances photosynthesis by elevating CO_2(aq) in the coral coelenteron, which prevents CO_2 limitation of the zooxanthellae; and that the relatively constant offset between light and dark calcification can be explained by increases in coelenteron pH associated with photosynthesis (that is, photosynthesis enhances calcification).

Several studies indicate that coral skeletal growth rates are more poorly correlated with environmental conditions than are tissue growth rates [Edinger et al., 2000], perhaps because energy allocation is prioritized for skeletal rather than tissue growth, at least in larger colonies [Anthony et al., 2002]. Nonetheless, some species show dramatic ranges in coral calcification rates, accommodated by changes in either skeletal extension or density. Calcification rates in massive *Porites* decrease five-fold from low to high latitudes, a change that is reflected in extension rather than skeletal density [Lough and Barnes, 2000]. In *Montastrea annularis*, extension rates appear to be conserved regardless of the calcification rate [Carricart-Ganivet, 2004]. In the branching species *Acropora formosa* from the high latitude Houtman Abrolhos reefs, extension rate and density varied considerably between environments, but overall skeletal mass did not [Harriott, 1998].

Such variability in skeletal growth strategies under natural environmental conditions makes it difficult to predict just how reef calcifiers will cope with any future reduction in calcification rates. It is reasonable to assume that some species will be affected more than others if their calcification rates decrease due to lowered saturation state. These effects could affect a reef calcifier at multiple stages of its life cycle, from larval settlement, to its ability to compete for space and light, to its ability to reproduce. These changes are likely to affect the composition of coral reef communities, and their cumulative calcification rates.

4.2. Calcification in Reef Communities and Reefs

Reef-building requires a net positive balance of calcium carbonate production. Even in areas where $CaCO_3$ production is high, if $CaCO_3$ removal is also high, reef-building will be low, and vice-versa. The direct removal of calcifiers, such as during a severe bleaching episode, affects reef-building by removing the main source of carbonate production. The net response of coral community calcification to changing seawater carbonate chemistry, for example a decrease in average pH from 8.2 to 7.8, will be the sum of many interrelated processes such as 1) the response of calcifying organisms, 2) changes in inorganic processes of carbonate precipitation and dissolution, and 3) the response of bioeroders to changes in community structure and perhaps in cementation patterns.

Reef building requires reef-builders. Once the calcifying community is removed, reef building ceases or reverses. The opposite notion – that reef-builders require reefs – is less certain [Kleypas et al., 2001]. Although a minimum amount of net carbonate production is required to build a reef, is there also a minimum amount necessary to support a coral reef community, regardless of whether it builds a reef or not? Many coral communities do not appear to be building a reef, yet they seem to function similarly to those that are building reefs. Some aspects of reef-building are obviously important to the reef community: 1) the ability to keep up with sea level rise, 2) the creation of spatial complexity that supports diversity, 3) the depth gradient that also supports biodiversity, and 4) the structural influence on the local hydrodynamic regime.

5. Knowledge Gaps and Future Research Directions

Finally, it is necessary to stress that the researches here described are but tentative early advances in a largely new field – so new that the greatest obstacle to progress is simply a lack of data. Faced with new problems we must go back to the beginning and patiently start again the task of describing and measuring before we can hope to make secure generalizations. – Bathurst, 1974.

Following the original warning that changing seawater carbonate chemistry might impact coral reefs of the future [Smith and Buddemeier, 1992], multiple studies have verified these concerns, both with respect to observations that the carbonate chemistry of the surface ocean is changing, and that calcification rates of major groups of organisms decline under increasing acidification. But many questions remain before we can make accurate predictions of how coral reef ecosystems will respond to calcium carbonate saturation states that are probably lower than have occurred for at least several hundred thousand years. We close this chapter by identifying major gaps in our scientific understanding of this problem, and a discussion of what approaches might best fill these gaps.

Most of the work that remains is an echo of that stated by Bathurst [1974] above, in his consideration of research needs within another aspect of carbonate chemistry, that of marine diagenesis. Even though the effects of changing seawater carbonate chemistry on coral reefs are potentially extremely important, there is resistance to understanding the problem, in part because the *greatest obstacle to progress is simply a lack of data.*

5.1. Gaps and the Need for a Comprehensive, Integrated Research Plan

5.1.1. Field measurements of seawater carbonate chemistry and calcification rates

One of the largest gaps in our understanding of coral reef calcification in relation to carbonate chemistry is our lack of observations of seawater carbonate chemistry in reef environments. There are very few measurements of seawater carbonate chemistry changes on coral reefs, and many of these are limited both spatially and temporally. If we are to understand the effects of changing ocean chemistry on coral reef ecosystems, many more measurements are needed in coral reef environments; perhaps not as routinely as temperature measurements, but at least as routine as measurements of light. The recent, large-scale observations of ocean chemistry associated with WOCE and JGOFS have provided invaluable data about oceanic responses to increasing atmospheric CO_2, but these observations were taken from open ocean waters, almost entirely away from carbonate platforms.

Field measurements of calcification on coral reefs are similarly limited, despite the fact that these observations attest to the tremendous impact of calcification on the surrounding seawater carbonate chemistry, and even fewer measurements have been made of dissolution. Can we design ways to monitor calcification and dissolution rates within reef-building organisms and within coral communities as a whole? And if so, can we find ways to effectively manipulate the carbonate chemistry on a coral reef similarly to the efforts in the terrestrial environment (e.g., Free Air CO_2 Enrichment (FACE) [Hendrey et al., [1999])?

These observational gaps exist for a good reason. There are tremendous difficulties associated with monitoring both the carbonate system in seawater and calcification rates. Characterization of the carbonate system requires measuring at least two of the parameters: pH, total CO_2, alkalinity, pCO_2. Although automated instrumentation is currently being developed to simultaneously measure these parameters [Kayanne et al., 2002a, 2002b; Watanabe et al., 2004], field measurements by-and-large still require sample collection and laboratory analysis. In addition, most previous studies relied on extension rates (length per unit time) as a measure of "growth" and did not measure true calcification rate (mass per unit time), and there is strong evidence that the two measurements are not interchangeable [Lough and Barnes, 1997].

Coral reefs occur across a range of environments that will affect both organic and inorganic carbon exchange on the reef (Figure 9) [Suzuki and Kawahata, 2003]; as well as the exchange between sediment pore-water and seawater. Even very good measurements in this environment must be coupled with good measurements of hydrodynamic exchanges,

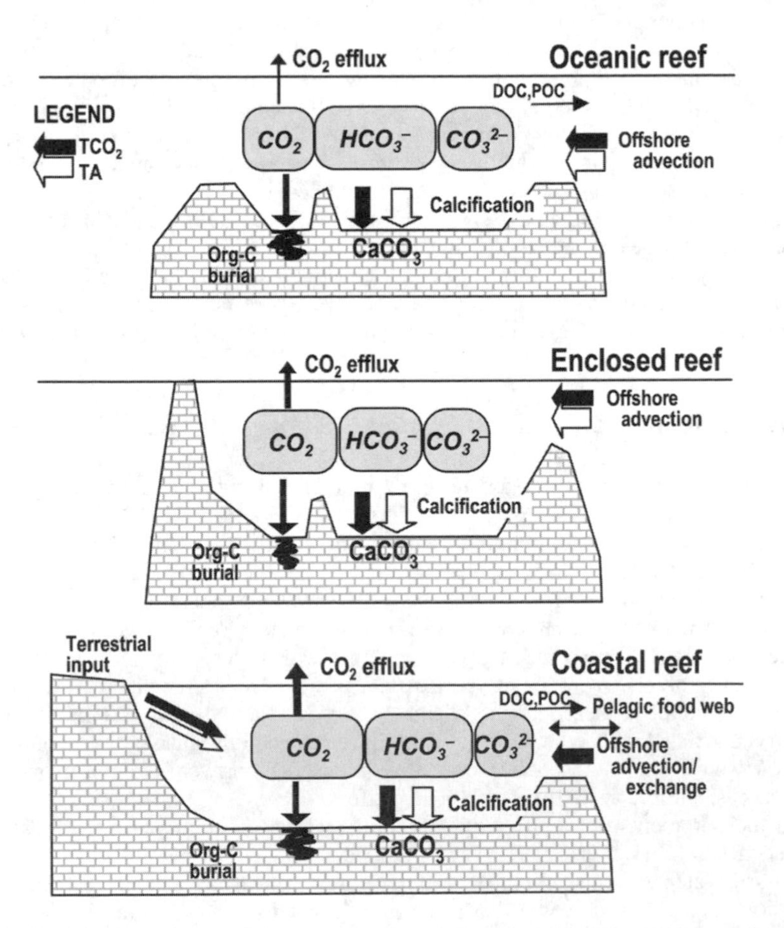

Figure 9. Variation in carbon cycling on coral reefs, as a function of reef morphology and proximity to land. AT = total alkalinity, DIC = dissolved inorganic carbon, DOC = dissolved organic carbon, POC = particulate organic carbon, DIN = dissolved inorganic nitrogen, DIP = dissolved inorganic phosphorus (from Suzuki and Kawahata [2003]; copyright Blackwell Publishing).

because it is likely the system is responding in multiple ways to CO_2 forcing, and in multiple parts of the system. Future collection of seawater carbonate chemistry and calcification data would ideally be complemented with measurements of those environmental variables that affect them. These include the obvious parameters of temperature, salinity and light, but also wind speed, the hydrodynamic regime, and the carbonate chemistry of the adjacent open ocean water mass. Installation of weather/ocean observing stations on reefs has increased in recent years, largely motivated by the need to monitor and research conditions that lead to coral bleaching, and many of these include measurements of temperature, salinity, light, and wind speeds (e.g., Berkelmans et al. [2002]). These stations are logical platforms for future collection of carbonate chemistry data.

5.1.2. Experimental needs

Essentially all published studies on the biocalcification response to increases in pCO_2 have come from laboratory experiments. Most of these have concentrated on measuring the response to saturation state alone, and most have focused on scleractinian corals. The next step in these experiments is to measure the combined effects of multiple variables known to affect coral calcification; e.g., temperature, light, and nutrients. In addition, there are many reef-building and reef-dwelling calcifying organisms that have not been tested in such experiments. One prime candidate for such studies is the calcifying green alga *Halimeda*, which is not only capable of high rates of calcium carbonate production [Hillis, 1997], but also appears to have little biological control on calcification. Other candidates include the main reef-building crustose coralline algae; but beyond the reef-builders many important members of the coral reef community, for example benthic foraminifera and echinoderms (both groups secrete high-Mg calcite), may also be affected by increases in pCO_2.

5.1.3. Standardization of measurements

Finally, there is a need to standardize methodologies. How can the various measurements of calcification – buoyant weight, alkalinity anomaly technique, ^{45}Ca uptake, densitometry, and tomography – be related to each other (Table 4)? Some of these techniques work better over short time scales and others over longer time scales; some require destruction of the coral while others do not. Many of these techniques have been qualitatively evaluated against other methods, but systematic comparisons of techniques within the same experiment are few. Understandably, these techniques do not necessarily measure the same aspect of calcification rate. Coral calcification rates taken over hours to weeks may not be comparable to calcification rates integrated over an entire year. In some measurements, dissolution and inorganic cementation must also be taken into account.

5.2. A Recommended Strategy

It is tempting to label as "controversial" the separate facts that 1) experimental results indicate that calcification rates should be decreasing in reef-building calcifiers due to rising CO_2, and 2) such a decrease has not been detected in massive corals. But differences between laboratory and field studies, particularly so few, should not be surprising. Nor is it surprising that coral reef calcification probably responds in complex ways to multiple

TABLE 4. Methods used to measure calcification rates in coral reef environments. G_{skel} = skeletal calcification; D_{skel} = skeletal dissolution; G_{inorg} = inorganic cementation; G_{sys} = system calcification; D_{sys} = system dissolution.

Technique	Parameter Measured	Timescale	Reference
Individual Organisms			
^{45}Ca labeling	G_{skel}	Duration of incubation (hours)	Clausen and Roth, 1975
Buoyant weight	$G_{skel} + D_{skel}$	Duration of experiment (weeks to years)	Jokiel et al., 1976 Dodge et al., 1984
Coral band increment (extension × density); density obtained via x-radiography tomography, densitometry	$G_{skel} + D_{skel} + G_{inorg}$	Integrated over time of band formation + post-depositional cementation	Lough and Barnes, 1997 Bessat and Buigues, 2001 Carricart-Ganivet, 2004
ΔAlk of monoculture	$G_{skel} + D_{skel}$	Discrete measurements over duration of experiment	Gattuso et al., 1998 Marubini et al., 2003
Systems			
ΔAlk of closed system	$G_{sys} + D_{sys}$	Discrete measurements over duration of experiment	Langdon et al., 2000 Leclercq et al., 2000
ΔAlk in open system	$G_{sys} + D_{sys}$ + mixing	Discrete measurements over duration of experiment – requires knowledge of mixing regime	Smith and Key, 1975
Geometric measurements (reef volume × reef density)	$G_{sys} + D_{sys}$ + imported $CaCO_3$ – exported $CaCO_3$	Integrated over geologic timescales; calculates volume of reef structure; $CaCO_3$ content based on assumptions of porosity of structure; requires radiometric dating to estimate time range of accumulation	Ryan et al., 2001

changes in a highly variable environment (temperature, saturation state, nutrients etc.). A successful research strategy to study the effects of changing seawater carbonate chemistry on calcification in coral reef ecosystems will require a combination of efforts from multiple disciplines and across multiple spatial and temporal scales. The following research needs provide a starting point toward developing such a strategy:

1. Increase measurements of the carbonate system on coral reefs, and commit to monitoring over periods adequate for detecting the response of the system to continued

increases in CO_2. Focus on collecting information from a variety of ocean settings that cover the important environmental ranges and seawater carbonate chemistry conditions; as well as the range of reef settings (e.g., well-mixed open ocean versus lagoonal) and environments (e.g., forereef, reef flat, lagoonal).

2. In concert with above, monitor *in situ* calcification and dissolution in organisms, with better characterization of the key environmental controls on calcification. Supplement and cross-check present-day measurements with coral skeletal records of calcification.
3. Continue experimental studies that combine multiple variables that affect calcification in organisms: saturation state, light, temperature, nutrients; and extend the range of calcifying groups tested.
4. Combine laboratory experiments with those in the natural environment. Develop and deploy technology for continuous field experiments analogous to the CO_2 enrichment experiments performed in terrestrial systems.
5. Incorporate ecological questions into observations and experiments; e.g., how does a change in calcification rate affect the ecology of an organism; at the ecosystem scale, what are the ecological differences between coral reefs and non-reef building coral communities.
6. Improve our accounting of coral reef carbonate budgets, through combined measurements of seawater carbonate chemistry, bioerosion, dissolution, and off-reef export of calcium carbonate.
7. Apply biogeochemical and ecological modeling to help quantify the mechanisms that contribute to the carbonate system, and to guide future sampling and experimental efforts.
8. Conduct physiological experiments to discriminate the various mechanisms of biocalcification within calcifying groups, and thus better understand the cross-taxa range of responses to changing seawater carbonate chemistry.
9. Develop protocols for the various methodologies used in seawater carbonate chemistry and calcification measurements. Establish the pros and cons of each procedure, and when possible, how each measurement can be related to the others.

5.3. Closing Comment

We now know that the carbonate chemistry of the surface ocean is changing in response to CO_2 forcing from fossil fuel emissions. We also know that the near-future changes in the CO_2 system in seawater will lie outside the range of conditions experienced by coral reefs for the past hundreds of thousands of years. Based on controlled experiments, calcification rates of reef organisms are expected to decrease under increased pCO_2 conditions; but coral records over the past century have not recorded a clear post-industrial signal of decreased calcification. The overarching question: "How will calcification rates of reef-building organisms and reefs themselves change in response to increased pCO_2 forcing" remains unanswered. Here, we have attempted to provide the background necessary for understanding the complexity of the question, as a first step toward answering it.

Acknowledgments. This chapter summarizes research from many in the scientific community, and in that sense, all of the names listed in the References are acknowledged for their contributions to this topic. Janice Lough, as usual, offered valuable comments on an early draft. We also thank Jonathan Erez and an anonymous reviewer for their very constructive inputs.

References

Agegian, C. R., The biogeochemical ecology of *Porolithon gardineri* (Foslie). Ph.D. Dissertation, University of Hawaii, pp. 178, 1985.

Andersson, A. J., F. T. Mackenzie, and L. M. Ver, Solution of shallow-water carbonates: An insignificant buffer against rising atmospheric CO_2, *Geology*, 31, 513-16, 2003.

Anthony, K. R. N., S. R. Connolly, and B. L. Willis, Comparative analysis of energy allocation to tissue and skeletal growth in corals, *Limnol. Oceanogr.*, 47, 1417-1429, 2002.

Archer, D., Biological fluxes in the ocean and atmospheric pCO_2, In *Treatise on Geochemistry,* vol. 6, *The Oceans and Marine Geochemistry*, H. Elderfield (ed.), pp. 275-291, Elsevier, 2003.

Archer, D., H. Kheshgi, and E. Maier-Reimer, Dynamics of fossil fuel CO_2 neutralization by marine $CaCO_3$, *Global Biogeochem. Cycles*, 12, 259-276, 1998.

Armstrong, R. A., C. Lee, J. I. Hedges, S. Honjo, and S. G. Wakeham, A new, mechanistic model for organic carbon fluxes in the ocean, based on the quantitative association of POC with ballast minerals. *Deep-Sea Res. II*, 49, 219-236, 2002.

Barnes, D. J., Light response curve for calcification in the staghorn coral, *Acropora acuminata*, *Comp. Biochem. Physiol.*, 73A, 41-45, 1982.

Bates, N. R., Interannual variability of oceanic CO_2 and biogeochemical properties in the Western North Atlantic subtropical gyre, *Deep-Sea Research II*, 48, 1507-1528, 2001.

Bates, N. R., L. Samuels, and L. Merlivat, Biogeochemical and physical factors influencing seawater fCO_2 and air-sea CO_2 exchange on the Bermuda coral reef, *Limnol. Oceanogr.*, 46, 833-846, 2001.

Bathurst, R. G. C., Marine diagenesis of shallow water calcium carbonate sediments, *Annual Review of Earth and Planetary Sciences*, 2, 257-274, 1974.

Berger, W. H., Increase of carbon dioxide in the atmosphere during deglaciation: The coral reef hypothesis, *Naturwissenschaften*, 69, 87-88, 1982.

Berkelmans, R., J. C. Hendee, P. A. Marshall, P. V. Ridd, A. R. Orpin, and D. Irvine, Automatic weather stations: Tools for managing and monitoring potential impacts to coral reefs. *Mar. Tech. Soc. Jour.*, 36, 29-38, 2002.

Bessat, F., and D. Buigues, Two centuries of variation in coral growth in a massive *Porites* colony from Moorea (French Polynesia): A response of ocean-atmosphere variability from south central Pacific, *Palaeogeogr., Palaeoclimatol., Palaeoecol.*, 175, 381-392, 2001.

Bischoff, W. D., M. A. Bertram, F. T. Mackenzie, and F. C. Bishop, Diagenetic stabilization pathways of magnesian calcites, *Carbonates and Evaporites*, 1, 82-89, 1993.

Bischoff, W. D., F. T. Mackenzie, and F. C. Bishop, Stabilities of synthetic magnesian calcites in aqueous solution: Comparison with biogenic materials, *Geochim. Cosmochim. Acta*, 51, 1413-1423, 1987.

Borowitzka, M. A., Photosynthesis and calcification in the articulated coralline alga *Amphiroa anceps* and *A. foliaceae*. *Marine Biology*, 62, 17-23, 1981.

Borowitzka, M. A., and A. W. D. Larkum, Calcification in the green alga *Halimeda*: III The sources of inorganic carbon for photosynthesis and calcification and a model of the mechanism of calcification, *J. Exp. Bot.*, 27, 879-893, 1976.

Bosence, D., Preservation of coralline-algal reef frameworks, *Proc. 5th Int. Coral Reef Cong.*, Tahiti, 1985, vol. 6, 623-628, 1985.

Boucher, G., J. Clavier, C. Hily, and J.-P. Gattuso, Contribution of soft-bottoms to the community metabolism (primary production and calcification) of a barrier reef flat (Moorea, French Polynesia), *J. Exp. Mar. Biol. Ecol.*, 225, 269-283, 1998.

Brennan, S. T., T. K. Lowenstein, and J. Horita, Seawater chemistry and the advent of biocalcification, *Geology*, 32, 473-476, 2004.
Broecker, W. S., and T. Takahashi, Calcium carbonate precipitation on the Bahama Banks, *J. Geophys. Res.*, 71, 1575-1602, 1966.
Broecker, W. S. C. Langdon, T. Takahashi, and T.-S. Peng, Factors controlling the rate of $CaCO_3$ precipitation on Grand Bahama Bank, *Global Biogeochemical Cycles*, 15, 589-596, 2001.
Carricart-Ganivet, J. P., Sea surface temperature and the growth of the west Atlantic reef-building coral *Montastraea annularis*, *J. Exp. Mar. Biol. Ecol.*, 302, 249-260, 2004.
Chalker, B. E., and D. L. Taylor, Rhythmic variations in calcification and photosynthesis associated with the coral *Acropora cervicornis* (Lamarck), *Proc. R. Soc. Lond. B*, 201, 179-189, 1978.
Chisholm, J. R. M., Calcification by crustose coralline algae on the Northern Great Barrier Reef, Australia, *Limnol. Oceanogr.*, 45, 1476-1484, 2000.
Chisholm, J. R. M., and J.-P. Gattuso, Validation of the alkalinity anomaly technique for investigating calcification and photosynthesis in coral reef communities. *Limnol. Oceanogr.*, 36, 1232-1239, 1991.
Clausen, C., Effects of temperature on the rate of ^{45}calcium uptake by *Pocillopora damicornis*, In: *Experimental Coelenterate Biology*, H. M. Lenhoff, L. Muscatine, and L. V. Davis (editors), University of Hawaii Press, Honolulu, pp. 246-259, 1971.
Clausen, C. D., and A. A. Roth, Effect of temperature and temperature adaptation on calcification rate in the hermatypic coral *Pocillopora damicornis*, *Marine Biology*, 33, 93-100, 1975.
Cohen, A. L., and T. A. McConnaughey, Geochemical Perspective on Coral Mineralization, In: *Reviews in Mineralogy and Geochemistry*, P. M. Dove, J. J. De Yoreo, and S. Weiner (editors), vol. 54, pp. 151-187, 2003.
Coles, S. L., and P. L. Jokiel, Synergistic effects of temperature, salinity and light on the hermatypic coral *Montipora verrucosa*, *Mar. Biol.*, 49, 187-195, 1978.
Cortés, J., Biology and geology of eastern Pacific coral reefs, *Coral Reefs*, 16 (suppl.), S39-S46, 1997.
Cruz-Pinon, G., J. P. Carricart-Ganivet, and J. Espinoza-Avalos, Monthly skeletal extension rates of the hermatypic corals *Montastraea annularis* and *Montastraea faveolata*: biological and environmental controls, *Marine Biology*, 143, 491-500, 2003.
De Beer, D., and A. W. D. Larkum, Photosynthesis and calcification in the calcifying algae *Halimeda discoidea* studied with microsensors, *Plant, Cell and Environment*, 24, 1209-1217, 2001.
Dodge, R. E., S. C. Wyers, H. R. Frith, A. H. Knap, S. R. Smith SR, C. B. Cook, and T. D. Sleeter, Coral calcification rates by the buoyant weight technique: effects of Alizarin staining, *J. Exp. Mar. Biol. Ecol.*, 75, 217-232, 1984.
DOE, *Handbook of methods for the analysis of the various parameters of the carbon dioxide system in sea water*, Version 2, A. G. Dickson and C. Goyet, eds, ORNL/CDIAC-74, 1994.
Eakin, C. M., Where have all the carbonates gone? A model comparison of calcium carbonate budgets before and after the 1982–1983 El Niño at Uva Island in the eastern Pacific, *Coral Reefs*, 15, 109-119, 1996.
Edinger, E. N., G. V. Limmon, J. Jompa, W. Widjatmoko, J. M. Keikoop, and M. J. Risk, Normal coral growth rates on dying reefs: Are coral growth rates good indicators of reef health? *Mar. Poll. Bull.*, 40, 404-425, 2000.
Enriquez, S., E. R. Mendez, and R. Iglesias-Prieto, Multiple scattering on coral skeletons enhances light absorption by symbiotic algae, *Limnol. Oceanogr.*, 50, 1025-1032, 2005.
Fautin, D. G., Reproduction of Cnidaria, *Can. J. Zool.,* 80, 1735-1754, 2002.
Feely, R. A., C. L. Sabine, K. Lee, W. Berelson, J. Kleypas, V. J. Fabry, and F. J. Millero, The impact of anthropogenic CO_2 on the $CaCO_3$ system in the ocean, *Science*, 305, 362-366, 2004.
Folk, R. L., and F. L. Lynch, Organic matter, putative nannobacteria and the formation of ooids and hardgrounds, *Sedimentology*, 48, 215-229, 2001.

Frankignoulle, M., J.-P. Gattuso, R. Biondo, I. Bourge, G. Copin-Montégut, and M. Pichon, Carbon fluxes in coral reefs. II. Eulerian study of inorganic carbon dynamics and measurement of air-sea CO_2 exchanges, *Mar. Ecol. Prog. Ser.*, 145, 123-132, 1996.

Friedman, G. M., A. J. Amiel, and N. Schneidermann, Submarine cementation in reefs: example from the Red Sea, *Journal of Sedimentary Petrology*, 44, 816-825, 1974.

Gao, K., Y. Aruga, K. Asada, T. Ishihara, T. Akano, and M. Kiyohara, Calcification in the articulated coralline alga *Corallina pilulifera*, with special reference to the effect of elevated CO_2 concentration. *Mar. Biol.*, 117, 129-132, 1993.

Gattuso, J.-P., D. Allemand, and M. Frankignoulle, Photosynthesis and calcification at cellular, organismal and community levels in coral reefs: a review on interactions and control by carbonate chemistry, *Am. Zool.*, 39, 160-183, 1999

Gattuso, J.-P., M. Frankignoulle, I. Bourge, S. Romaine and R. W. Buddemeier, Effect of calcium carbonate saturation of seawater on coral calcification, *Glob. Planet. Changes*, 18, 37-46, 1998.

Gattuso, J.-P., M. Pichon, B. Delesalle, C. Canon, and M. Frankignoulle, Carbon fluxes in coral reefs. I. Lagrangian measurement of community metabolism and resulting air-sea CO_2 disequilibrium, *Mar. Ecol. Prog. Ser.*, 145, 109-121, 1996.

Gattuso, J.-P., S. Reynaud-Vaganay, P. Furla, S. Romaine-Lioud, and J. Jaubert, Calcification does not stimulate photosynthesis in the zooxanthellate scleractinian coral *Stylophora pistillata*, *Limnol. Oceanogr*, 45, 246-250, 2000.

Glynn, P. W., State of coral reefs in the Galápagos Islands: natural vs anthropogenic impacts, *Mar. Poll. Bull.*, 29, 131-140, 1994.

Glynn, P. W., and G. M. Wellington, *Corals and Coral Reefs of the Galapagos Islands.* Univ. California, 1983.

Halley, R. B., and K. K. Yates, Will reef sediments buffer corals from increased global CO_2, *in* Hopley, D. et al., eds., *Proc.9th Int. Coral Reef Sym.*, Abstracts: Indonesia, State Ministry for the Environment, 2000.

Harriott, V. J., Growth of the staghorn coral *Acropora formosa* at Houtman Abrolhos, Western Australia, *Mar. Biol.*, 132, 319-325, 1998.

Hendrey, G. R., D. S. Ellsworth, K. F. Lewin, and J. Nagy, A free-air enrichment system for exposing tall forest vegetation to elevated atmospheric CO_2, *Global Change Biol.*, 5, 293-309, 1999.

Hillis, L., Coralgal reefs from a calcareous green alga perspective, and a first carbonate budget, *Proc. 8th Int. Coral Reef Sym.*, 1, 761-766, Panama, 1997.

Hoegh-Guldberg, O., Climate change, coral bleaching and the future of the world's coral reefs, *Mar. Freshw. Res.*, 50, 839-866, 1999.

Horst, G. P., and P. J. Edmunds, Effects of temperature and pH on calcification and quantum-yield efficiency of *Madracis mirabilis*, *Limnol. Oceanogr.*, in press, 2006.

Hossain, M. M., and S. Ohde, Calcification of cultured *Porites* and *Fungia* under different aragonite saturation states of seawater, *Proc. 10th Int. Coral Reef Sym.*, 597-606, Okinawa, Japan, 2006.

Houck, J. E., R. W. Buddemeier, S. V. Smith, and P. L. Jokiel, The response of coral growth rate and skeletal strontium content to light intensity and water temperature, *Proc. 3rd Int. Coral Reef Sym. 2*, 425-431, Miami, 1987.

James, N. P., R. M. Ginsburg, D. S. Marszalek, and P. W. Choquette, Facies and fabric specificity of early subsea cements in shallow Belize (British Honduras) reefs, *J. Sedim. Petrol.*, 46, 523-544, 1976.

Jokiel, P. L., and S. L. Coles, Effects of temperature on the mortality and growth of Hawaiian reef corals, *Marine Biology*, 43, 201-208, 1997.

Jokiel, P. L., J. W. Maragos, and L. Franzisket, Coral growth buoyant weight technique. In: *Coral reefs: research methods*, D. R. Stoddart and R. E Johannes (editors), Monographs on oceanographic methodology. UNESCO, *Paris.* pp. 529-542, 1978.

Kawahata, H., A. Suzuki, T. Ayukai, and K. Goto, Distribution of the fugacity of carbon dioxide in the surface seawater of the Great Barrier Reef, *Mar. Chem.*, 72, 257-272, 2000.

Kayanne H., H. Hata, K. Nozaki, K. Kato, A. Negishi, H. Saito, H. Yamano, T. Isamu, H. Kimoto, M. Tsuda, F. Akimoto, K. Kawate, and I. Iwata, Submergible system to measure seawater pCO_2 on a shallow sea floor, *Marine Technology Society Journal*, 36, 23-28, 2002a.

Kayanne, H., S. Kudo, H. Hata, H. Yamano, K. Nozaki, K. Kato, A. Negishi, H. Saito, F. Akimoto, and H. Kimoto, Integrated monitoring system for coral reef water pCO_2, carbonate system and physical parameters, *Proc. 9th Int. Coral Reef Sym.*, 2, 1079-1084, Bali, Indonesia, 2002b.

Keeling, C. D., H. Brix, and N. Gruber, Seasonal and long-term dynamics of the upper ocean carbon cycle at Station ALOHA near Hawaii, *Global Biogeochem. Cycles*, 18, GB4006, doi: 10.1029/2004GB002227, 2004.

Kleypas, J. A., Geological development of fringing reefs in the Southern Great Barrier Reef, Australia, Unpublished Ph.D. thesis, James Cook University of North Queensland, Australia, pp. 199, 1991.

Kleypas, J. A., R. W. Buddemeier, and J.-P. Gattuso, The future of coral reefs in an age of global change, *Int. J. Earth* Sci., 90, 426-437, 2001.

Kleypas, J. A., R. W. Buddemeier, D. Archer, J.-P. Gattuso, C. Langdon, and B. N. Opdyke, Geochemical consequences of increased atmospheric CO_2 on coral reefs, *Science*, 284, 118-120, 1999.

Kleypas, J. A., R. W. Buddemeier, C. M. Eakin, J.-P Gattuso, J. Guinotte, O. Hoegh-Guldberg, R. Iglesias-Prieto, P. L. Jokiel, C. Langdon, W. Skirving, and A. E. Strong, Comment on "Coral reef calcification and climate change: The effect of ocean warming," *Geophys. Res. Lett.*, 32, L08601, doi:10.1029/2004GL022329, 2005.

Kump, L. R., J. F. Kasting, and R. G. Crane, *The Earth System*, Prentice Hall, Inc., New Jersey.

Lang, J., and E. A. Chornesky, Competition between scleractinian reef corals – a review of mechanisms and effects, In Z. Dubinsky, editor, *Ecosystems of the World, Vol. 25, Coral Reefs*, Elsevier, pp. 209-257, 1990.

Langdon, C., Review of experimental evidence for effects of CO_2 on calcification of reef builders, *Proc. 9th Int. Coral Reef Sym.*, 2, 1091-1098, Bali, Indonesia, 2002.

Langdon, C., and M. J. Atkinson, Effect of elevated pCO_2 on photosynthesis and calcification of corals and interactions with seasonal change in temperature/irradiance and nutrient enrichment, *J. Geophys. Res.*, 110, C09S07, doi:10.1029/2004JC002576, 2005.

Langdon, C., T. Takahashi, C. Sweeney, D. Chipman, J. Goddard, F. Marubini, H. Aceves, H. Barnett, and M. J. Atkinson, Effect of calcium carbonate saturation state on the calcification rate of an experimental coral reef, *Global Biogeochemical Cycles*, 14, 639-654, 2000.

Leclercq, N., J.-P. Gattuso, and J. Jaubert, CO_2 partial pressure controls the calcification rate of a coral community, *Global Change Biology*, 6, 329-334, 2000.

Leclercq, N., J.-P. Gattuso, and J. Jaubert, Primary production, respiration, and calcification of a coral reef mesocosm under increased CO_2 partial pressure, *Limnol. Oceanogr.*, 47, 558-564, 2002.

Leder, J. J., A. M. Szmant, and P. K. Swart, The effect of prolonged "bleaching" on skeletal banding and stable isotopic composition in *Montastrea annularis*, *Coral Reefs*, 10, 19-27, 1991.

Lighty, R. G., Preservation of internal reef porosity and diagenetic sealing of submerged early Holocene barrier reef, southeast Florida shelf, in *Carbonate Cements*, pp. 123-151, SEPM Special Publication No. 36, 1985.

Lirman, D., Fragmentation in the branching coral *Acropora palmata* (Lamarck): growth, survivorship, and reproduction of colonies and fragments, *J. Exp. Mar. Biol. Ecol.*, 251, 41-57, 2000.

Lough, J. M., and D. J. Barnes, Several centuries of variation in skeletal extension, density and calcification in massive *Porites* colonies from the Great Barrier Reef: A proxy for seawater temperature and a background of variability against which to identify unnatural change, *J. Exp. Mar. Biol. Ecol.*, 211, 29-67, 1997.

Lough, J. M., and D. J. Barnes, Environmental controls on growth of the massive coral *Porites*, *J. Exp. Mar. Biol. Ecol.*, 245, 225-243, 2000.

Marshall, A. T., and P. Clode, Effect of increased calcium concentration in sea water on calcification and photosynthesis in the scleractinian coral *Galaxea fascicularis*, *J. Exp. Biol.*, 205, 2107-2113, 2002.

Marshall, A. T., and P. Clode, Calcification rate and the effect of temperature in a zooxanthellate and an azooxanthellate scleractinian reef coral, *Coral Reefs*, 23, 218-224, 2004.

Marshall, J. F., Cross-shelf and facies related variations in submarine cementation in the Central Great Barrier Reef, *Proc. 5th Int. Coral Reef Sym.*, 3, 509-512, Townsville, Australia, 1985.

Marubini, F., and P. S. Davies, Nitrate increases zooxanthellae population density and reduces skeletogenesis in corals, *Mar. Biol.*, 127, 319-328, 1996.

Marubini, F., and B. Thake, Bicarbonate addition promotes coral growth, *Limnol. Oceanogr.*, 44, 716-720, 1999

Marubini, F., H. Barnett, C. Langdon, and M. J. Atkinson, Interaction of light and carbonate ion on calcification of the hermatypic coral *Porites compressa, Marine Ecology Progress Series*, 220, 153-162, 2001.

Marubini, F., C. Ferrier-Pages, and J.-P. Cuif, Suppression of growth in scleractinian corals by decreasing ambient carbonate ion concentration: a cross-family comparison. *Proceedings of the Royal Society* B, 270, 179-184, 2003.

Matsumoto, K., J. L. Sarmiento, R. M. Key, O. Aumont, J. L. Bullister, K. Caldeira, J. M. Campin, S. C. Doney, H. Drange, J. C. Dutay, M. Follows, Y. Gao, A. Gnanadesikan, N. Gruber, A. Ishida, F. Joos, K. Lindsay, E. Maier-Reimer, J. C. Marshall, R. J. Matear, P. Monfray, A. Mouchet, R. Najjar, G. K. Plattner, R. Schlitzer, R. Slater, P. S. Swathi, I. J. Totterdell, M. F. Weirig, Y. Yamanaka, A. Yool, and J. C. Orr, Evaluation of ocean carbon cycle models with data-based metrics, *Geophys. Res. Lett.*, 31, L07303, doi:10.1029/2003GL018970, 2004.

McConnaughey, T., and J. F. Whelan, Calcification generates protons for nutrient and bicarbonate uptake. *Earth Sci. Rev.*, 42, 95-117, 1997.

McNeil, B. I., R. J. Matear, and D. J. Barnes, Coral reef calcification and climate change: The effect of ocean warming, *Geophys. Res. Lett.*, 31, L22309, doi:10.1029/2004GL021541, 2004.

Milliman, J. D., *Marine Carbonates*, pp. 375, Springer-Verlag, New York, 1974.

Morse, J. W., Formation and Diagenesis of Carbonate Sediments, In *Treatise on Geochemistry*, vol. 7, pp. 67-85, F. Mackenzie (ed.), Elsevier, 2003.

Morse, J. W., D. K. Gledhill, and F. J. Millero, $CaCO_3$ precipitation kinetics in waters from the Great Bahama Bank: Implications for the relationship between Bank hydrochemistry and whitings, *Geochim. Cosmochim. Acta*, 67, 2819-2826, 2003.

Morse, J. W., and F. T. Mackenzie, *Geochemistry of Sedimentary Carbonates*, pp. 707, Elsevier, Amsterdam.

Morse, J. W., D. K. Gledhill, and F. J. Millero, $CaCO_3$ precipitation kinetics in waters from the Great Bahama Bank: Implications for the relationship between Bank hydrochemistry and whitings, *Geochim. Cosmochim. Acta*, 67, 2819-2826, 2003.

Mucci, A., and J. W. Morse, The solubility of calcite in seawater solutions of various magnesium concentration, It = 0.679 m and 25C and one atmosphere total pressure, *Geochim. Cosmochim. Acta*, 48, 815-822, 1984.

Müller, A., M. K. Gagan, and M. T. McCulloch, Early marine diagenesis in corals and geochemical consequences for paleoceanographic reconstructions, *Geophysical Research Letters*, 28, 4471-4474, 2001.

Munhoven, G., Glacial-interglacial changes of continental weathering: estimates of the related CO_2 and HCO_3^- flux variations and their uncertainties, *Global Planetary Change*, 33, 155-176, 2002.

Oberdorfer, J. A., and R. W. Buddemeier, Coral-reef hydrology: field studies of water movement within a barrier reef, *Coral Reefs*, 5, 7-12, 1986.
Ohde, S., and M. M. Hossain, Effect of $CaCO_3$ (aragonite) saturation state of seawater on calcification of *Porites* coral, *Geochemical Journal*, 38, 613-621, 2004.
Ohde, S., and R. van Woesik, Carbon dioxide flux and metabolic processes of a coral reef, Okinawa, *Bull. Mar. Sci.*, 65, 559-576, 1999.
Opdyke, B. N., and J. C. G. Walker, Return of the coral reef hypothesis: Basin to shelf partitioning of $CaCO_3$ and its effect on atmospheric CO_2, *Geology*, 20, 730-736, 1992.
Orr, J.C., E. Maier-Reimer, U. Mikolajewicz, P. Monfray, J. L. Sarmiento, J. R. Toggweiler, N. K. Taylor, J. Palmer, N. Gruber, C. L. Sabine, C. Le Quéré, R. M. Key, and J. Boutin, Estimates of anthropogenic carbon uptake from four three-dimensional global ocean models, *Global Biogeochem. Cycles*, 15, 43-60, 2001.
Quay, P. D., B. Tilbrook, and C. S. Wong, Oceanic uptake of fossil fuel CO_2: Carbon-13 evidence, *Science*, 256, 74-79, 1992.
Reaka-Kudla, M. L., J. S. Feingold, and W. Glynn, Experimental studies of rapid bioerosion of coral reefs in the Galápagos Islands, *Coral Reefs*, 15, 101-107, 1996.
Renegar, D. A., and B. M. Riegl, Effect of nutrient enrichment and elevated CO_2 partial pressure on growth rate of Atlantic scleractinian coral *Acropora cervicornis*, *Mar. Ecol.-Prog. Ser.*, 293, 69-76, 2005.
Reynaud, S., N. Leclercq, S. Romaine-Lioud, C. Ferrier-Pages, J. Jaubert, and J.-P. Gattuso, Interacting effects of CO_2 partial pressure and temperature on photosynthesis and calcification in a scleractinian coral, *Global Change Biol.*, 9, 1660-1668, 2003.
Ridgwell, A., A. J. Watson, M. A. Maslin, and J. O. Kaplan, Implications of coral reef buildup for the controls on atmospheric CO_2 since the Last Glacial Maximum, *Paleoceanography*, 18, 1083, doi:10.1029/2003PA000893, 2003.
Robbins, L. L., and P. L. Blackwelder, Biochemical and ultrastructural evidence for the origin of whitings: A biological induced calcium-carbonate precipitation mechanism, *Geology*, 20, 464-468, 1992.
Ryan, D. A, B. N. Opdyke, and J. S. Jell, Holocene sediments of Wistari Reef: towards a global quantification of coral reef related neritic sedimentation in the Holocene, *Palaeogeogr., Palaeoclim., Palaeoecol.*, 175, 173-184, 2001.
Sabine, C. L., R. A. Feely, N. Gruber, R. M. Key, K. Lee, J. L. Bullister, R. Wanninkhof, C. S. Wong, D. W. R. Wallace, B. Tilbrook, F. J. Millero, T. -H. Peng, A. Kozyr, T. Ono, and A. F. Rios, The oceanic sink for anthropogenic CO_2, *Science*, 305, 367-371, 2004.
Sakai, K., Effect of colony size, polyp size, and budding mode on egg production in a colonial coral, *Biol. Bull.*, 195, 319-325, 1998.
Schneider, K., and J. Erez, The effect of carbonate chemistry on calcification and photosynthesis in the hermatypic coral *Acropora eurystoma*, *Limnol. Oceanogr.*, 51, 1284-1293, 2006.
Smith, S. V., and R. W. Buddemeier, Global change and coral reef ecosystems, *Ann. Rev. Ecol. Syst.*, 23, 89-118, 1992.
Smith, S. V., and G. S. Key, Carbon dioxide metabolism in marine environments, *Limnol. Oceanogr.*, 20, 493-495, 1975.
Smith, S. V., and D. W. Kinsey, Calcification and organic carbon metabolism as indicated by carbon dioxide, In *Coral reefs: research methods*, D. R. Stoddart and R. E Johannes (eds), pp. 469-484, Monographs on oceanographic methodology. UNESCO, Paris, 1978.
Stanley, S. M., and L. A. Hardie, Secular oscillations in the carbonate mineralogy of reef-building and sediment-producing organisms driven by tectonically forced shifts in seawater chemistry, *Paleogeography, Paleoclimatology, Palaeoecology*, 144, 3-19, 1998.
Stanley, S. M., J. B. Ries, and L. A. Hardie, Low-magnesium calcite produced by coralline algae in seawater of Late Cretaceous composition, *Proc. Nat. Acad. Sci.*, 99, 15323-15326, 2002.

Suzuki, A., and H. Kawahata, Partial pressure of carbon dioxide in coral reef lagoon waters: Comparative study of atolls and barrier reefs in the Indo-Pacific oceans, *Jour. Oceanogr.*, 55, 731-745, 1999.

Suzuki, A., and H. Kawahata, Carbon budget of coral reef systems: an overview of observations in fringing reefs, barrier reefs and atolls in the Indo-Pacific regions, *Tellus*, 55B, 428-444, 2003.

Walter, L. M., and J. W. Morse, Magnesium calcite stabilities: A reevaluation, *Geochim. Cosmochim. Acta*, 48, 1059-1070, 1984.

Watanabe, A., H. Kayanne, K. Nozaki, K. Kato, A. Negishi, S. Kudo, H. Kimoto, M. Tsuda, and A. G. Dickson, A rapid, precise potentiometric determination of total alkalinity in seawater by a newly developed flow-through analyzer designed for coastal regions, *Marine Chemistry*, 85, 75-87, 2004.

Weiner, S., and P. M. Dove, An overview of biomineralization processes and the problem of vital effect, In *Biomineralization, Reviews in Mineralogy and Geochemistry*, Volume 54, edited by P. M. Dove, J. J. De Yoreo, and S. Weiner, pp. 1-29, Mineralogical Society of America, Geochemical Society, 2003.

Zeebe, R. E., and D. Wolf-Gladrow, *CO_2 in Seawater: Equilibrium, Kinetics, Isotopes*, pp. 346, Elsevier Science B.V., Amsterdam, 2001.

6

Analyzing the Relationship Between Ocean Temperature Anomalies and Coral Disease Outbreaks at Broad Spatial Scales

Elizabeth R. Selig, C. Drew Harvell, John F. Bruno, Bette L. Willis, Cathie A. Page, Kenneth S. Casey, and Hugh Sweatman

Abstract

Ocean warming due to climate change could increase the frequency and severity of infectious coral disease outbreaks by increasing pathogen virulence or host susceptibility. However, little is known about how temperature anomalies may affect disease severity over broad spatial scales. We hypothesized that the frequency of warm temperature anomalies increased the frequency of white syndrome, a common scleractinian disease in the Indo-Pacific. We created a novel 4 km satellite temperature anomaly dataset using data from NOAA's Pathfinder program and developed four different temperature anomaly metrics, which we correlated with white syndrome frequency at 47 reefs spread across 1500 km of the Great Barrier Reef. This cross-sectional epidemiological analysis used data from disease field surveys conducted by the Australian Institute of Marine Science six to twelve months after the summer of 2002, a year of extensive coral bleaching. We found a highly significant positive relationship between the frequency of warm temperature anomalies and the frequency of white syndrome. There was also a highly significant, nearly exponential relationship between total coral cover and the number of disease cases. Furthermore, coral cover modified the effect of temperature on disease frequency. Both high coral cover (>50%) and anomalously warm water appear to be necessary for white syndrome outbreaks to occur and these two risk factors explained nearly 75% of the variance in disease cases. These results suggest that rising ocean temperatures could exacerbate the effects of infectious diseases on coral reef ecosystems.

1. Introduction

Over the last four decades, coral cover has declined dramatically on reefs worldwide [Gardner et al., 2003; Bellwood et al., 2004]. Several factors are thought to be responsible for this decline including overfishing [Jackson, 1997; Pandolfi et al., 2003], terrestrial run-off [Fabricius, 2005], climate change [Hoegh-Guldberg, 1999; Hughes et al., 2003], and infectious disease [Aronson and Precht, 2001]. There is growing recognition that we

Coral Reefs and Climate Change: Science and Management
Coastal and Estuarine Studies 61
Copyright 2006 by the American Geophysical Union.
10.1029/61CE07

need to focus on possible synergisms among these and other stressors [Hughes and Connell, 1999; Lenihan et al., 1999]. In the last several years, the relationship between climate conditions and disease has received more attention as researchers have connected factors such as temperature and precipitation with increases in human and wildlife diseases [Pascual et al., 2000; Patz, 2002; Kutz et al., 2005; Pounds et al., 2006]. Yet, few studies have focused on the effects of climate change on diseases in the ocean, particularly at broad spatial scales.

Disease has already had significant effects on coral reef ecosystems. Several diseases have altered the landscape of Caribbean reefs, causing the near extirpation of the keystone herbivore *Diadema antillarum* [Lessios, 1988] and dramatic losses of *Acropora cervicornis* and *Acropora palmata* [Aronson and Precht, 2001; Aronson et al., 2002]. These disease outbreaks mediated a shift from coral- to algal-dominated communities in the Caribbean [Aronson and Precht, 2001]. The scale and severity of coral loss on many Caribbean reefs is unprecedented in the paleontological record [Aronson and Precht, 2001; Wapnick et al., 2004] of many reefs and indicative of the emergence of a novel stressor [Aronson and Precht, 2001]. Recent studies quantifying both disease reports [Ward and Lafferty, 2004] and the number of described coral diseases [Sutherland et al., 2004] suggest that the frequency of marine diseases is increasing and many of these diseases are the result of previously unknown pathogens. Reports from the Pacific indicate that diseases of reef-building corals may be far more widespread than previously believed [Sutherland et al., 2004; Willis et al., 2004; Aeby, 2005; Raymundo et al., 2005]. Although coral disease is likely underreported in the Pacific due to a lack of disease research, increases in coral disease cases have been detected since the late 1990s [Willis et al., 2004]. These lines of evidence provide strong support for the hypothesis that there has been a real increase in the number of coral disease reports over the last three decades [Harvell et al., 1999; Ward and Lafferty, 2004].

The causes underlying recent increases in coral disease outbreaks are complex and poorly understood, in part because of a paucity of knowledge about the identity and sources of most coral pathogens [Sutherland et al., 2004]. Pathogens have a variety of purported vectors and reservoirs including algae [Nugues et al., 2004], invertebrates [Rosenberg and Falkovitz, 2004; Williams and Miller, 2005], sediment transported from the Sahel region of northern Africa [Shinn et al., 2000], and sewage effluent [Patterson et al., 2002]. In addition, some diseases are hypothesized to be associated with changes in corals' microbial communities, rather than the result of an external infectious pathogen.

Abiotic factors like nutrients and temperature can exacerbate disease severity, but the scale of their effects may vary. For example, nutrients increased the severity of two coral diseases in experimental manipulations [Bruno et al., 2003]. However, inputs of terrestrial pollution, including nutrients, do not always result in elevated disease levels on affected reefs [Weil, 2004; Willis et al., 2004]. Although localized inputs of nutrients may play some role in disease outbreaks, the regional scale of most outbreaks [Aronson and Precht, 2001; Kim and Harvell, 2004] indicates that a climatic variable like temperature may be a critical driver of disease dynamics. For example, disease prevalence, or the proportion of the total population that is diseased, may be related to the frequency and magnitude of warm temperature anomalies. Extensive work clearly links these anomalies to coral bleaching events [Glynn et al., 1988; Glynn and D'croz, 1990; Hoegh-Guldberg, 1999; Bruno et al., 2001; Fitt et al., 2001; Liu et al., 2003; Strong et al., 2004]. Since bleaching is a sign of physiological stress in corals [Glynn, 1993; Brown, 1997], it is expected that bleached or thermally-stressed corals would be more susceptible to opportunistic and residential pathogens [Hayes et al., 2001; Rosenberg and Ben-Haim, 2002].

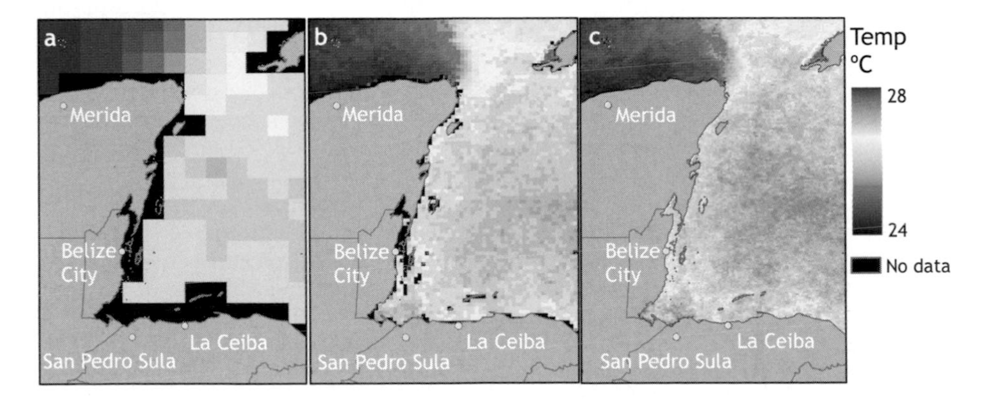

Plate 1. An example of the benefits of increasing resolution on the Yucatan peninsula and Caribbean Sea from; (a) 50 km HotSpot data to (b) 9 km Pathfinder data to (c) 4 km Pathfinder data. In the 4 km data, there is less missing data, allowing for greater coverage of coastal areas where many reefs occur. In addition, the 4 km data displays more spatial structure and precision in the temperature values. Data are from January climatological averages, monthly for the 50 km data and from the first week of January for the 9 km and 4 km data.

Forecast sea surface temperature (SST) models predict that the frequency and severity of warm temperature anomalies will increase with climate change [Hoegh-Guldberg, 1999; Sheppard, 2003; Sheppard and Rioja-Nieto, 2005]. Ocean temperature has already increased, on average, 0.4-0.8°C [Folland et al., 2001] from 1861 to 2000, with some regional variation [Casey and Cornillon, 2001]. These anomalies and the general warming of the ocean could have several effects on disease. One possible outcome of global warming is that the summer "disease season" may become more severe, as summer temperature maxima increase, and longer, as these elevated thermal regimes start earlier and persist later in the season. Seasonal variability in coral disease abundance has been found in multiple field studies in different regions. In the Caribbean, several coral diseases, including black band [Edmunds, 1991; Kuta and Richardson, 2002], white pox [Patterson et al., 2002], and dark spots disease [Gil-Agudelo and Garzon-Ferreira, 2001], are more prevalent or spread across colonies more rapidly during summertime than during cooler seasons. On the Great Barrier Reef (GBR), white syndrome frequency was greater in summer than winter on surveyed reefs [Willis et al., 2004]. Similarly, in the summer of 2002 on the GBR, Jones et al. [2004] documented a localized outbreak of atrementous necrosis. In addition, shorter or warmer winters may release some infectious diseases from the low-temperature control that provides hosts with a seasonal escape from disease [Harvell et al., 2002]. Climate warming has also been predicted to alter the geographic distribution of infectious disease by shifting host and pathogen latitudinal ranges pole-ward [Marcogliese, 2001; Harvell et al., 2002].

Testing the hypothesis that the frequency or intensity of temperature anomalies can influence coral disease dynamics requires high quality data on temperature and disease frequency. Here we discuss how analysis of these broad scale questions is now possible using a newly developed high-resolution satellite temperature anomaly dataset. We then present a case study where we examine the effects of the warm temperature anomalies on white syndrome, an emergent sign of disease on Great Barrier Reef corals. We also discuss future research directions for testing the relationship between temperature and disease.

2. Using Remote Sensing Data to Explore the Climate Warming Disease Outbreak Hypothesis

Most documented coral disease outbreaks have been at the scale of ocean basins [Lessios, 1988; Aronson and Precht, 2001; Willis et al., 2004]. Previously, correlating these outbreaks with ocean temperature was complicated by a scale mismatch. *In situ* temperature loggers can be highly effective at capturing small-scale variability (1-100 m) [Leichter and Miller, 1999; Castillo and Helmuth, 2005; Leichter et al., 2005], but are limited in their spatial extent. On the other hand, the 50 km HotSpot mapping (Plate 1a) [Strong et al., 2004], although effective for predicting bleaching at broad scales [Bruno et al., 2001; Berkelmans et al., 2004; Strong et al., 2004], is too coarse to accurately represent the temperature of waters surrounding many reefs and may not sufficiently capture local variability [Toscano et al., 2000]. Therefore, the development of new, higher resolution remote sensing products is required to detect correlations between temperature anomalies and disease outbreaks.

We developed a novel satellite temperature anomaly product to investigate the relationship between sea surface temperature and disease at broad spatial scales. Our dataset used the 4 km Advanced Very High Resolution Radiometer (AVHRR) Pathfinder Version 5.0 SST dataset produced by NOAA's National Oceanographic Data Center and the University of Miami's Rosenstiel School of Marine and Atmospheric Science (http://pathfinder.nodc.noaa.gov). These data now provide the longest sea surface temperature record at the highest resolution of any global satellite dataset. The Pathfinder Version 4.2 dataset had a 9 km resolution [Kilpatrick et al., 2001], a marked improvement from the 50 km data. However, inaccuracies in the land mask, which defines the extent of the data, resulted in coverage of only 60% of reef areas (Plate 1b). Using a more refined land mask and other algorithm enhancements, the 4 km Pathfinder Version 5.0 dataset reprocessed the full AVHRR record from 1985-2005 to create an improved SST climate data record. With these improvements, temperature records are now available for more than 98% of reefs worldwide (Plate 1c). Appropriate use of satellite temperature data requires validation with *in situ* loggers [Reynolds et al., 2002]. In theory, an infrared observing satellite like AVHRR only measures an integrated temperature over approximately the top 10 micrometers of the ocean surface. Therefore, we validated the assumption that the Pathfinder 4 km Pathfinder Version 5.0 SST values reflect temperatures on shallow reefs, where the majority of reef-building corals are found, by comparing Pathfinder SST estimates to *in situ* temperature logger data collected by the Australian Institute of Marine Science (www.reeffutures.org) on nine shallow reefs on the GBR (5 m to 9 m depth). We fit a linear multi-level model and found a highly significant relationship ($p < 0.0001$) with a common value for the slope of all the reefs of 0.96 (SE = 0.017). We found the difference between satellite-derived and benthic temperatures was generally less than the 0.2°C error of most *in situ* loggers. We also ran linear regression analyses for each reef to investigate the generality of this relationship at all nine reefs (Table 1) and found that the satellite temperature measurements were a very good predictor of benthic water temperature.

One key advantage of satellite data is their ability to provide a long-term temporal record of temperature values, which enables users to create accurate climatologies, or average long-term patterns (Plate 2a). These climatologies are then used as the basis for calculating deviations from typical weekly, monthly, seasonal, or annual temperatures. For disease analyses, relevant data include the frequency of deviations or temperature anomalies, the duration of anomaly events, occurrence of wintertime anomalies, and the geographic extent of the anomaly.

TABLE 1. Relationship between weekly averaged satellite and *in situ* temperatures at nine reefs on the GBR reefs at 5-9 m depth. Field data were collected by the Australian Institute of Marine Science in cooperation with CRC Reef Research Centre and the Great Barrier Reef Marine Park Authority.

Reef	Latitude	Longitude	Period	n	P	R^2
Agincourt	−16.0384	145.8688	1996-2004	173	$p < .0001$	0.91
Chicken	−18.6521	147.7217	1996-2004	244	$p < .0001$	0.94
Davies	−18.8060	147.6686	1996-2004	200	$p < .0001$	0.96
Dip	−18.3999	147.4519	1997-2004	197	$p < .0001$	0.92
East Cay	−21.4698	152.5665	1995-2004	248	$p < .0001$	0.96
John Brewer	−18.6188	147.0815	1996-2004	239	$p < .0001$	0.94
Lizard Island	−14.6915	145.4692	1996-2004	134	$p < .0001$	0.90
Myrmidon	−18.2572	147.3813	1995-2004	205	$p < .0001$	0.93
Turner Cay	−21.7031	152.5601	1997-2004	204	$p < .0001$	0.96

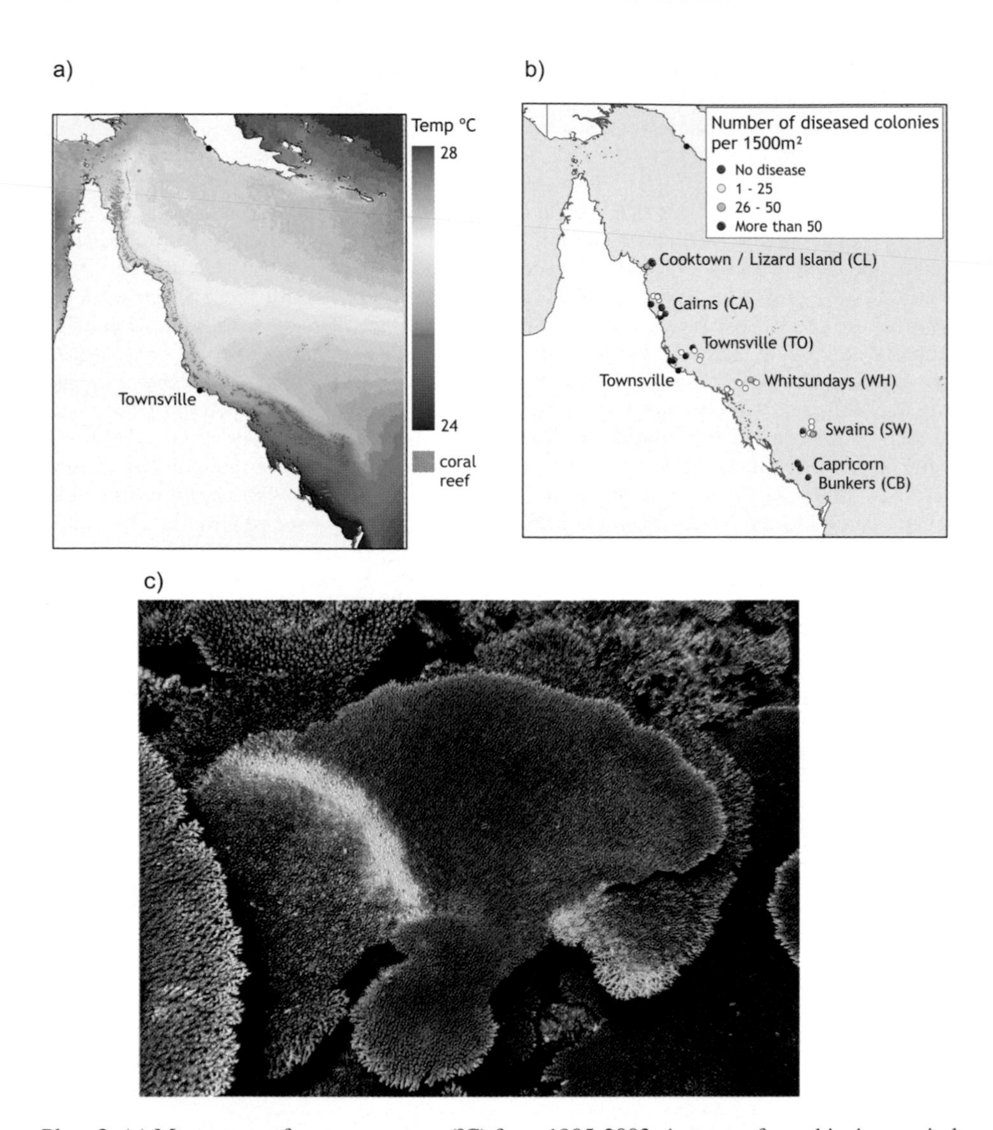

Plate 2. (a) Mean sea surface temperature (°C) from 1985-2003. Averages from this time period were used to create weekly climatologies for each surveyed reef. (b) Number of white syndrome cases in 2002-2003 based on data collected by the Australian Institute of Marine Science's Long-term Monitoring Program surveys. Each of the six latitudinal areas or sectors are labeled with their names and abbreviations. Cooktown/Lizard Island (CL) has nine surveyed reefs, Cairns (CA) has ten, Townsville (TO) has eight, Whitsundays (WH) has nine reefs, Swains (SW) has seven, and Capricorn Bunkers (CB) has four. For each reef, three different sites were surveyed. Disease cases were highest in the Cooktown/Lizard Island and Capricorn Bunkers sectors. (c) Example of white syndrome spreading across *Acropora cytherea*. White areas are coral skeleton that have been recently exposed following die-off behind the disease front.

3. Case Study on the Great Barrier Reef

The Great Barrier Reef (GBR) is the largest barrier reef system in the world and one of the most highly managed, with more than a third of reef area in marine reserves, or no-take areas [Fernandes et al., 2005]. Recent surveys suggest that disease is more prevalent than previously believed on the GBR [Willis et al., 2004]. Black band, white syndrome, brown band, and skeletal eroding band have all been reported on the GBR [Willis et al., 2004]. In spite of the well-documented relationship with increased temperatures and black band disease in the Caribbean [Edmunds, 1991; Kuta and Richardson, 2002], it did not vary in frequency over the course of our study [Willis et al., 2004]. We focused our analysis on white syndrome, which increased dramatically in 2002-2003 on some reefs. We tested specific hypotheses about how temperature might affect disease frequency by using an information-theoretic approach to evaluate different thermal stress metrics.

Disease Surveys

The disease data used for this analysis were collected by the Australian Institute of Marine Science's Long-term Monitoring Program (AIMS LTMP) during their 2002-2003 surveys. Surveys were performed at 47 reefs in a stratified design. Five permanent 50 m × 2 m belt transects were surveyed at three sites on each reef at approximately 6-9 m depth (full methods in *Sweatman et al.*, 2003). For each transect, the number of colonies infected with white syndrome was quantified. Reefs were grouped by latitudinal sectors, which together cover more than 1500 km and a variety of different oceanographic regimes (Plate 2b). White syndrome is characterized by a band of recently-exposed white skeleton, sometimes preceded by a band of bleached tissue at the tissue-skeleton interface. The white band moves across the colony as the front of tissue mortality progresses, potentially resulting in mortality of the whole colony (Plate 2c). White syndrome has been recorded on at least 17 species from 4 families, including Acroporidae, Pocilloporidae and Faviidae, families that constitute a significant percentage of overall coral cover on the GBR [Willis et al., 2004]. White syndrome is similar to Caribbean white diseases such as white band I, white band II, white plague I, and white plague II in its disease signs. Because the pathogen(s) that cause white disease in the Pacific are not known, we cannot state with certainty that the syndrome does not represent more than one distinct disease [Willis et al., 2004] or that the underlying etiology is infectious.

Temperature Data

Consistent with previous analyses of satellite data in coral studies, we used nighttime weekly temperature averages [Liu et al., 2003; Strong et al., 2004]. Nighttime daily averaged data had too many gaps and would have required extensive interpolation. Although they may not capture short duration events, weekly data provide substantially more continuity in the record and still represent a time scale short enough to capture most thermal stress events that negatively impact corals [Glynn, 1993; Podesta and Glynn, 2001]. To measure anomalies, we first calculated weekly climatologies, or the mean values at each calendar week from 1985-2003 for each 4 × 4 km pixel (Plate 2a). Missing data in the climatologies were interpolated using a Piecewise Cubic Hermite Interpolating Polynomial (PCHIP) function in MATLAB [The Mathworks Inc., 2005]. The climatologies were then smoothed using a five-week running mean to minimize unusual fluctuations from periods of limited data availability. Gaps in weekly temperature observations were also interpolated using the PCHIP function without modifying the original data.

Thermal Stress Metrics

To test the hypothesis that temperature affects disease frequency, we first designed temperature metrics relevant to disease. Infectious diseases are interactions between hosts and pathogens. Temperature can increase host susceptibility, but it can also increase pathogen growth rate, transmission rate, and over-wintering survival [Harvell et al., 2002]. Because of the potential complexity of this relationship and the paucity of data about most pathogens, no specific algorithm exists for disease prediction. However, data on how corals respond to stress and epidemiological theory provide a general guide to temperature thresholds that may be applicable to disease dynamics. We developed a series of temperature metrics for our case study based on metrics known to have a physiological effect on coral health. Three of our four thermal stress metrics are based on an anomaly threshold of 1°C, because this is widely assumed to estimate the point at which a warm temperature anomaly induces a measurable physiological stress in a coral host, and in general, increases of ≥1°C above normal summertime temperatures are thought to induce bleaching [Glynn, 1993; Glynn, 1996; Winter et al., 1998; Hoegh-Guldberg, 1999; Berkelmans, 2002]. All of the metrics measured deviations from the location-specific climatologies we created. Because disease surveys were conducted at different times of year (November-March), we standardized all of our metrics to include the number of anomalies during the 52 weeks prior to each disease survey. We used the latitude and longitude of each reef to match it with its corresponding pixel in the satellite data.

We developed each temperature metric to investigate a specific hypothesis related to the relationship between thermal stress and disease. The first three metrics are location-specific, which assumes that corals are acclimated to the thermal regime at their location. Work by Berkelmans and Willis [1999] suggests that corals exhibit some degree of local acclimation or adaptation. The fourth metric assumes that temperature is affecting disease rates at regional scales. If thermal anomalies are acting on the pathogen itself, by increasing growth or reproductive rates, they would likely be acting at a regional scale due to high pathogen mobility [McCallum et al., 2003].

Metric Descriptions

1) *WSSTA = Weekly Sea Surface Temperature Anomalies* = deviations of 1°C or greater from the mean climatology during a particular week at a given location from 1985-2003. This metric is designed to be both location and season-specific by determining whether the temperature is unusual for that location at a particular week of the year.
2) *TSA = Thermal Stress Anomalies* = deviations of 1°C or greater from the mean maximum climatological weekly temperature from 1985-2003. The mean maximum climatological week is the warmest week of the 52 weekly climatologies. This metric is also site-specific but designed to detect deviations from typical summertime highs and is similar to metrics used by the Coral Reef Watch program [Liu et al., 2003; Strong et al., 2004].
3) *LTSA = Local Temperature Stress Anomalies* = deviations from the upper 2.5% of all weekly measurements taken at that location from 1985-2003. This metric is site-specific. It is designed to detect whether temperatures are unusual based on the distribution and extremes of all measured temperatures, regardless of calendar week.
4) *RTSA = Regional Thermal Stress Anomalies* = deviations from the upper 2.5% of all weekly measurements taken at all reefs. For this metric, local temperatures are compared to regional average values.

TABLE 2. For each thermal metric we calculated: the AIC, a measure of expected relative Kullback Leibler information; the AIC_C, AIC corrected for small sample size; the difference, Δ_i between each model and the best model in the set; and the Akaike weight, w_i. The sum of all the Akaike weights is equal to 1.0 and each weight represents the approximate likelihood that a specific model is the best model of those compared, in a Kullback-Leibler information sense. For the models that involve a transformed response, likelihoods were adjusted with the Jacobian of the transformation to make the AIC_C values comparable.

Metric	AIC	AICc	Δ_i	w_i
WSSTA	1340	1342	0	1
TSA	1464	1465	123	0
LTSA	1479	1481	138	0
RTSA	1476	1477	135	0

Metric Selection and Statistical Analysis

We evaluated the different thermal stress metrics (Table 2) using Akaike Information Criterion (AIC) [Akaike, 1973; R 2.1.1, 2005]. Identifying which thermal stress metric best explains the relationship between temperature and disease can facilitate the development of an appropriate model for a specific coral-pathogen system. Different diseases may be affected by different temperature characteristics. For example, TSAs may be highly correlated with an increase of one disease while another disease may be more correlated with WSSTAs. AIC is an estimate of expected relative Kullback-Leibler information, an information-theoretic measure of the distance between models where smaller AIC values should be preferred. In practice, only the relative differences in AIC between models are meaningful. When these differences are normalized as Akaike weights, (w_i), they can also be given a probabilistic interpretation. Each w_i is the weight of evidence favoring model i as the best model among the models under consideration [Anderson et al., 2000; Anderson and Burnham, 2002].

Coral cover is clearly related to the number of white syndrome cases on a reef (Figure 1A) and was also included in the analysis. We used multiple, nonlinear regression analysis [StataCorp LP, 2006] to analyze the relationship between the two independent variables, thermal stress and coral cover, and the dependent variable, the total number of white syndrome cases at each reef (i.e., the number/1500m^2). We used a Poisson model and a quadratic function for the thermal stress metrics and a linear function for coral cover.

Results

The austral summer in 2002 triggered the most extensive bleaching event ever documented on the GBR [Berkelmans et al., 2004]. Surveys conducted after this summer found a major increase in disease frequency with some sampled reef areas having more than 300 cases of disease [Willis et al., 2004]. AIC_C, a small-sample (second order) bias-adjusted variant of AIC [Burnham and Anderson, 2002], indicated that the number of WSSTAs provided the best model fit (Table 2) and was used in the main analysis as the thermal

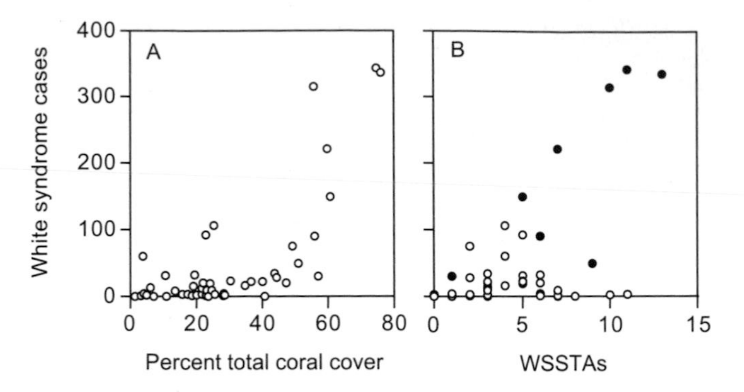

Figure 1. Relationships between total percent coral cover (A) and WSSTA frequency (B) and the number of white syndrome cases. Solid points in B represent reefs with >50% coral cover. Each point represents the values from a single reef (n = 47).

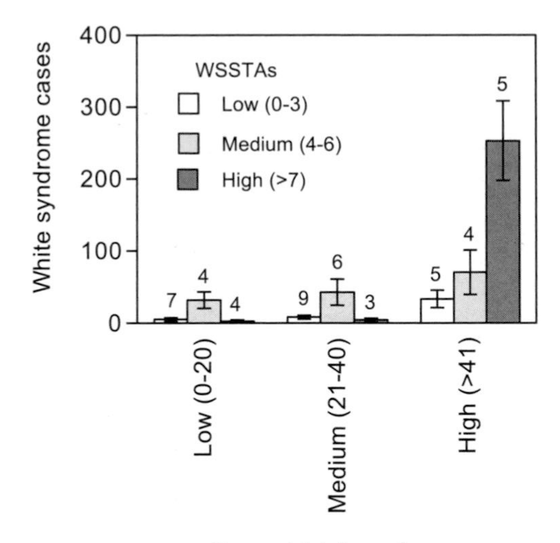

Figure 2. The effect of total coral cover and the number of WSSTAs on the number of white syndrome cases. Values are mean ±1 SE. Values above error bars are the number of surveyed reefs in each of the nine categories.

stress metric. The whole Poisson regression model ($R^2 = 0.73$) and both main effects (WSSTAs and coral cover) were all highly statistically significant at the reef scale (all $P < 0.0001$, n = 47). There was also a highly significant interaction between coral cover and WSSTAs ($P < 0.0001$). At low and intermediate cover (0-40%), the number of white syndrome cases was greatest when annual WSSTAs were 4-6 and declined slightly with increasing WSSTA (Figure 2). However, when cover was high (>41%), white syndrome frequency was more than 3x greater when the frequency of WSSTAs was >7 than when it was 4-6 (Figure 2).

Discussion

Temperature anomalies are known to be the underlying cause of mass coral bleaching [Podesta and Glynn, 2001; Liu et al., 2003; Berkelmans et al., 2004; Strong et al., 2004], but their relationship with infectious disease dynamics is not well understood. Our results suggest that warm temperature anomalies can significantly affect the frequency of white syndrome on the GBR, especially where coral cover is high. Total coral cover was clearly related to the number of white syndrome cases on a reef (Figure 1A) [Willis et al., 2004]. With few exceptions white syndrome frequency was relatively low (<30 cases/1500 m^2, the area surveyed at each reef) when coral cover was <50% (Figure 1A) and was very high, 192 cases ± 46 (mean ± 1 SE) at the eight reefs where coral cover was >50%. The nearly exponential relationship between coral cover and disease cases (Figure 1A) suggests there is a threshold coral cover of approximately 50% that is generally required to for an outbreak of white syndrome to occur. This could be due to higher host densities on high coral cover reefs. Host density is widely known to influence disease dynamics [Anderson and May, 1986]. White syndrome has been documented in the major coral families on the GBR, including the abundant staghorn and tabular species of *Acropora* [Willis et al., 2004]. Therefore total coral cover is likely to be directly related to white syndrome host cover. However, the host density-prevalence relationship is not always clear for infectious coral diseases such as sea fan aspergillosis [Kim and Harvell, 2004] possibly because secondary transmission is rare among host colonies [Edmunds, 1991]. Additionally, several other aspects of total coral cover could also influence the dynamics of white syndrome and other infectious coral diseases. For example, coral cover could be positively related to animal vectors (i.e., coral predators and mutualists), which could increase rates secondary transmission, although Willis et al. [2004] found that the density of the corallivorous snail *Drupella* spp. was unrelated to white syndrome frequency on the GBR.

Temperature measured as WSSTA frequency also had a strong effect on the number of disease cases (Figure 1B). The selection of the WSSTA metric as the best model using an AIC approach suggests that the effect of thermal stress on white syndrome is both seasonal and location-specific. The WSSTA metric is the only metric we tested that incorporates temperature anomalies throughout the year. These findings are consistent with coral physiology studies, which have found that at the beginning of cooler months, zooxanthellae densities increase and coral tissue is built up [Brown et al., 1999]. With warmer or longer than usual summers or warmer winters these accumulations may not occur, increasing corals' vulnerability to future stress [Fitt et al., 2001]. Higher WSSTA frequencies may lead to chronic stress, which could increase host susceptibility and disease prevalence. However, WSSTAs could also have influenced disease frequency by increasing pathogen virulence. The relationship between WSSTA and white syndrome cases (Figure 1B) is also suggestive of a weak threshold response when the number of annual WSSTAs exceeds approximately seven. There were several reefs with >7 WSSTAs and a low number of cases (<10), however, all were low coral cover reefs (mean cover 15.1 ± 3.0, n = 7). In fact, the only reefs with >200 cases had >50% cover and a WSSTA frequency >7 (Figure 1B), suggesting that both conditions are necessary for white syndrome outbreaks to occur. Furthermore, these two risk factors explained nearly 75% of the variance in disease cases.

White syndrome frequency varied substantially among the six sectors, possibly due in part to regional temperature variation. Average WSSTA frequency within sectors was positively related to average disease frequency, largely because Capricorn Bunkers, the sector with by far the greatest number of cases also had the greatest number of WSSTAs (Figure 3) and the highest average coral cover [Willis et al., 2004]. Although other studies have found that higher nutrient levels are associated with increased disease severity [Kuta and Richardson,

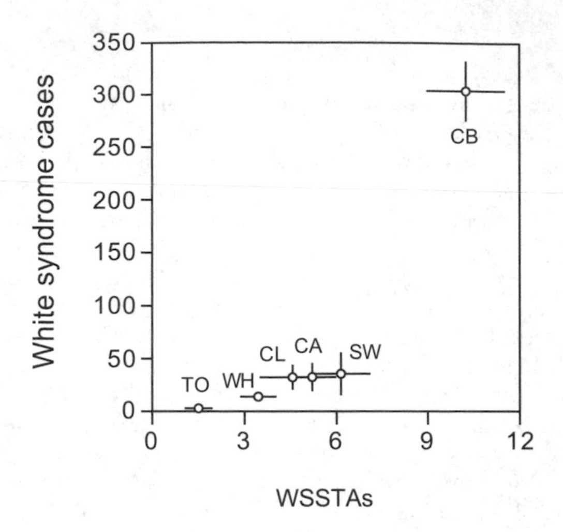

Figure 3. Relationship between mean thermal stress (WSSTAs) and the mean number of white syndrome cases at the sector scale. See Plate 2b for sector delineations and abbreviations. Values are mean ±1 SE.

2002; Bruno et al., 2003], Willis et al. [2004] found that outer shelf reefs had higher levels of white syndrome than inner shelf reefs. Assuming that distance from shore is a proxy for nutrient availability, these findings suggest that nutrient and sediment input may not be primary contributors to increases in white syndrome [Willis et al., 2004]. In fact, several inner shelf reefs relatively close to shore in the Cairns sector had high WSSTA frequencies but very few disease cases in 2002, possibly because host density was also low or because other abiotic or biotic conditions inhibited white syndrome [Willis et al., 2004].

The importance of different abiotic and biotic factors driving white syndrome and other diseases is likely to vary with scale and the host-pathogen system [Bruno et al., 2003]. Ocean currents may facilitate spread or isolation of different pathogens [McCallum et al., 2003], but no empirical studies have yet quantified potential dispersal patterns. Biotic factors like host age or size structure are also likely to have significant effects on disease prevalence [Anderson and May, 1986; Dube et al., 2002; Lafferty and Gerber, 2002]. Older or larger hosts may be more vulnerable to disease [Dube et al., 2002; Borger and Steiner, 2005], which could be particularly devastating for coral populations where older or larger individuals are likely to have higher reproductive output [Hall and Hughes, 1996; Sakai, 1998; Dube et al., 2002].

Uncertainty about the identity and source of most pathogens represents a major challenge in understanding the factors that determine pathogen survival and development. Of the currently described coral diseases, only 5 of 18 [Sutherland et al., 2004] have been identified through fulfillment of Koch's postulates, which require a putative pathogen to be 1) isolated from a diseased individual, 2) grown in pure culture, and 3) transferred to a healthy organism where it induces the disease state [Koch, 1882]. An additional postulate, that the pathogen be reisolated from the infected organism, was not formulated by Koch, but is also typically recommended [Fredricks and Relman, 1996; Richardson, 1998; Sutherland et al., 2004]. Fulfilling Koch's postulates for coral pathogens has been challenging, in part due to the complex nature of the host-pathogen relationship, the possibility of multiple disease agents, and the difficulty of natural inoculation [Richardson, 1998;

Sutherland et al., 2004]. Identifying pathogens through Koch's postulates is not essential for epidemiological study [Fredricks and Relman, 1996], but without isolating them, it has been difficult to determine the mechanisms behind disease dynamics. For example, with a known, cultured pathogen, manipulative experiments on both the isolated pathogen and the host-pathogen system could yield insights into whether temperature is affecting disease by increasing expression of pathogen virulence factors, increasing pathogen growth or reproductive rates, or increasing host susceptibility [Harvell et al., 1999; Harvell et al., 2002].

Although the mechanisms are not known, this study found a strong correlative relationship between white syndrome and warm temperature anomalies. This relationship could have several implications for coral communities on the GBR. White syndrome affects key reef-building species on the GBR including the competitively dominant tabular acroporid corals that constitute a substantial percentage of total coral cover [Baird and Hughes, 2000; Connell et al., 2004]. Reductions in the abundance of these corals could cause shifts in species assemblages or abundances [Baird and Hughes, 2000]. In the Caribbean, loss of acroporid corals due to white band disease has led to a shift in dominance to *Agaricia* sp. on some reefs [Aronson et al., 2002] and precipitated a shift to macroalgal dominance on others. Predicted increases in the frequency and severity of thermal stress anomalies with global climate change [Hoegh-Guldberg, 1999] could exacerbate these kinds of disease effects. Our results and the balance of the published evidence from field studies comparing coral disease prevalence among seasons [Edmunds, 1991], years [Willis et al., 2004] and sites [Kuta and Richardson, 2002] suggest that water temperature plays a substantial role in coral disease dynamics. Because reef building corals are irreplaceable as marine foundation species [Bruno and Bertness, 2001], the synergism between temperature and disease could have cascading effects throughout reef ecosystems [Bruno et al., 2003].

4. Future Research Directions

The development of 20 years of consistently processed satellite sea surface temperature and anomaly data for the GBR region represents a meaningful advancement in understanding the effects of temperature on several parameters of coral health including disease. Further validation of the satellite estimates will provide a better understanding of its limitations and enable more productive and effective use of the dataset. For example, in areas with persistent cloud cover, low data availability may decrease accuracy [Kilpatrick et al., 2001]. The relationship of the satellite-measured surface temperatures to temperatures at different depths will depend on bottom topography and the presence of oceanographic features like internal tidal bores [Leichter and Miller, 1999], which can alter the temperature regime experienced by corals. Assessing patterns of frequency, extent, and intensity of anomalies over the full time period of the dataset could also identify areas that are more or less vulnerable to thermal stress. The presence or absence of these correlations could help determine whether marine protected areas can be used to protect areas of greater resilience or resistance to thermal stress [West and Salm, 2003; Obura, 2005; Wooldridge et al., 2005].

To facilitate regional scale investigations of coral disease risk factors and dynamics, refinement of remote sensing tools must be complemented with rigorous disease monitoring protocols. Much of the current monitoring is idiosyncratic, often in response to a disease outbreak with little baseline or long-term monitoring data [but see Kim and Harvell, 2004; Willis et al., 2004; Santavy et al., 2005]. Long-term longitudinal and cross-sectional epidemiological studies are an essential component of elucidating density-dependence in disease dynamics, susceptible age classes, possible pathogens or modes of transmission, and potential effects on reproductive output and population dynamics. Monitoring programs intended for use in conjunction with satellite temperature data should sample from different 4 km grid cells as defined by the Pathfinder dataset so that there is adequate replication for analysis. In addition, surveys should be conducted within a relatively close time frame so that the data do not covary with other temporal patterns. Finally, manipulative laboratory experiments are also an essential complement to these correlative studies to identify mechanisms driving correlations between temperature and rates of infection and spread.

5. Conclusions

Until very recently it was impossible to correlate satellite-derived temperature anomalies with *in situ* disease surveys. Temporally-consistent satellite temperature data at 4 km resolution enabled us to observe a positive relationship between temperature anomalies and outbreaks of white syndrome at broad spatial scales. A continuing challenge in assessing the importance of temperature anomalies as factors in disease dynamics will be to measure the role of other variables such as proximity to pathogen sources, local water quality, and physical oceanographic patterns. In some cases, these variables could be more important risk factors than temperature. A priority for coral reef scientists is the implementation of disease surveys on a global scale to adequately test the hypotheses that temperature anomalies and other factors drive coral disease. It is likely that there will be different thermal stress thresholds in different biogeographic regions for each disease. Work is underway to describe disease syndromes worldwide and develop a repeatable method for surveying coral disease [Willis et al., 2004] that can be implemented globally. In combination with satellite temperature data and other data sources, we can then begin to understand disease dynamics at regional scales.

Acknowledgments. We thank A. Alker, A. Barton, K. France, S. Lee, A. Melendy Bruno, S. Neale, M. O'Connor, N. O'Connor, L. Stearns, all past and present members of the Australian Institute of Marine Science's Long-term Monitoring Program involved in collecting the disease data, and two anonymous reviewers for their comments. This research was funded in part by the National Science Foundation (OCE-0326705 to J.F.B), an EPA STAR Fellowship to E.R.S, the NOAA Coral Reef Conservation Program and its NESDIS Coral Reef Watch project, and the University of North Carolina at Chapel Hill.

References

Aeby, G. S., Outbreak of coral disease in the Northwestern Hawaiian Islands, *Coral Reefs*, 24(3), 481-481, 2005.

Akaike, H., Information theory and an extension of the maximum likelihood principle, In *Second International Symposium on Information Theory*, edited by B. N. Petrov, and F. Csaki, pp. 267-281, Akademiai Kaido, Budapest, 1973.

Anderson, D. R., K. P. Burnham, and W. L. Thompson, Null hypothesis testing: Problems, prevalence, and an alternative, *Journal of Wildlife Management*, 64(4), 912-923, 2000.

Anderson, D. R., and K. R. Burnham, Avoiding pitfalls when using information-theoretic methods, *Journal of Wildlife Management*, 66(3), 912-918, 2002.

Anderson, R. M., and R. M. May, The invasion, persistence and spread of infectious diseases within animal and plant communities, *Philosophical Transactions of the Royal Society of London Series B-Biological Sciences*, 314(1167), 533-570, 1986.

Aronson, R. B., I. G. MacIntyre, W. F. Precht, T. J. T. Murdoch, and C. M. Wapnick, The expanding scale of species turnover events on coral reefs in Belize, *Ecological Monographs*, 72(2), 233-249, 2002.

Aronson, R. B., and W. F. Precht, White-band disease and the changing face of Caribbean coral reefs, *Hydrobiologia*, 460, 25-38, 2001.

Baird, A. H., and T. P. Hughes, Competitive dominance by tabular corals: an experimental analysis of recruitment and survival of understorey assemblages, *Journal of Experimental Marine Biology and Ecology*, 251(1), 117-132, 2000.

Bellwood, D. R., T. P. Hughes, C. Folke, and M. Nystrom, Confronting the coral reef crisis, *Nature*, 429(6994), 827-833, 2004.

Berkelmans, R., Time-integrated thermal bleaching thresholds of reefs and their variation on the Great Barrier Reef, *Marine Ecology Progress Series*, 229, 73-82, 2002.

Berkelmans, R., G. De'ath, S. Kininmonth, and W. J. Skirving, A comparison of the 1998 and 2002 coral bleaching events on the Great Barrier Reef: spatial correlation, patterns, and predictions, *Coral Reefs*, 23(1), 74-83, 2004.

Berkelmans, R., and B. L. Willis, Seasonal and local spatial patterns in the upper thermal limits of corals on the inshore Central Great Barrier Reef, *Coral Reefs*, 18, 219-228, 1999.

Borger, J. L., and S. C. C. Steiner, The spatial and temporal dynamics of coral diseases in Dominica, West Indies, *Bulletin of Marine Science*, 77(1), 137-154, 2005.

Brown, B. E., Coral bleaching: causes and consequences, *Coral Reefs*, 16, S129-S138, 1997.

Brown, B. E., R. P. Dunne, I. Ambarsari, M. D. A. Le Tissier, and U. Satapoomin, Seasonal fluctuations in environmental factors and variations in symbiotic algae and chlorophyll pigments in four Indo-Pacific coral species, *Marine Ecology Progress Series*, 191, 53-69, 1999.

Bruno, J. F., and M. D. Bertness, Habitat modification and facilitation in benthic marine communities, In *Marine Community Ecology*, edited by M. D. Bertness, M. E. Hay, and S. D. Gaines, pp. 201-218, Sinauer, Sunderland, MA, 2001.

Bruno, J. F., L. E. Petes, C. D. Harvell, and A. Hettinger, Nutrient enrichment can increase the severity of coral diseases, *Ecology Letters*, 6, 1056-1061, 2003.

Bruno, J. F., C. E. Siddon, J. D. Witman, P. L. Colin, and M. A. Toscano, El Niño related coral bleaching in Palau, Western Caroline Islands, *Coral Reefs*, 20(2), 127-136, 2001.

Burnham, K. P., and D. R. Anderson, *Model Selection and Multimodel Inference*, Springer-Verlag, New York, 2002.

Casey, K. S., and P. Cornillon, Global and regional sea surface temperature trends, *Journal of Climate*, 14(18), 3801-3818, 2001.

Castillo, K. D., and B. S. T. Helmuth, Influence of thermal history on the response of Montastraea annularis to short-term temperature exposure, *Marine Biology*, 148(2), 261-270, 2005.

Connell, J. H., T. E. Hughes, C. C. Wallace, J. E. Tanner, K. E. Harms, and A. M. Kerr, A long-term study of competition and diversity of corals, *Ecological Monographs*, 74(2), 179-210, 2004.

Dube, D., K. Kim, A. P. Alker, and C. D. Harvell, Size structure and geographic variation in chemical resistance of sea fan corals Gorgonia ventalina to a fungal pathogen, *Marine Ecology Progress Series*, 231, 139-150, 2002.

Edmunds, P. J., Extent and effect of black band disease on a Caribbean reef, *Coral Reefs*, 10(3), 161-165, 1991.

Fabricius, K. E., Effects of terrestrial runoff on the ecology of corals and coral reefs: review and synthesis, *Marine Pollution Bulletin*, 50(2), 125-146, 2005.

Fernandes, L., J. Day, A. Lewis, S. Slegers, B. Kerrigan, D. Breen, D. Cameron, B. Jago, J. Hall, D. Lowe, J. Innes, J. Tanzer, V. Chadwick, L. Thompson, K. Gorman, M. Simmons, B. Barnett, K. Sampson, G. De'ath, B. Mapstone, H. Marsh, H. Possingham, I. Ball, T. Ward, K. Dobbs, J. Aumend, D. Slater, and K. Stapleton, Establishing representative no-take areas in the Great Barrier Reef: Large-scale implementation of theory on marine protected areas, *Conservation Biology*, 19(6), 1733-1744, 2005.

Fitt, W. K., B. E. Brown, M. E. Warner, and R. P. Dunne, Coral bleaching: interpretation of thermal tolerance limits and thermal thresholds in tropical corals, *Coral Reefs*, 20(1), 51-65, 2001.

Folland, C. K., N. A. Rayner, S. J. Brown, T. M. Smith, S. S. P. Shen, D. E. Parker, I. Macadam, P. D. Jones, R. N. Jones, N. Nicholls, and D. M. H. Sexton, Global temperature change and its uncertainties since 1861, *Geophysical Research Letters*, 28(13), 2621-2624, 2001.

Fredricks, D. N., and D. A. Relman, Sequence-based identification of microbial pathogens: A reconsideration of Koch's postulates, *Clinical Microbiology Reviews*, 9(1), 18-33, 1996.

Gardner, T. A., I. M. Côté, J. A. Gill, A. Grant, and A. R. Watkinson, Long-term region-wide declines in Caribbean corals, *Science*, 301, 958-960, 2003.

Gil-Agudelo, D. L., and J. Garzon-Ferreira, Spatial and seasonal variation of Dark Spots Disease in coral communities of the Santa Marta area (Colombian Caribbean), *Bulletin of Marine Science*, 69(2), 619-629, 2001.

Glynn, P. W., Coral-Reef Bleaching - Ecological Perspectives, *Coral Reefs*, 12(1), 1-17, 1993.

Glynn, P. W., Coral reef bleaching: Facts, hypotheses and implications, *Global Change Biology*, 2(6), 495-509, 1996.

Glynn, P. W., J. Cortes, H. M. Guzman, and R. H. Richmond, El Niño (1982-1983) associated coral mortality and relationship to sea surface temperature deviations in the tropical eastern Pacific, *Proceedings of the Sixth International Coral Reef Symposium*, 2, 693-698, 1988.

Glynn, P. W., and L. D'croz, Experimental-evidence for high-temperature stress as the cause of El-Niño-coincident coral mortality, *Coral Reefs*, 8(4), 181-191, 1990.

Hall, V. R., and T. P. Hughes, Reproductive strategies of modular organisms: Comparative studies of reef-building corals, *Ecology*, 77(3), 950-963, 1996.

Harvell, C. D., K. Kim, J. M. Burkholder, R. R. Colwell, P. R. Epstein, D. J. Grimes, E. E. Hofmann, E. K. Lipp, A. D. M. E. Osterhaus, R. M. Overstreet, J. W. Porter, G. W. Smith, and G. R. Vasta, Emerging marine diseases–climate links and anthropogenic factors, *Science*, 285, 1505-1510, 1999.

Harvell, C. D., C. E. Mitchell, J. R. Ward, S. Altizer, A. P. Dobson, R. S. Ostfeld, and M. D. Samuel, Climate warming and disease risks for terrestrial and marine biota, *Science*, 296, 2158-2162, 2002.

Hayes, M. L., J. Bonaventura, T. P. Mitchell, J. M. Prospero, E. A. Shinn, F. Van Dolah, and R. T. Barber, How are climate and marine biological outbreaks functionally linked?, *Hydrobiologia*, 460, 213-220, 2001.

Hoegh-Guldberg, O., Climate change, coral bleaching and the future of the world's coral reefs, *Marine Freshwater Research*, 50, 839-866, 1999.

Hughes, T. P., A. H. Baird, D. R. Bellwood, M. Card, S. R. Connolly, C. Folke, R. Grosberg, O. Hoegh-Guldberg, J. B. C. Jackson, J. Kleypas, J. M. Lough, P. Marshall, M. Nystrom, S. R. Palumbi, J. M. Pandolfi, B. Rosen, and J. Roughgarden, Climate change, human impacts, and the resilience of coral reefs, *Science*, 301, 929-933, 2003.

Hughes, T. P., and J. H. Connell, Multiple stressors on coral reefs: a long-term perspective, *Limnology and Oceanography*, 44 (3, part 2), 932-940, 1999.

Jackson, J. B. C., Reefs since Columbus, *Coral Reefs*, 16, S23-S32, 1997.

Jones, R. J., J. Bowyer, O. Hoegh-Guldberg, and L. L. Blackall, Dynamics of a temperature-related coral disease outbreak, *Marine Ecology Progress Series*, 281, 63-77, 2004.

Kilpatrick, K. A., G. P. Podesta, and R. Evans, Overview of the NOAA/NASA advanced very high resolution radiometer Pathfinder algorithm for sea surface temperature and associated matchup database, *Journal of Geophysical Research-Oceans*, 106(C5), 9179-9197, 2001.

Kim, K., and C. D. Harvell, The rise and fall of a six-year coral-fungal epizootic, *American Naturalist*, 164(5), S52-S63, 2004.

Koch, R., Die aetiologie der tuberculose, *Berl Klinische Wochenschrift*, 19, 221-230, 1882.

Kuta, K. G., and L. L. Richardson, Ecological aspects of black band disease of corals: relationships between disease incidence and environmental factors, *Coral Reefs*, 21(4), 393-398, 2002.

Kutz, S. J., E. P. Hoberg, L. Polley, and E. J. Jenkins, Global warming is changing the dynamics of Arctic host-parasite systems, *Proceedings of the Royal Society B-Biological Sciences*, 272(1581), 2571-2576, 2005.

Lafferty, K. D., and L. R. Gerber, Good medicine for conservation biology: The intersection of epidemiology and conservation theory, *Conservation Biology*, 16(3), 593-604, 2002.

Leichter, J. J., G. B. Deane, and M. D. Stokes, Spatial and temporal variability of internal wave forcing on a coral reef, *Journal of Physical Oceanography*, 35(11), 1945-1962, 2005.

Leichter, J. J., and S. L. Miller, Predicting high-frequency upwelling: Spatial and temporal patterns of temperature anomalies on a Florida coral reef, *Continental Shelf Research*, 19(7), 911-928, 1999.

Lenihan, H. S., F. Micheli, S. W. Shelton, and C. H. Peterson, The influence of multiple environmental stressors on susceptibility to parasites: an experimental determination with oysters, *Limnology and Oceanography*, 44 (3, part 2), 910-924, 1999.

Lessios, H. A., Mass mortality of Diadema-Antillarum in the Caribbean - what have we learned, *Annual Review of Ecology and Systematics*, 19, 371-393, 1988.

Liu, G., W. Skirving, and A. E. Strong, Remote sensing of sea surface temperatures during 2002 Barrier Reef coral bleaching, *EOS*, 84(15), 137-144, 2003.

Marcogliese, D. J., Implications of climate change for parasitism of animals in the aquatic environment, *Canadian Journal of Zoology-Revue Canadienne De Zoologie*, 79(8), 1331-1352, 2001.

McCallum, H., C. D. Harvell, and A. P. Dobson, Rates of spread in marine pathogens, *Ecology Letters*, 6, 1062-1067, 2003.

Nugues, M. M., G. W. Smith, R. J. Hooidonk, M. I. Seabra, and R. P. M. Bak, Algal contact as a trigger for coral disease, *Ecology Letters*, 7(10), 919-923, 2004.

Obura, D. O., Resilience and climate change: lessons from coral reefs and bleaching in the Western Indian Ocean, *Estuarine Coastal and Shelf Science*, 63(3), 353-372, 2005.

Pandolfi, J. M., R. H. Bradbury, E. Sala, T. P. Hughes, K. A. Bjorndal, R. G. Cooke, D. McArdle, L. McClenachan, M. J. H. Newman, G. Paredes, R. R. Warner, and J. B. C. Jackson, Global trajectories of the long-term decline of coral reef ecosystems, *Science*, 301, 955-958, 2003.

Pascual, M., X. Rodo, S. P. Ellner, R. R. Colwell, and M. J. Bouma, Cholera dynamics and El Niño-Southern Oscillation, *Science*, 289, 1766-1769, 2000.

Patterson, K. L., J. W. Porter, K. E. Ritchie, S. W. Polson, E. Mueller, E. C. Peters, D. L. Santavy, and G. W. Smiths, The etiology of white pox, a lethal disease of the Caribbean elkhorn coral, Acropora palmata, *Proceedings of the National Academy of Sciences of the United States of America*, 99(13), 8725-8730, 2002.

Patz, J. A., A human disease indicator for the effects of recent global climate change, *Proceedings of the National Academy of Sciences of the United States of America*, 99(20), 12506-12508, 2002.

Podesta, G. P., and P. W. Glynn, The 1997-98 El Niño event in Panama and Galapagos: An update of thermal stress indices relative to coral bleaching, *Bulletin of Marine Science*, 69(1), 43-59, 2001.

Pounds, J. A., M. R. Bustamante, L. A. Coloma, J. A. Consuegra, M. P. L. Fogden, P. N. Foster, E. La Marca, K. L. Masters, A. Merino-Viteri, R. Puschendorf, S. R. Ron, G. A. Sanchez-Azofeifa, C. J. Still, and B. E. Young, Widespread amphibian extinctions from epidemic disease driven by global warming, *Nature*, 439(7073), 161-167, 2006.

R Development Core Team, R: A language and environment for statistical computing, R Foundation for Statistical Computing, Vienna, Austria, 2005.

Raymundo, L. J., K. B. Rosell, C. T. Reboton, and L. Kaczmarsky, Coral diseases on Philippine reefs: genus Porites is a dominant host, *Diseases of Aquatic Organisms*, 64(3), 181-191, 2005.

Reynolds, R. W., N. A. Rayner, T. M. Smith, D. C. Stokes, and W. Q. Wang, An improved in situ and satellite SST analysis for climate, *Journal of Climate*, 15(13), 1609-1625, 2002.

Richardson, L. L., Coral diseases: what is really known?, *Trends in Ecology & Evolution*, 13(11), 438-443, 1998.

Rosenberg, E., and Y. Ben-Haim, Microbial diseases of corals and global warming, *Environmental Microbiology*, 4(6), 318-326, 2002.

Rosenberg, E., and L. Falkovitz, The Vibrio shiloi/Oculina patagonica model system of coral bleaching, *Annual Review of Microbiology*, 58, 143-159, 2004.

Sakai, K., Effect of colony size, polyp size, and budding mode on egg production in a colonial coral, *Biological Bulletin*, 195(3), 319-325, 1998.

Santavy, D. L., J. K. Summers, V. D. Engle, and L. C. Harwell, The condition of coral reefs in South Florida (2000) using coral disease and bleaching as indicators, *Environmental Monitoring and Assessment*, 100(1-3), 129-152, 2005.

Sheppard, C., and R. Rioja-Nieto, Sea surface temperature 1871-2099 in 38 cells in the Caribbean region, *Marine Environmental Research*, 60(3), 389-396, 2005.

Sheppard, C. R. C., Predicted recurrences of mass coral mortality in the Indian Ocean, *Nature*, 425(6955), 294-297, 2003.

Shinn, E. A., G. W. Smith, J. M. Prospero, P. Betzer, M.L. Hayes, V. Garrison, and R.T. Barber, African dust and the demise of Caribbean coral reefs, *Geophysical Research Letters*, 27(19), 3029-3032, 2000.

StataCorp LP, Stata 9, StataCorp LP, College Station, TX, 2006.

Strong, A. E., G. Liu, J. Meyer, J. C. Hendee, and D. Sasko, Coral Reef Watch 2002, *Bulletin of Marine Science*, 75(2), 259-268, 2004.

Sutherland, K. P., J. W. Porter, and C. Torres, Disease and immunity in Caribbean and Indo-Pacific zooxanthellate corals, *Marine Ecology Progress Series*, 266, 273-302, 2004.

Sweatman, H., D. Abdo, S. Burgess, A. Cheal, G. Coleman, S. Delean, M. J. Emslie, I. Miller, K. Osborne, C. A. Page, and A. Thompson, *Long-Term Monitoring of the Great Barrier Reef: Status Report Number* 6, Australian Institute of Marine Science, Townsville, 2003.

The Mathworks Inc., Matlab, The Mathworks Inc., Natick, MA, 2005.

Toscano, M. A., G. Liu, G. I. C., K. S. Casey, A. Strong, and J. E. Meyer, Improved prediction of coral bleaching using high-resolution HotSpot anomaly mapping, in *Ninth International Coral Reef Symposium*, edited by M. K. Moosa, S. Soemodihardjo, A. Nontji, A. Soegiarto, K. Romimohtarto, Sukarno, and Suharsono, pp. 1143-1148, Ministry of the Environment, the Indonesian Institute of Sciences, and the International Society for Reef Studies, Bali, Indonesia, 2000.

Wapnick, C. M., W. F. Precht, and R. B. Aronson, Millennial-scale dynamics of staghorn coral in Discovery Bay, Jamaica, *Ecology Letters*, 7(4), 354-361, 2004.

Ward, J. R., and K. D. Lafferty, The elusive baseline of marine disease: Are diseases in ocean ecosystems increasing?, *PloS Biology*, 2(4), 542-547, 2004.

Weil, E., Coral Reef Diseases in the Wider Caribbean, in *Coral Health and Disease*, edited by E. Rosenberg, and Y. Loya, Springer, Berlin, 2004.

West, J. M., and R. V. Salm, Resistance and resilience to coral bleaching: Implications for coral reef conservation and management, *Conservation Biology*, 17(4), 956-967, 2003.

Williams, D. E., and M. W. Miller, Coral disease outbreak: pattern, prevalence and transmission in Acropora cervicornis, *Marine Ecology-Progress Series*, 301, 119-128, 2005.

Willis, B. L., C. A. Page, and E. A. Dinsdale, Coral disease on the Great Barrier Reef, in *Coral Health and Disease*, edited by E. Rosenberg, and Y. Loya, pp. 69-104, Springer-Verlag, Berlin, 2004.

Winter, A., R. S. Appeldoorn, A. Bruckner, E. H. Williams, and C. Goenaga, Sea surface temperatures and coral reef bleaching off La Parguera, Puerto Rico (northeastern Caribbean Sea), *Coral Reefs*, 17(4), 377-382, 1998.

Wooldridge, S., T. Done, R. Berkelmans, R. Jones, and P. Marshall, Precursors for resilience in coral communities in a warming climate: a belief network approach, *Marine Ecology-Progress Series*, 295, 157-169, 2005.

7

A Coral Population Response (CPR) Model for Thermal Stress

R. van Woesik and S. Koksal

Abstract

More than two decades of higher than average sea surface temperatures in the tropics have led to unprecedented thermal stress, coral bleaching and extensive coral mortality. Field and experimental evidence shows that thermal stress is a function of high-water temperature and high irradiance, while water flow dampens thermal stress effects. Coral populations clearly differ in their susceptibility to thermal stress, some are simply more physiologically tolerant, yet other differences are influenced by colony morphology, colony size and depth distribution. Although, physiological responses are measured in hours and days population changes take weeks, months or years, yet ultimately the latter is a time-integrated response of the former. Few studies to date have suggested ways to accurately translate coral physiology data into coral population responses. Our Coral Population Response (CPR) model is a system of non-linear ordinary differential equations that links the physiological responses of corals to water temperature, irradiance and water-flow rates to accurately predict thermal stress and subsequent population change. This model allows hindcasting and forecasting of changes in coral populations based on current and predicted environmental scenarios involving global climate change.

Introduction

Coral Bleaching and Global Effects

After 200,000 years of comparative stasis in coral community composition [Pandolfi and Jackson, 2001], the last two decades has seen sudden changes to coral reefs worldwide [Jackson et al., 2001]. While we think of reef building corals and their symbiotic dinoflagellates as warmth loving, they have an upper temperature tolerance, and clearly temperature stress has caused shifts in coral community structure in the Indian Ocean [McClanahan et al., 2001; Sheppard, 2003], southern Japan [Loya et al., 2001], the Caribbean [Aronson et al., 2000] and on the Great Barrier Reef [Baird and Marshall, 1998; Berkelmans and Oliver, 1999]. Contemporary research efforts are focused on whether these

Coral Reefs and Climate Change: Science and Management
Coastal and Estuarine Studies 61
Copyright 2006 by the American Geophysical Union.
10.1029/61CE08

community shifts are permanent, which species are most resilient, and which species can acclimatize and adapt to forthcoming thermal stress events [Van Woesik et al., 2004].

Coral bleaching is generally defined as the loss of symbiotic algae (i.e., zooxanthellae) and/or pigment because of stress at and above the corals' acclimation capacity [Glynn, 1993; Fitt and Warner, 1995; Brown, 1997; Fitt et al., 2001]. Higher than average Sea Surface Temperatures (SST) under high solar radiation is reflected in the high frequency of coral bleaching events over the last two decades [Glynn, 1991; Glynn, 1993; Brown, 1997; Hoegh-Guldberg, 1999; Wilkinson, 1999]. While extreme events are elevated end points of seasonal cycles, where paling and photoinhibition is common [Warner et al., 2002], the global coral-bleaching event witnessed in 1997-98 coincided with the highest amplitude and duration of Sea Surface Temperatures (SSTs) on record in the Indian and Pacific Oceans [Berkelmans and Oliver, 1999; Strong et al., 2000; Marshall and Baird, 2000; McClanahan, 2000; Edwards et al., 2001; McClanahan et al., 2001]. High temperatures also appear to onset the virulence of primary infectious bacteria (i.e., *Vibrio shiloi*, *Vibrio coralliitycus* [Kushmaro et al., 1996]) that reduce symbiont populations and produce an immuno-compromised host [Rosenberg, 2004]. The global nature of up-regulation of coral-associated bacteria and the increase in susceptibility of corals to other potential stresses, including diseases, need further examination.

Physiological Responses

Most evidence suggests that photosynthetic impairment of symbiotic zooxanthellae in corals (which average 10^6 algae per cm^2 of coral tissue) when subjected to elevated temperature and high irradiance leads to loss of pigment and loss of symbionts (i.e., bleaching). Such conditions may be temporary or lead to coral mortality [Gates et al., 1992; Fitt et al., 2000; Fitt et al., 2001; Jones and Hoegh-Guldberg, 2001]. Symbiont reshuffling and shifts in relative abundance of symbionts in colonies may lead to photo-acclimation and coral survival [Buddemeier and Fautin, 1993; Baker, 2001], yet this adjustment is not an adaptation in a Darwinian sense, because that involves differential-reproductive rates on different individuals within populations and any adverse effects on coral reproduction is counterintuitive to any adaptive process [Van Woesik, 2001; Hoegh-Guldberg et al., 2002].

The photosynthetic capacity of coral symbionts change in accordance with the daily course of irradiance. Increasing morning irradiance leads to increased rates of photosynthesis, to a point where light becomes saturating and photosynthetic rates plateau, or even decrease [Chalker, 1981]. Prolonged high irradiance leads to the over reduction of the light reaction centers and the production of harmful photosynthetic byproducts. Such conditions can lead to photoinhibition, which can commence around mid-morning on cloudless days and continues until mid-afternoon [Brown et al., 1999]. Indeed, under high irradiance, a diel decrease in photosynthetic capacity, or reduction in the quantum yield of photosystem II (PSII) photochemistry, in corals closely follows the diel flux of photosynthetically active radiation [Gorbunov et al., 2001]. This reduction agrees with the photoprotective dynamics of xanthophyll in coral symbionts [Brown et al., 1999] and the down regulation of damaged reaction centers of PSII complexes [Gorbunov et al., 2001].

Warner et al. [1999] argued that the sensitivity of the D1 protein in PSII characterizes a symbionts susceptibility to photoinhibition, and Takahashi et al. [2004] concurred, but suggested that the repair process of D1 protein is the limiting factor, and rapid repair rates of D1 protein equate with high tolerance. Takahashi et al. [2004] also showed that coral species differ considerable in their ability to recover from photoinhibition; while they only measured 5 different coral species, the physiological response of each species agreed with field observations [Loya et al., 2001]. Implicit in this argument is the fact that coral species house different strains of symbionts [LaJeunesse et al., 2003]. It is becoming clear that symbiont strains differ in susceptibility to heat and irradiance stress, with D-type zooxanthellae potentially showing greatest (global) tolerance [Baker, 2004]. Yet, juvenile corals infected with D-type zooxanthellae show poor holobiont [i.e., the coral and its symbionts) growth, suggesting that the emergent holobiont physiology differs in accordance with zooxanthellae strain [Little et al., 2004]. But there are still few studies linking zooxanthellae genetic constitution with holobiont performance.

Dynamic photoinhibition is, however, a reversible process, which redirects excitation energy away from the reaction centers. Yet, daily time courses of photosynthetic activity show that high irradiance has a sustained effect on the symbiont's photochemistry, and quantum yield will remain suppressed even after 12 hours of darkness, but only if the corals are subjected to high irradiance prior to the onset of darkness [Jones and Hoegh-Guldberg, 2001]. Irreversible, photo-damage to the PS II reaction center can occur as chronic photoinhibition and the photosystem can only resume efficient photosynthesis when the primary reaction centers recover [Warner et al., 1999; Takahashi et al., 2004]. Therefore, sustained levels of dynamic photoinhibition can lead to chronic photoinhibition, which in turn often leads to symbiont damage, loss of pigment and expulsion of symbionts through a bleaching event [Jones and Hoegh-Guldberg, 2001]. Such is the case when water temperatures are elevated above ambient because the enzymes within PSII are particularly sensitive to heat stress [Iglesias-Prieto et al., 1992], and corals perceive heat stress as a further increase in excitation pressure over PS II [Iglesias-Prieto et al., 2004]. Therefore, low or reduced irradiance during times of temperature stress reduces photoinhibition, coral bleaching and coral mortality [Mumby et al., 2001] as does the reciprocal involving moderate water temperature at high irradiance [Takahashi et al., 2004].

Environmental Reciprocity

While environmental reciprocity between temperature and irradiance are clearly recognized [Iglesias-Prieto et al., 1992; Maxwell et al., 1994; Warner et al., 1996; Takahashi et al., 2004], few studies to date have focused on other physical variables influencing bleaching susceptibility; for example the dynamics of water flow may explain some patterns of bleaching and post-bleaching recovery [Nakamura and Van Woesik, 2001]. Water flowing across marine organisms exerts forces that thin boundary layers which increases the potential for mass transfer across the solid-liquid interface [Patterson and Sebens, 1989]. Flux of dissolved gases and metabolites, between an organism and the surrounding water, are also largely dependent on the size and morphology of the organism [Lesser et al., 1994]. Elevated flow rates typically lead to significantly higher rates of photosynthesis, colony respiration and calcification [Dennison and Barnes, 1988; Patterson, 1992; Atkinson et al., 1994; Lesser et al., 1994]. Furthermore, Atkinson and colleagues demonstrated that nutrient uptake by corals is mass transfer limited [Atkinson and Bilger, 1992; Baird and Atkinson, 1997; Thomas and Atkinson, 1997].

Recent experimental evidence has shown that the prevention of, and rapid recovery from, bleaching by enhancing water flow rates strongly implies that mass-transfer-limited processes are involved in coral bleaching [Nakamura and Van Woesik, 2001; Nakamura et al., 2003; Nakamura et al., 2005; Van Woesik et al., 2005; Fabricius, 2006]. Nakamura and Van Woesik [2001] suggested that water flow reduces stress by increasing the efflux of oxygen radicals from the corals under temperature and irradiance stress. This hypothesis has recently been confirmed by Finelli et al. [2006], who undertook experiments on two coral species, the massive *Montastrea annularis* and the complex foliose, bifacial fronded *Agaricia agaricites*. Finelli et al. showed a significant increase in photosynthetic efficiency with an increase in water flow, but only for *Agaricia agaricites*. They also showed that by increasing oxygen concentrations the effect of flow on *Agaricia agaricites* was reversed, which clearly suggests that water flow directly removes accumulated, harmful photosynthetic oxygen species from corals. Also, there are species specific and morphological differences, whereby massive coral colonies may be less sensitive to an increase in water flow compared with more complex morphologies. Indeed, Nakamura and colleagues also used a complex *Acropora* morphology in past experiments [*Acropora digitifera* was used by Nakamura and Van Woesik, 2001; Nakamura et al., 2003; Nakamura et al., 2005].

Linking Physiology to Populations

While the above physiological responses are measured in hours and days [Jones and Hoegh-Gulberg, 2001; Gorbunov et al., 2001; Brown et al., 2002] population changes take weeks, months or years [Loya et al., 2001; Van Woesik et al., 2004], yet ultimately the latter is a time-integrated response of the former, just as evolutionary change is an integrated response of changes that occur over ecological time. Few studies to date have suggested a way to translate coral physiology data into coral population responses. Baker [2001] inferred that the genetic constitution of symbionts produced differential bleaching and survival (at the population level), but Brown et al. [2002] showed that genetically identical symbionts can also produce differential bleaching within coral holobionts; they suggested differential bleaching within colonies with identical symbiont strains was more a consequence of recent (irradiance) experience. Iglesias-Prieto et al. [2004] linked the molecular response of two coral species' symbiotic populations with their photosynthetic capacity, and then in turn linked that physiology with their vertical distribution along the reef slope. Here we develop a model to link coral photo-physiology to population response with the underlying assumption that coral populations that are less likely to suffer photoinhibition, measurable at the physiological level, are also less likely to bleach and suffer mortality losses.

Models

We consider models as merely simplifications of nature; we do not necessarily view them as new sources of knowledge showing emergent properties. Rather, we treat them as organizing tools that express ideas and perspectives and capture the essence of patterns, mechanisms and regulatory processes. This organizing framework allows synthesis and integration that may lead to extrapolation across spatial and temporal scales. Predictions may include specific and even hypothetical scenarios involving climate change. Within this philosophical framework we present a novel model that links the physiological response of corals, in an experimental setting, with the predicted population response in the field.

Our Coral Population Response (CPR) model is a system of non-linear ordinary differential equations that links the physiological and population responses of corals to water temperature, irradiance and water-flow rates to accurately predict thermal stress and subsequent population change. This model may pave the way for translating data from physiological scales to populations, and eventually to complex coral reef-landscape scales. Yet, continued observations and time-series analyses of coral populations in the field are necessary because they provide fuel for validating predictive models. This model also allows hindcasting and forecasting of coral-population changes based on current and predicted global climate change scenarios.

Methods

Thermal Stress Functions

A proxy for the overall photosynthetic capacity of PSII in coral symbionts is a function of the ratio of dark adapted variable fluorescence to maximal fluorescence (Fv/Fm) [Jones et al., 1998; Warner et al., 1999]. A recent study by Takahashi et al. [2004] examined the response of *Acropora digitifera* colonies to different temperature and irradiance treatments. The maximum potential quantum yield of PSII (Fv/Fm) of dark adapted colonies was measured using a DIVING-PAM Fluorometer [Walz, Germany] after 3 hours of irradiance ranging from 0, 100, 250, 500, 1000 and 1500 μmol photons m^{-2} s^{-1}, under both 28 and 32°C [Takahashi et al., 2004]. Just as the photochemical efficiency of PS II (Fv/Fm) describes the overall capacity of the photosystem, we argue that the reciprocal $\left(\frac{1}{Fv/Fm}\right)$ should describe the overall level of dysfunction of the corals' endosymbionts. The Fv/Fm end-points of the empirical data were treated as reciprocals to derive a stress response and plotted against irradiance (Figure 1). We note that all notations and abbreviations used in the following text and equations are provided in Table 1.

Mathematical Model

From above we understand that dynamic photoinhibition is a function of the duration and intensity of light [Takahashi et al., 2004]. As a function of time, $l(t)$, the amount of irradiance ($\mu\, mol$ photons $m^{-2}s^{-1}$) at time t, is given by the periodic function

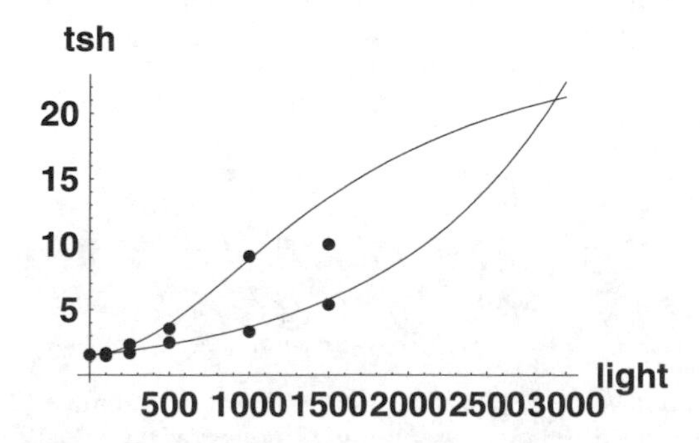

Figure 1. Thermal stress as a function of irradiance under high (32°C, top-line) and moderate SST (28°C, bottom-line).

TABLE 1. Notations and abbreviations used in the model.

Notation	Unit	Interpretation
t	hour	time
$N(t)$	number	Coral population function
N_{max}	"	Carrying capacity
N_0	"	Initial population
l_0	μ mol photons $m^{-2}s^{-1}$	Maximum irradiance
ω	$(time)^{-1}$	frequency
$t_{sh}(t)$	temperature . μ mol photons $m^{-2}s^{-1}$	Thermal stress function
t^0_{sh}	"	Initial value of thermal stress
$l(t)$	μ mol photons m^{-2} s^{-1}	Light function
m	$(time)^{-1}$	Migration rate
p	$(time)^{-1}$	Mortality rate
r_1	$(time)^{-1}$	Intrinsic growth rate of population
k_1	$(time)^{-1}$(temperature . μ mol photons m^{-2} $s^{-1})^{-1}$	The rate at which t_{sh} *effects the population*
k_2	$(time)^{-1}$	The cumulative rate of change of t_{sh} wrt. *Time during the day*
k_3	dimensionless	Adjustment parameter
$\tilde{k}_2$	$(time)^{-1}$	Proportionality constant of the rate of change of t_{sh} wrt time during the night
$f_{mod}(l)$	temperature . μ mol photons m^{-2} s^{-1}	Thermal stress as a function of irradiance under moderate SST
$f_{high}(l)$	temperature . μ mol photons m^{-2} s^{-1}	Thermal stress as a function of irradiance under high SST
a_1,b_1	temperature . μ mol photons m^{-2} s^{-1}	Parameters estimated for f_{high}(l) by curve fitting
c_1	μ mol photons m^{-2} s^{-1}	Parameter estimated by curve fitting for f_{high}(l)
a_2	temperature . μ mol photons m^{-2} s^{-1}	Parameter estimated for f_{mod}(l) by curve fitting
b_2	(μ mol photons m^{-2} $s^{-1})^{-1}$	Parameter estimated for f_{mod}(l) by curve fitting
F_0	cm . s^{-1}	Flow rate of water
k_4	$(cm)^{-1}$	Adjustment coefficient

$$l(t) = \begin{cases} I_0 \sin\left(\frac{\pi t}{12}\right), & 0 \le t \le 12 \\ 0, & 12 \le t \le 24 \end{cases} \tag{1}$$

with $l(t + 24) = l(t)$, and I_0 being the maximum value for irradiance (2000 μmol photons m^{-2} s^{-1}), and t is measured in hours, assuming 12 hours of day and 12 hours of night. We assume that there is minimal cloud cover over the 15 days of running the model, which is realistic for a bleaching event. For example in the western Pacific Ocean, during the 1998 and the 2001-bleaching event, cloud cover was minimal. We therefore assume maximum irradiance while running the model.

Photoinhibition is exacerbated by elevated water temperature [Jones and Hoegh-Guldberg, 2001; Takahashi et al., 2004]; in combination we define high irradiance and high water temperature as a thermal stress variable, and denote it by t_{sh} whose unit is °C μmol photons m^{-2} s^{-1}. The method of least squared approximation is used to represent thermal stress as a function of irradiance. Under the high SST (32°C), thermal stress is defined by

$$t_{sh} = f_{high}(l) = a_1 + \frac{b_1 l^2}{c_1^2 + l^2} \tag{2}$$

where the constant coefficients are determined by fitting the curve given in Figure 1 to the empirical data. In Equation 2, the units of a_1 and b_1 are the same as the unit of t_{sh} and the unit of c_1 is μmol photons m^{-2} s^{-1}. On the other hand, when SST is moderate (28°C), thermal stress is given by

$$t_{sh} = f_{\text{mod}}(l) = a_2 2^{b_2 l} \tag{3}$$

As before, the constant coefficients a_2 and b_2 are determined by curve fitting (Figure 1), and have the units of t_{sh} and (μmol photons m^{-2} s^{-1})$^{-1}$, respectively. In our model, the parameters in $f_{high}(l)$ and $f_{\text{mod}}(l)$ were estimated with high precision having the mean squared error values of 12.92% and 0.37%, respectively.

In what follows we used SST, irradiance and water flow rates to model the interactions between thermal stress and coral populations in term of a system of nonlinear ODEs. In this system we consider a population of corals, N, and thermal stress, t_{sh}, as functions of time, t. Here we assume that the rate of change of thermal stress t_{sh} with respect to time is directly proportional to the amount of thermal stress at present and the total rate of change of thermal stress function, $f_{\text{mod}}(l)$, given in Equation 3. We also assume that the coral population increases logistically with the carrying capacity, $N_{\max}$, and decreases at a rate proportional to the interaction between thermal stress and population. Hence, in the case of lower (28°C) than critical temperatures (30°C), during the day irradiance, the dynamics of $N(t)$ and $t_{sh}(t)$ are governed by the system of ODEs:

$$\frac{dN}{dt} = mN + r_1 N\left(1 - \frac{N}{N_{\max}}\right) - k_1 t_{sh} N - pN; \quad N(0) = N_0 \tag{4}$$

$$\frac{dt_{sh}}{dt} = k_2 t_{sh} + k_3 f(t,l); \qquad t_{sh}(0) = t_{sh}^0 \tag{5}$$

where r_1 is the intrinsic rate of population increase, m and p are the migration and mortality rates, respectively, $N_{\max}$ is the carrying capacity, $f(t,l) = \frac{df_{\text{mod}}}{dl}\frac{dl}{dt}$, the total rate of change of f_{mod} given in Equation 3, k_1, k_2 are proportionality constants, and k_3 is the unitless adjustment parameter. The constants m, p, r_1 and k_2 have the units (time)$^{-1}$ and k_1 has the units (temperature · μmol photons · m^{-2} $s^{-1})^{-1}$ $(time)^{-1}$. When the water temperature is high (32°C), in Equation 5 $f(t, l)$ is replaced by the total rate of change of f_{high} (Equation 2), $g(t,l) = \frac{df_{high}}{dl}\frac{dl}{dt}$.

Thermal Stress and Population Dynamics

In the Equation (4), we assumed that the terms representing migration and mortality (independent of thermal stress) are significant only over long periods of time; therefore we made these terms equal over the 15-day iterations of the model system. We ran the models for 3 months, yet they showed greatest change in the first 15 days and little change thereafter. We used the system (4-5) to study the daytime dynamics of thermal stress and the coral population. To obtain the response of the system during the night, we replaced Equation 5 by

$$\frac{dt_{sh}}{dt} = -\tilde{k}_2 t_{sh} \qquad (6)$$

with the proportionality constant $\tilde{k}_2$ having the unit of $(time)^{-1}$. The system (4-6) does not include water flow. Water flow is incorporated into Equations (5) and (6) by adding the term $-k_4 F_0 t_{sh}$, where F_0 (*cm/s*) is the flow rate of water, and k_4 is the proportionality constant having the unit $(cm)^{-1}$. Similar adjustments were made to the system when SST was high (32°C).

Model Simulations

To illustrate the model system and determine the significance of the parameters and the initial value of thermal stress, we considered one coral species *Acropora digitifera*, which is a common Indo-Pacific coral for which we have plenty of physiological and ecological data [Loya et al., 2001; Nakamura and Van Woesik, 2001; Nakamura et al., 2003; Nakamura et al., 2005]. We note that all parameters used in the following simulations are provided in Tables 2 and 3. The values 1500 and .05 were used as initial values of population and thermal stress, respectively. The model was solved numerically using the 4th order Runge-Kutta method for four cases: High SST (32°C) with and without water flow and moderate SST (28°C) with and without water flow. We tested the model with different values of initial thermal stress and the parameters $\tilde{k}_2$ (the rate at which thermal stress decreases at night) and k_2 (the rate of increase in thermal stress during the day time). We kept the values of the adjustment parameter, k_3, and the intrinsic rate of change of population, r_1, the same for all cases. Over the 15-day period, we estimated population and thermal stress values. By curve fitting, we also obtained functional relations for the values of thermal stress before sunset and sunrise. We observed that the values of thermal stress before sunset and sunrise could be estimated by the exponential functions of the forms $h_1(x) = c_1 e^{c_2 x}$ when there was no water flow and $h_2(x) = c_1 e^{-c_2 x}$ when there was water flow. In these functions, c_1 and c_2 were determined for each case by fitting a curve to the data obtained by the numerical values of thermal stress before sunset and sunrise. The mean squared error values of these estimations varied between $5.59 \bullet 10^{-11}\%$ and 2.7%. The values of all parameters used in testing the models are given in Tables 1 and 2.

TABLE 2. Parameter values used in the simulations to derive the graphs.

Parameters	High temp	Moderate temp
k_2	.034	.015
k_3	.0025	.0025
k_1	.01	.000001
m	.02	.02
p	.02	.02
r_1	.00125	.00125
$\tilde{k}_2$	.0001	.01
F_0	.03	.03
k_4	.95	.95

TABLE 3. Parameter values used in the thermal stress functions.

Parameters	$f_{high}(l)$	$f_{mod}(l)$
a_1	1.6	
a_2		1.5
b_1	25	
b_2		.0013
c_1	2450000	

Results and Discussion

Linking Physiology to Population Response

The models show that thermal stress follows a periodic function similar to available irradiance, with maximum thermal stress at noon (Figures 2 and 3). This agrees with the strong inverse relationships shown for studies on diel flux of photosynthetically active radiation and diurnal fluctuations in the quantum yield of PSII photochemistry in corals [Gorbunov et al., 2001]. The modeled periodicity also corresponds closely with the photoprotective dynamics of xanthophyll in coral symbionts [Brown et al., 1999].

Daily time courses of photosynthetic activity show that high irradiance has a sustained effect on the symbiont's photochemistry since quantum yield may remain suppressed even after 12 hours of darkness [Jones and Hoegh-Guldberg, 2001]. Similarly, the modeled time courses show a steady rise in thermal stress after daily iterations. This accumulative response is directly proportional to SST – showing reduced rates of thermal stress accumulation at lower SST (Figures 2 and 3). Our model also shows that thermal stress is considerably reduced under moderate water flow while thermal stress is high under still conditions (Figures 2 and 3). These results agree with experimental manipulations [Nakamura and Van Woesik, 2001; Nakamura et al., 2003; Fabricius, 2006], and highlights the recent experimental evidence that suggests that rates of down regulation in PS II activity during acute light stress as well as the degree of subsequent recovery in dim light are directly affected by the flow regime [Nakamura et al., 2005].

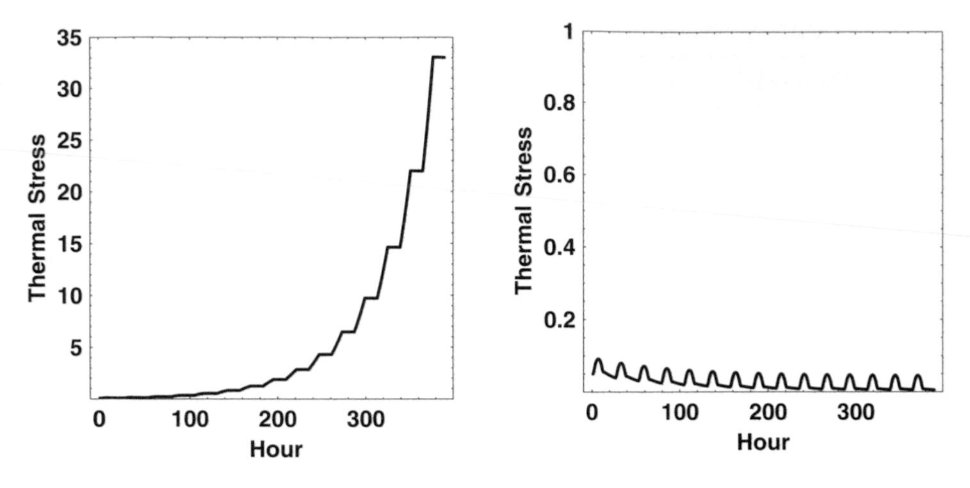

Figure 2. Thermal stress under high SST (32°C) (a) with no flow, (b) with flow.

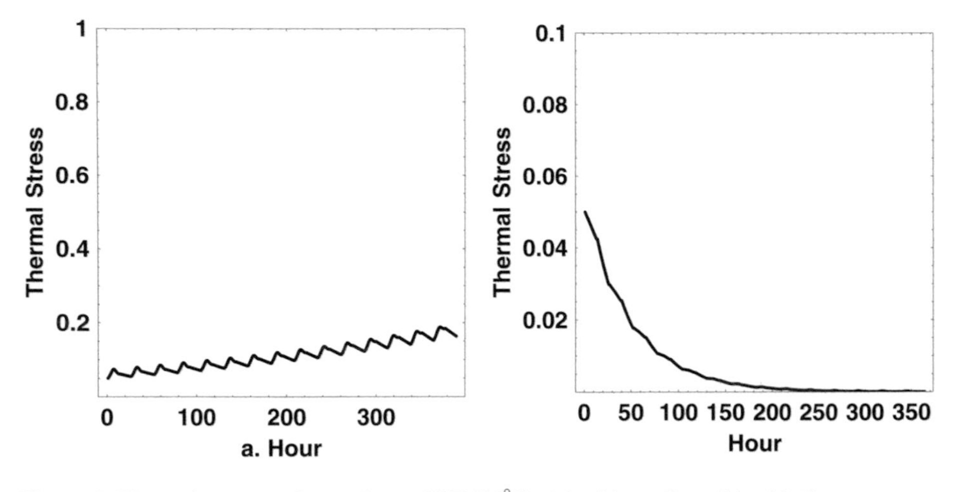

Figure 3. Thermal stress under moderate SST (28°C): (a) with no flow, (b) with flow.

Our model demonstrates a strong dependency on the initial value of thermal stress. When we tested both temperature systems, we observed that they both responded similarly at an initial thermal stress of 0.5 or larger (°C μmol photons m^{-2} s^{-1}) – we suspect that this value corresponds to critical temperature, which in turn is locality dependent, however temperature is implicitly embedded in the ODE model (in this case the critical temperature was close to 30°C, based on experiments and field data from southern Japan). It is appealing to use the initial thermal stress conditions of the models as proxies for SST because coral colonies closer to a critical temperature threshold, which again will vary geographically and among species, are effectively set at higher initial conditions, and are more likely to become chronically photoinhibited. Under sustained thermal stress, photosynthetic dysfunction leads to detrimental cascading effects at the population level (Figure 4).

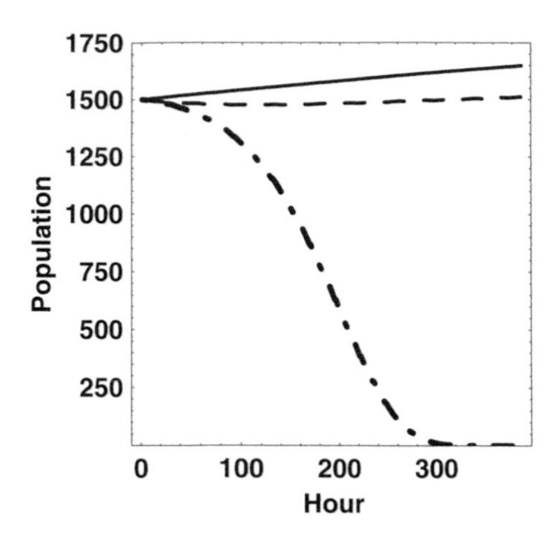

Figure 4. Population change relative to thermal stress and water flow for 15 days; Dashed-dot is the population function, $N(t)$, under high SST (32°C) no water flow; Dashed is the population function, $N(t)$, under high SST (32°C) with 3 cm/s water flow; Line is the population function, $N(t)$, under moderate SST (28°C) with 3 cm/s water flow; Dots is the population function, $N(t)$, under moderate SST (28°C) no water flow. (Note that the graphs of population under high SST with water flow and population under moderate SST with no water flow almost coincide). All four populations had initial populations, N_o of 1500.

Our model also showed that under moderate SST, when $\frac{k_2}{r_1} > 12$ and $\frac{\tilde{k}_2}{r_1} < 8$, coral populations declined, although the rate at which they declined was less than for high temperature. Yet, both models behaved similarly. In the figures shown here, we took $\frac{\tilde{k}_2}{r_1} = 8$ and $\frac{k_2}{r_1} = 12$. We also consider that the parameter k_1 (the rate at which thermal stress effects the coral population) also plays an essential role in the model, yet we ran the tests for only different values of k_2 and $\tilde{k}_2$ since these values can be easily estimated from empirical data.

Spatial Differences

Our model links the physiological mechanisms of thermal stress with the population response. At high SST we show a dramatic population decline compared with moderate declines at lower SSTs, under the same irradiance flux (Figures 4 and 5). These results concur with field observations for the Indian Ocean [McClanahan et al., 2001; Sheppard, 2003] and the southern Japanese islands in the 1998 and 2001 summer [Loya et al., 2001; Van Woesik et al., 2004], which displayed major population changes under high thermal stress, whereas under moderate SST anomalies on the Great Barrier Reef in 1998 [Baird and Marshall, 1998] and 2002 [Van Woesik, personal observations] coral bleaching occurred but mortality was relatively low.

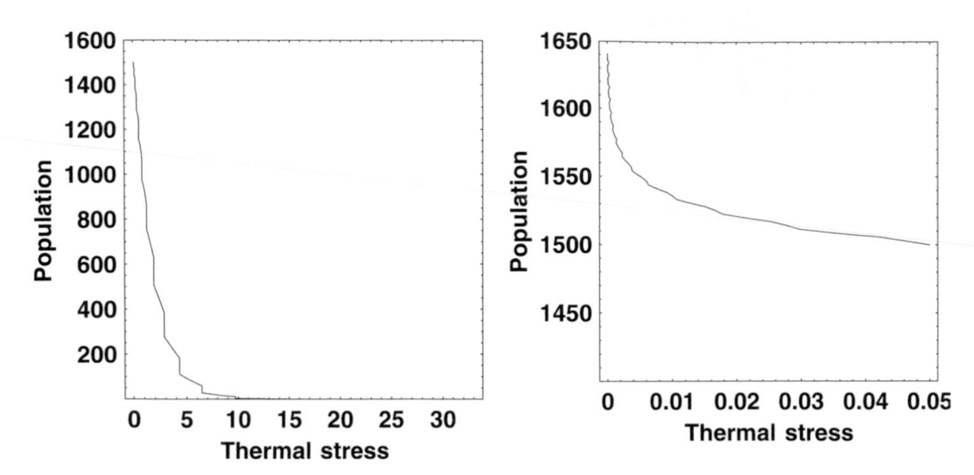

Figure 5. Coral population responses to thermal stress, (a) under high SST with no water flow, and (b) under moderate SST with water flow.

Undoubtedly coral populations decline once thermal stress reaches a critical level; the corals get forced toward irreversible, chronic photoinhibition of photosynthesis, which inevitably leads to bleaching and to mortality. Critical levels of thermal stress however differ in accordance with regional setting [McClanahan et al., 2001; Loya et al., 2001; Aronson et al., 2000; Baird and Marshall, 1998], thermal history [Castillo and Helmuth, 2005], recent experience [Brown et al., 2002], and species composition [Loya et al., 2001]. It is conceivable that when an entire region is experiencing high temperatures, the localities with low-seasonal ranges in temperature, for example windward reefs, may be more susceptible than leeward reefs, since the latter experience high seasonal temperature fluctuations. Clearly, low water flow may lead to high local temperature, but as far as we know the low flow *per se* does not induce tolerance [Nakamura and Van Woesik, 2001], rather the adjustment, or acclimatization, stems from the high range in local water temperatures. Clearly, the next generation of experiments needs to treat for water flow rates and temperature range (i.e., variance) under controlled conditions, because low flow and high temperatures are often not separable in the field [Van Woesik et al., 2005].

Under projected climate change scenarios, which plainly suggests increases in intensities and frequencies of thermal stress events [Hough-Guldberg, 1999], some thermally tolerant corals are destined to become 'the winners' [sensu Loya et al., 2001], while other thermally intolerant coral species are destined to become 'the losers'. But does a short-term winning strategy relay into long-term population success? In other words is 'tolerance-strategy' time invariant? What may appear to be a winning strategy in the short term, through survival of small colonies or the apparent short-term survival of a 'winning' growth form, may turn out to be detrimental in the long term, especially if thermal stress events increase in frequency and the time period for colony growth is reduced and reproduction is affected.

Model Applications

The dynamics of winners and losers need serious consideration, as do local thermal histories and subsequent responses to regional thermal anomalies [Van Woesik et al., 2005].

Yet, physiological tolerances of different holobionts are easily measured under controlled conditions [Iglesias-Prieto et al., 1992; Warner et al., 1999; Jones et al., 1998; Jones and Hoegh-Guldberg, 2001] and input as corresponding functions into Equations 2 and 3. In other words, short-term experimental studies on different coral species, under known and controlled conditions are easily input into the CPR model, allowing sensitivity studies and hindcasting to assess accuracy, and forecasting to examine long-term population responses. For example, experimental work shows that massive colonies are less sensitive to an increase in water flow and thermal stress compared with more complex morphologies [Finelli et al., 2006], which may be related to the general tolerance of massive colonies to thermal stress reported from all oceans.

Conclusions

Coral populations are indeed highly diverse and complex biological entities, with capacities to acclimatize and adapt to thermal stresses. We suspect that thermal tolerance may be related to symbiont composition [LaJeunesse et al., 2003; Baker, 2004] and the inherent diversity of the holobiont system. Thermal sensitivity of symbionts has been related to membrane lipid composition [Tchernov et al., 2004], which is highly plausible since the saturation state of fatty acids influences temperature tolerance and PSII recovery [Wada et al., 1994], which in turn relates closely to differential sensitivity of the D1 protein in PSII [Warner et al., 1999], characterizing a symbionts susceptibility to photoinhibition and differential repair rates [Takahashi et al., 2004]. Our model does not make these links, yet. The above model is merely the beginning of a formalized system that links coral physiology to subsequent population change by capturing the essence of processes involved in coral bleaching – taking measurements of photochemical efficiency of the coral holobiont and converting them to thermal stress functions, under influences of irradiance, temperature and water flow – and projecting those processes through time to estimate proportions of populations that survive under different stress scenarios. Further physiology and population studies will allow model validation and adjustment that will allow hindcasting and forecasting of changes in coral populations based on current and predicted environmental scenarios involving global climate change, which is surely one of the most pertinent problems facing coral reefs globally.

Acknowledgments. We would like to thank Peter Mumby, Bill Fitt, Sandra van Woesik, Ove Hoegh-Guldberg, Simon Levin and anonymous reviewers for comments on the manuscript. This research is supported by the National Science Foundation, Grant # 0308960.

References

Atkinson M. J., and R. W. Bilger, Effects of water velocity on phosphate uptake in coral reef-flat communities. *Limnol Oceanogr*, 37: 273-279, 1992.

Aronson, R. B., W. F. Precht, I. G. Macintyre, and T. J. T. Murdoch, Coral bleach-out in Belize. *Nature*, 405, 36, 2000.

Baker, A. C., Reef corals bleach to survive change. *Nature*, 411, 765-766, 2001.

Baker A. C., Symbiont diversity on coral reefs and its relationship to bleaching resistance and resilience. In *Coral Health and Disease*, E. Rosenberg and Y. Loya eds, Springer, 177-194, 2004.

Baird, A. H., and P. A. Marshall, Mass bleaching of corals on the Great Barrier Reef, *Coral Reefs*, 17, 376, 1998.

Baird M. E., and M. J. Atkinson, Measurement and prediction of mass transfer to experimental coral reef communities. *Limnol and Oceanogr.*, 42(8): 1685-1693, 1997.
Berkelmans, R., and Oliver, J.K., Large-scale bleaching of corals on the great barrier reef. *Coral Reefs*, 18, 55-60, 1999.
Brown B. E., Coral Bleaching: causes and consequences. *Coral Reefs*, 16, 129-138, 1997.
Brown, B. E., I. Ambarsi, M. E. Warner, W. K. Fitt, R. P. Dunne, S. W. Gibb, and D. G. Cummings, Diurnal changes in photochemical efficiency and xanthophylls concentrations in shallow water reef corals: evidence for photoinhibition and photoprotection. *Coral Reefs*, 18: 99-105, 1999.
Brown, B. E., R. P. Dunne, M. S. Goodson, and A. E. Douglas, Experience shapes the susceptibility of a reef coral to bleaching. *Coral Reefs*, 119-126, 2002.
Buddemeier R. W., and D. G. Fautin, Coral bleaching as an adaptive mechanism: a testable hypothesis. *Bioscience*, 43: 320-326, 1993.
Castillo K. D., and B. S. T. Helmuth, Influence of thermal history on the response of *Montastraea annularis* to short-term temperature exposure. Marine Biology, 148: 261-270, 2005.
Chalker B. E., Simulating light-saturation curves for photosynthesis and calcification by reef-building corals. *Marine Biology*, 63: 135-141, 1981.
Dennison W. C., and D. J. Barnes, Effect of water motion on coral photosynthesis and calcification. *J Exp Mar Biol Ecol.*, 115: 67-77, 1988.
Edwards A. J., S. Clark, H. Zahir, A. Rajasuriya, A. Naseer, and J. Rubens, Coral bleaching and mortality on artificial and natural reefs in Maldives in 1998, sea surface temperature anomalies and initial recovery. *Mar Poll Bull*, 42: 7-15, 2001.
Fabricius K. E., Effects of irradiance, flow, and colony pigmentation on the temperature microenvironment around corals: implications for coral bleaching? *Limno Oceanog.*, 51(1): 30-37, 2006.
Finelli, C. M., Helmuth, B. S. T., Pentcheff, N. D., Wethey, D. S., Water flow influences oxygen transport and photosynthetic efficiency in corals. *Coral Reefs*, 25: 47-57, 2005.
Fitt, W. K., M. E. Warner, Bleaching patterns of four species of Caribbean reef corals. *Biol Bull*, 189: 298-307, 1995.
Fitt, W. K., McFarland, F. K., Warner, M. E., and G. C. Chilcoat, Seasonal patterns of tissue biomass and densities of symbiotic dinoflagellates in reef corals and relation to coral bleaching. *Limnology and Oceanography*, 45(3), 677-685, 2000.
Fitt, W. K., Brown, B. E., Warner M. E., and R. P. Dunne, Coral Bleaching: Interpretation of thermal tolerance limits and thermal thresholds in tropical corals. *Coral Reefs*, 20: 1-27, 2001.
Gates, R. D., G. Baghdasarian, and L. Muscatine, Temperature stress causes host-cell Detachment in Symbiotic Cnidarians - Implications for Coral Bleaching, *Biological Bulletin*, 182, 324-332, 1992.
Glynn, P. W., Coral reef bleaching in the 1980s and possible connections with global warming. *Trends Ecol. Evol.*, 6, 175-179, 1991.
Glynn, P. W., Coral reef bleaching: Ecological perspectives. *Coral Reefs*, 12, 1-17, 1993.
Gorbunov, M. Y., Z. S. Kolber, M. P. Lesser, and P. G. Falkowski, Photosynthesis and photoprotection of symbiotic corals. *Limno and Oceanogr.*, 46(1): 75-85, 2001.
Hoegh-Guldberg, O., Climate change, coral bleaching and the future of the world's coral reefs. *Marine and Freshwater Research*, 50, 839-866, 1999.
Hoegh-Guldberg, O., R. J. Jones, S. Ward, and W. K. Loh, Is coral bleaching really adaptive? *Nature*, 415: 601-602, 2002.
Iglesias-Preito, R., J. L. Matta, W. A. Robins, and R. K. Trench, Photosynthetic response to elevated temperature in the symbiotic dinoflagellate *Symbiodinium microadriaticum* in culture. *Proc. Natl. Acad. Sci.*, USA 89: 10302-10305, 1992.
Iglesias-Prieto, R., V. H. Beltran, T. LaJeunesse, H. Reyes-Bonilla, P. E. Thome, Different algal symbionts explain the vertical distribution of dominant reef corals in the eastern Pacific. *Proc Royal Soc Lond.*, B doi:10.1098/rspb.2004.2757, 2004.

Jackson, J. B. C., et al., Historical overfishing and the recent collapse of coastal ecosystems. *Science*, 293: 629-638, 2001.
Jones, R. J., O. Hoegh-Guldberg, A. W. L. Larkum, and U. Schreiber, Temperature-induced bleaching of corals begins with impairment of the CO_2 fixation mechanism in zooxanthellae. *Plant Cell and Environment*, 21, 1219-30, 1998.
Jones, R. J., and O. Hoegh-Guldberg, Diurnal changes in the photochemical efficiency of the symbiotic dinoflagellates (Dinophyceae) of corals: photoprotection, photoinactivation and the relationship to coral bleaching. *Plant Cell and Environment*, 24, 89-99, 2001.
Kushmaro, A., Y. Loya, M. Fine, and E. Rosenberg, Bacterial infection and coral bleaching. *Nature*, 380, 396, 1996.
LaJeunesse, T. C., W. K. W. Loh, R. Van Woesik, O. Hoegh-Guldberg, G. W. Schmidt, W. K. Fitt, Low symbiont diversity in southern Great Barrier Reef corals relative to those of the Caribbean. *Limnol Oceanogr.*, 48: 2046-2054, 2003.
Lesser, M. P., V. M. Weis, M. R. Patterson, P. L. Jokiel, Effects of morphology and water motion on carbon delivery and productivity in the reef coral, *Pocillopora damicornis* (Linnaeus): Diffusion barriers, inorganic carbon limitation, and biochemical plasticity. *J Exp Mar Biol Ecol.*, 178: 153-179, 1994.
Lesser, M. P., Oxidative stress causes coral bleaching during exposure to elevated temperatures. *Coral Reefs*, 16: 187-192, 1997.
Little, A. F., M. J. H. van Oppen, B. L. Willis, Flexibility in algal endosymbioses shapes growth in reef corals. *Science*, 304: 1492-1494, 2004.
Loya, Y., K. Sakai, K. Yamazato, Y. Nakano, H. Sambali, and R. van Woesik, Coral bleaching: the winners and the losers. *Ecology Letters*, 4: 122-131, 2001.
McClanahan, T. R. Bleaching damage and recovery potential of Maldivian coral reefs. *Mar Poll Bull*, 40: 587-597, 2000.
McClanahan, T. R., Muthiga, N. A., and S. Mangi, Coral and algal changes after the 1998 coral bleaching: interaction with reef management and herbivores on Kenyan reefs. *Coral Reefs*, 19: 380-391, 2001.
McClanahan, T. R., N. A. Muthiga, S. Mangi, Coral and algal changes after the 1998 coral bleaching: interaction with reef management and herbivores on kenyan reefs. *coral reefs*, 19: 380-391, 2001.
Marshall, P. A., and A. H. Baird, Bleaching of corals on the Great Barrier Reef: differential susceptibilities among taxa. *Coral Reefs*, 19: 155-163, 2000.
Maxwell, D. P., S. Falk, C. G. Trick, and N. P. A. Huner, Growth at low temperature mimics high-light acclimation in *Chlorella Vulgaris*. *Plant Physiol.*, 105: 535-543, 1994.
Mumby, P. J., J. R. M. Chisholm, A. J. Edwards, S. Andrefouet, and J. Jaubert, Cloudy weather may have saved Society Island reef corals during the 1998 ENSO event. *Marine Ecology Progress Series*, 222: 209-216, 2001.
Murray, J. D., *Mathematical Biology*. I: An Introduction. Springer.
Nakamura, T., and Van Woesik, R., 2001. Water-flow rates and passive diffusion partially explain differential survival of corals during the 1998 bleaching event. *Marine Ecology Progress Series*, 212, 301-304, 2001.
Nakamura, T., H. Yamasaki, and R. Van Woesik, Water flow facilitates recovery from bleaching in the coral *Stylophora pistillata*. *Marine Ecology Progress Series*, 256: 287-291, 2003.
Nakamura, T., R. Van Woesik, H. Yamasaki, Photoinhibition of photosynthesis is reduced by water flow in the reef-building coral *Acropora digitifera, Marine Ecology Progress Series*, 301: 109-118, 2005.
Patterson, M. R., and K. P. Sebens, Forced convection modulates gas exchange in cnidarians. *Proc Natl Acad Sci.*, (USA) 86: 8833-8836, 1989.
Patterson, M. R., A mass-transfer explanation of metabolic scaling relations in some aquatic invertebrates and algae. *Science*, 255: 1421-1423, 1992.
Pandolfi, J. M., and J. B. C. Jackson, Community structure of Pleistocene coral reefs of Curacao, Netherlands Antilles. *Ecological Monographs*, 71(1), 49-67, 2001.

Rosenberg, E., The bacterial disease hypothesis of coral bleaching. In *Coral Health and Disease*, Eugene Rosenberg, and Yossi Loya, eds., Springer, 445-462, 2004.

Sheppard, C. R., Predicted recurrences of mass coral mortality in the Indian Ocean. *Nature*, 425: 294-297, 2003.

Strong, A. E., E. Kearns, K. K. Gjovig, Sea surface temperature signals from satellites-an update. *Geophysical Research Letters*, 27(11): 1667-1670, 2000.

Takahashi, S., T. Nakamura, M. Sakamizu, R. van Woesik, and H. Yamasaki, Repair machinery of symbiotic photosynthesis as the primary target of heat stress for reef-building corals. *Plant and Cell Physiology*, 45(2): 251-255, 2004.

Tchernov, D., M. Y. Gorbunov, C. de Vargas, S. N. Yadav, A. J. Milligan, M. Haggblom, P. G. Falkowski, Membrane lipids of symbiotic algae are diagnostic of sensitivity to thermal bleaching in corals. PNAS 101: 13531-13535, 2004.

Thomas, F. I. M., and M. L. Atkinson, Ammonium uptake by coral reefs: effects of water velocity and surface roughness on mass transfer. *Limnol Oceanogr.*, 42(1): 81-88, 1997.

Van Woesik, R., Coral bleaching: transcending spatial and temporal scales. Trends in Ecology & Evolution, 16: 119-121, 2001.

Van Woesik, R., A. Irikawa, Y. Loya, Coral bleaching: signs of change in southern Japan. In *Coral Health and Disease*, Eugene Rosenberg, and Yossi Loya, eds., Springer, 119-141, 2004.

Van Woesik, R., T. Nakamura, H. Yamasaki, C. Sheppard, Comment on ‘ Effects of geography, taxa, water flow, and temperature variation on coral bleaching intensity in Mauritius’ by McClanahan et al [2005]. *Marine Ecology Progress Series*, 305: 297-299, 2005.

Wada, H., Contribution of membrane lipids to the ability of the photosynthetic machinery to tolerate temperature stress. *Proc Nat Acad Sci.*, USA 91: 4273-4277, 1994.

Warner, M. E., W. K. Fitt, and G. W. Schmidt, The effects of elevated temperature on the photosynthetic efficiency of zooxanthellae *in hospite* from four different species of reef coral: a novel approach. *Plant Cell Environ.*, 19: 291-299, 1996.

Warner, M. E., W. K. Fitt, and G. W. Schmidt, Damage to photosystem II in symbiotic dinoflagellates: A determinant of coral bleaching, *Proc. Natl. Acad. Sci.*, USA 96, 8007-8012, 1999.

Warner, M. E., G. C. Chilcoat, F. K. McFarland, W. K. Fitt, Seasonal fluctuations in the photosynthetic capacity of photosystem II in symbiotic dinoflagellates in the Caribbean reef-building coral *Montastraea*, *Mar Biol.*, 141: 31-38, 2002.

Wilkinson, C. R., Global and local threats to coral reef functioning and existence: review and predictions. *Marine & Freshwater Research*, 50, 867-878, 1999.

8

The Hydrodynamics of a Bleaching Event: Implications for Management and Monitoring

William Skirving, Mal Heron and Scott Heron

Abstract

This chapter examines the hydrodynamic conditions that are present during a coral bleaching event. Meteorological and climate parameters and influences are discussed. The physics of mixing and its influence on the horizontal and vertical variations of sea temperature are examined. A specialized hydrodynamic model for Palau is then presented as a case study to demonstrate the utility of these models for understanding spatial variations during bleaching events. This case study along with the other sections of this chapter provide the foundation for concluding that hydrodynamic modeling can provide us with a relatively accurate glimpse of the spatial variation of thermal stress and, therefore, what future stress events may hold for corals. Although the timing of a coral bleaching event is unknown and cannot be predicted with current technology, the relative patterns of sea surface temperature during individual bleaching events can be predicted using current modeling techniques. However, improvements in our understanding of coral physiology and higher spatial-resolution climate models are necessary before the full potential of these predictions can be utilized in management decisions.

Introduction

Coral bleaching is a generalized stress response by the coral-zooxanthellae symbiosis and is not necessarily related to any one stressor [Glynn, 1993]. To date, mass coral bleaching events have been correlated with thermal stress [e.g., Dennis and Wicklund, 1993; Drollet et al., 1994; Winter et al., 1998; Hoegh-Guldberg, 1999; McField, 1999; Berkelmans, 2002]. The physiological mechanism is that high temperatures damage the photosynthetic pathway, which leads to a breakdown of the photosynthetic process [Jones et al., 1998]. After the thermal threshold is surpassed, the normally robust photo system can be overwhelmed by significant amounts of light, eventually causing the formation of reactive oxygen molecules that eventually destabilizes the relationship between corals and their symbionts [Hoegh-Guldberg, 1999; Downs et al., 2002]. Therefore, although light is an important factor in the coral bleaching story, it is not normally a stressor until water temperatures have exceeded certain limits [Berkelmans, 2002].

Coral Reefs and Climate Change: Science and Management
Coastal and Estuarine Studies 61
Copyright 2006 by the American Geophysical Union.
10.1029/61CE09

At present, coral bleaching conditions are monitored in near real-time using satellite-based Advanced Very High Resolution Radiometer data. The U.S. National Oceanic and Atmospheric Administration (NOAA) Coral Reef Watch (CRW) program produces half-weekly, 50-km resolution sea surface temperature (SST) and various derivative products with global coverage (see http://coralreefwatch.noaa.gov). These products provide reef managers and stakeholders with up-to-date information on their jurisdiction. However, physical characteristics of a reef-site that can enhance or mitigate thermal stress are identifiable and, as such, can be used to provide managers with a map defining which regions are more susceptible to thermal stress. This chapter will investigate the origin and spatial variation of the warm water that is known to be a major factor in coral bleaching.

Mass Coral Bleaching: Climate or Weather?

Links between El Niño and Bleaching

There is much talk about El Niño being the "cause" of the 1998 coral bleaching event. If this link exists then we would expect a correlation between El Niño and bleaching around the world. Arzayus and Skirving [2004] took the CRW satellite-derived Degree Heating Week (DHW) product and using the suggested value of DHW = 4 to indicate bleaching, they hindcast bleaching conditions back to 1985 for the entire world. A comparison was then made between the hindcast bleaching conditions and El Niño, La Niña and Neutral states (here referred as ENSO states) as defined by the National Centers for Environmental Prediction's Oceanic Niño Index. (http://www.cpc.noaa.gov/products/analysis_monitoring/ensostuff/ensoyears.shtml).

Arzayus and Skirving [2004] defined that a 50km-square region is correlated with an ENSO state if 70% or more of bleaching events occur during that ENSO state. They found that only 0.2% of bleaching events on reefs are correlated with an ENSO state (0.05% with El Niño, 0.14% with La Niña, and 0.01% with Neutral conditions). Although the effects of different ENSO states on bleaching severity were not examined, it was clearly shown that the onset of bleaching is not correlated with ENSO for the vast majority of world reefs. In fact, the variability of local weather conditions is greater than the climatological means which are used to characterize ENSO states.

Bleaching Weather

Skirving and Guinotte [2001] investigated the origin of the warm water that caused parts of the Great Barrier Reef (GBR) to bleach during 1998. They noted that a combination of low wind speed and neap tides was correlated with high SST. They also noted that during these warm periods there was another correlation between shallow bathymetry and relatively cooler SST.

These correlations led them to conclude that the warm water was a result of local heating from solar radiation in conditions where there was a lack of hydrodynamic mixing. The idea that SST anomalies leading to coral bleaching are mostly a result of local heating has since been supported by many field observations [Wilkinson, 1998; Wilkinson, 2000; Berkelmans et al., 2004; Bird et al., 2004; Skirving et al., 2004].

Very few mass coral bleaching events in the world are a result of advected warm water [Skirving, 2004]. Little to no wind, clear sunny skies and weak ocean currents characterize these events and, as such, local heating is the cause of almost all thermally-induced mass coral bleaching events. It would therefore be more accurate to describe mass coral bleaching as a weather phenomenon rather than the result of climate, as is currently popular. Climate is likely to modulate the frequency and intensity of these weather events, but more research is necessary before direct links between climate states [e.g., El Niño] and coral bleaching can be understood.

El Niño and Weather: A GBR Case Study

The record 1998 GBR bleaching event occurred during the intense 1998 El Niño event and, as a result, many scientists and managers believed that El Niño may be a key component to significant bleaching events on the GBR.

The 2002 coral bleaching event in the GBR was more significant than the 1998 bleaching event in every measurable aspect; SSTs were generally higher, bleaching was more extensive and there was higher mortality [Wilkinson, 2002]. The puzzling thing was that the El Niño did not begin until a number of months after the GBR bleached. This cast doubt on the causal link between El Niño and GBR bleaching events.

In 2003, the Coral Reef Watch team within NOAA/NESDIS used their DHW and HotSpot satellite products to examine the intensity and accumulated heat stress for the GBR during both bleaching events [Liu et al., 2003]. They noticed that each bleaching event was accompanied by a significant pool of anomalously warm water that covered thousands of square kilometers. The key difference between the 1998 and 2002 bleaching events was not in the intensity of the SST anomaly that caused each event, but in the proximity of the anomaly to the GBR [Liu et al., 2003]. The fact that the 1998 bleaching occurred during a strong El Niño, while the 2002 event occurred outside an El Niño, may be related to the position of the centre of the anomaly. Every significant El Niño event since 1985 has been accompanied by a significant SST anomaly situated off the east coast of Australia, well south of the GBR (see hindcast SST anomaly products on http://coralreefwatch.noaa.gov/satellite/). The 2002 bleaching event had a smaller, less intense anomaly than that of 1998, but it was situated in the Coral Sea directly off the central GBR. It is this proximity that allowed it to have a far greater influence over local GBR conditions and hence the more significant levels of bleaching observed during 2002. [Liu et al., 2003]

Clearly short- and long-term effects are intertwined. There is a need to monitor local conditions in order to forecast coral bleaching events and also to expect some modulation of the frequency and intensity of bleaching events due to the changing climate.

Spatial Variability of SST During a Bleaching Event

During a bleaching event, spatial patterns of SST are quite complex and have a scale of hundreds to tens of thousands of meters. Plate 1a is taken from Skirving and Guinotte [2001] and is an SST image of the southern GBR during the 1998 bleaching event. It clearly shows the high complexity that existed in the spatial patterns of SST during this event. Skirving and Guinotte [2001] also point out that this bleaching event (like most others around the world) was characterized by bright sunny skies and very low winds. This is generally accepted as a crucial part of the formula for a thermally induced mass coral bleaching event.

One problem with this is that the spatial scales of bright sunny skies (i.e., no cloud) are much larger than the observed variability in SST. This means that we need to look for local

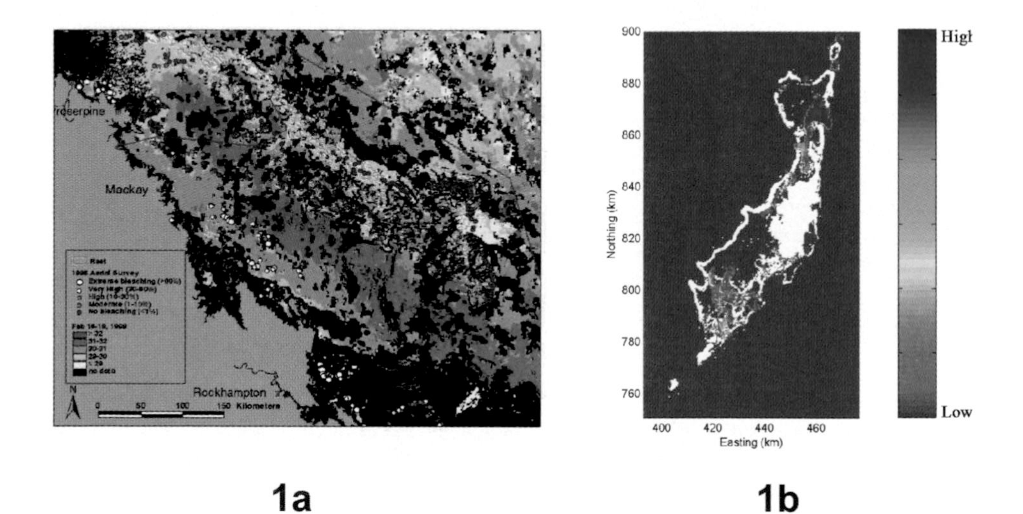

Plate 1. (a) Average SST for 16-18th February for the Southern GBR region. Reefs and bleaching are also depicted. (from Skirving and Guinotte, 2001). (b) Map of thermal capacitance for Palau, ranging from low (red) to high (blue). (modified from Heron and Skirving, 2004)

conditions which intervene in the causal link between insolation and sea surface temperature and which impose the observed spatial variability in the SST. It has been suggested that, in the absence of wind, hydrodynamic mixing is the only mechanism that could create such a complex SST pattern [Skirving and Guinotte, 2001; Skirving, 2004; Skirving et al., 2004].

Solar energy is absorbed mainly within the top few meters of the water column and, without any vertical mixing, tends to form a stable stratified layer with the warmer water at the top. If there is no vertical mixing then the warm surface layer has the potential to cause coral bleaching. If there is vertical mixing then the temperature of the surface water is reduced and approaches the average temperature of the water column, and the SST condition for bleaching is less likely to occur.

There are four different mechanisms that can vertically mix the water column: wind, low frequency currents [e.g., East Australian Current, Gulf Stream, etc], high frequency currents (e.g., tides) and swell waves.

The effect of winds on the sea surface is to cause surface stress which in turn can form surface gravity waves and also drive surface currents. To a fairly good approximation, low amplitude surface gravity waves are linear and do not lose energy as they propagate. As the wind becomes stronger and the wave heights increase the non-linear effects grow and energy is lost at the surface in micro-breakers, whitecaps and breaking waves. Energy which is lost from the waves ends up as turbulent mixing in the water column. There is little research on the effects of low wind speeds on coral bleaching conditions, and in particular on the effects of current shear caused near the surface by these winds.

Low- and high-frequency currents on continental shelves experience frictional stress at the sea floor, and this is projected through the vertical column as turbulent kinetic energy and vertical mixing. There are also inherent instabilities in water flow when the current speeds are high. Since the water at the surface is warmer than the water below it, complete mixing in the vertical will result in a cooling at the surface and a warming at depth. Further to this, deeper locations would experience a larger reduction in the surface temperature by this mechanism. Therefore, during a bleaching event, there is a relationship between the patterns of SST and a combination of the depth of water and strength of the currents [Skirving et al., 2004; Heron and Skirving, 2004].

Swell can affect the local vertical mixing in the water column. Swell is defined as long wavelength gravity waves generated outside of the local area. Thus, it is quite possible to have swell when there are no local wind waves. The most dramatic effect of swell is on the exposed side of a reef where swell breaks. Some of the energy is transferred into an overflowing bore which carries water onto the reef flat, but most of it is dispersed into turbulence at the reef front. There is a zone at the outer edge of a reef on the exposed side where significant vertical mixing can occur even in the absence of winds and currents.

The Physics of Vertical Mixing

During the daylight hours there is generally a net flow of heat into the ocean. The major part of this flux is in the form of electromagnetic radiation which is absorbed in the upper layers of the ocean. The resulting warm layer may then be mixed down by dynamical processes driven by currents and waves. The outflow of heat through the sea surface dominates at night. The combination of these fluxes gives a net diurnal variation whose long-term mean sets the climatological mean for the sea surface temperature.

Insolation

The solar radiation spectrum at the bottom of the atmosphere peaks in the visible range of wavelengths and has absorption lines and bands due to the composition of the atmosphere. Figure 1 (upper panel) shows typical spectra for the incident solar radiation at the top of the atmosphere and at the sea surface. This a conceptual sketch and for accurate calculations we need to use detailed estimates which take into account the variations in the solar constant and the latitude of the reef. The atmospheric model for absorption in this case changes the total energy from 1353 W/m^2 at the top of the atmosphere [Thekaekara and Drummond, 1971] to 933 W/m^2 at the sea surface. To calculate the atmospheric absorption for a particular site an atmospheric model, such as MODTRAN [Bernstein et al., 1996], can be used. The main control on the energy arriving at the sea surface is exerted by the aerosol and water vapor content of the atmosphere. Therefore, the type of air mass above a coral reef will have considerable influence over the amount of surface solar radiation. A clear sky and low aerosol content, which occur in regions such as the Red Sea and parts of the GBR, will provide maximum insolation. This is evident in the extreme SSTs often recorded in parts of these regions.

Conversely, regions that experience dust storms, high humidity and cloudy conditions will experience lower amounts of insolation, with the same sun angles, resulting in lower SSTs.

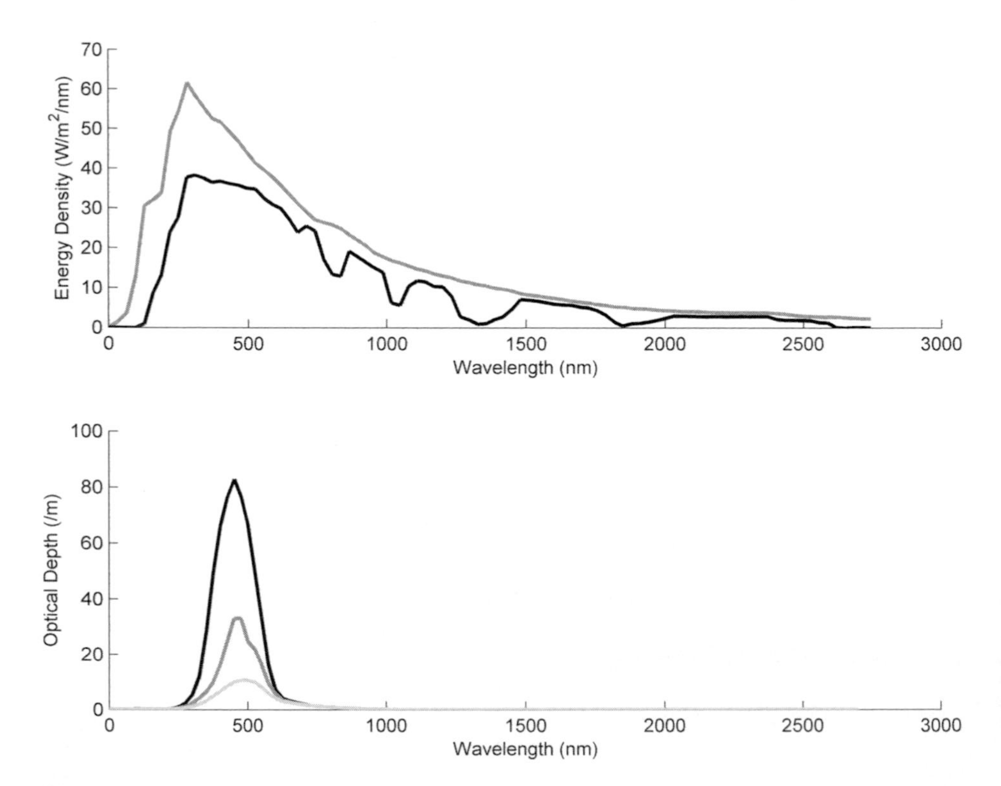

Figure 1. (Upper panel) Stylized graph of solar energy density at the top of the atmosphere (dark grey) and at the sea surface (black). (Lower panel) Stylized optical depth for case I water (black), case II water (dark grey) and case III water (light grey).

Absorption

For a single radiation wavelength and homogeneous water the amount of radiant energy absorbed, ΔI, in an interval of depth, Δz, is assumed to depend on the amount incident on that interval. This can be written as

$$\Delta I = -\alpha I \Delta z \tag{1}$$

where α is an absorption coefficient equal to the inverse of the optical depth, I is the incident energy, and z is the depth. Equation (1) leads directly to the expression for the radiant energy, $I(z)$, at depth z in the water column in terms of the energy I_0 incident at the surface.

$$I(z) = I_0 \exp(-\alpha z) \tag{2}$$

Equations (1) and (2) work for a single absorbing constituent in the water, and for an absorption coefficient which has no variation with depth. This is often not the case. Figure 1 (lower panel) shows how the optical depth ($1/\alpha$) varies with wavelength for three different types of water (case I – extremely pure ocean water, case II – turbid tropical-subtropical water, case III – mid-latitude water [Stewart, 2005; Jerlov, 1976]). This schematic illustrates the point that radiation in the visible band penetrates much further into the water than infra-red radiation which is absorbed very close to the surface, even in the clearest of case I waters. Under these conditions the absorption has to be calculated by summation across all wavelengths for each layer of water. To express this we recognize that the incident radiation varies with wavelength as shown in the upper panel of Figure 1

$$I_0 = \sum_{\lambda} E(\lambda)\Delta\lambda \tag{3}$$

where $E(\lambda)$ is the energy density at the sea surface. Because each wavelength has its own value of optical depth ($1/\alpha$) we have to calculate the attenuation through the water column to depth z for each wavelength, and then add them up to find the total remaining energy at that depth.

$$I(z) = \sum_{z}\sum_{\lambda} E(\lambda)\exp(-\alpha z)\Delta\lambda\Delta z \tag{4}$$

This is the value plotted in Figure 2 for the insolation and optical depths shown in Figure 1. Note that the curves shown in Figure 2 are not exponential; they are the sum of exponentials with differing optical depths. The most striking feature of Figure 2 is that 90% of the insolation energy is absorbed above 2.32 m, 0.81 m and 0.35 m for case I, case II and case III water respectively. This is juxtaposed with the fact that case I water has an optical depth of up to 80 m at about 450 nm.

For the purpose of discussing the heating effect of insolation energy in the water column, the heating effect is restricted to the top (approx) 1 m in case II water. This clearly demonstrates the importance of vertical mixing if this heat energy is to be removed from the surface layer.

Equation (4) assumes that the optical depth of the water is constant through the water column. If there is any layering of the biomass or sediments then the optical depth ($1/\alpha$) may vary with depth and this could be put into a modified form of Equation (4).

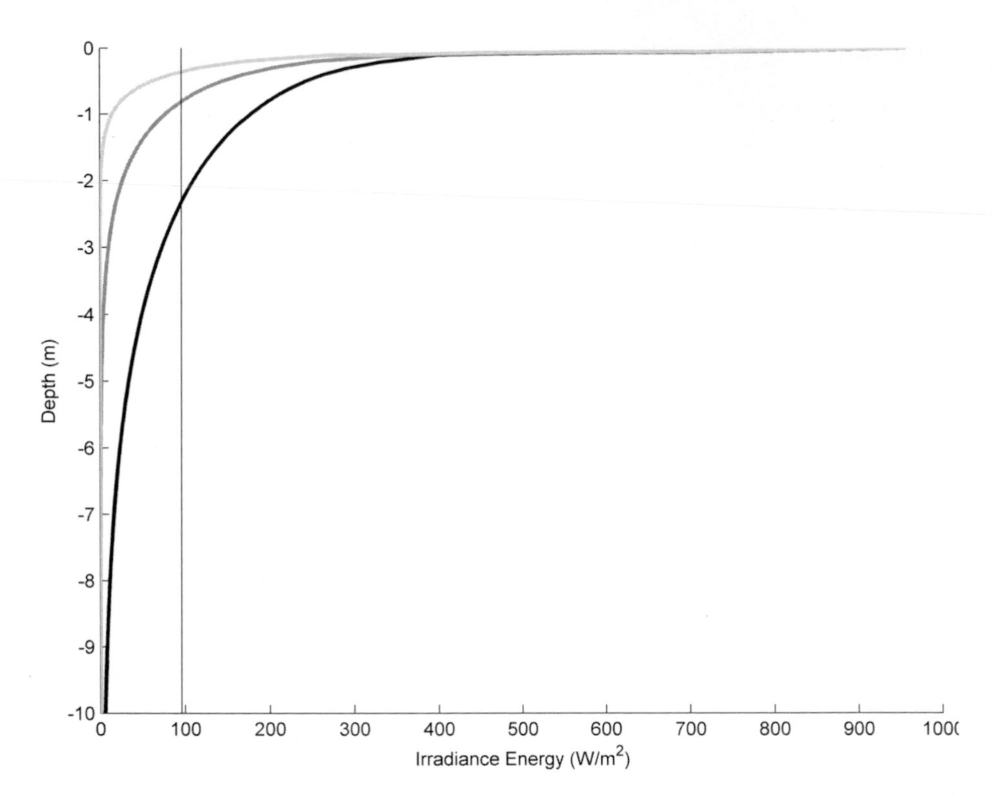

Figure 2. Total radiation energy, integrated over all wavelengths, decreases with depth for the case I water (black), case II water (dark grey) and case III water (light grey) data shown in Figure 1. The thin vertical line can be used to find the depths at which the energy density in the radiation has fallen to 10% of the value at the surface.

Dynamical mixing by currents

Ocean currents have a tendency to induce mixing under most conditions. In shallow water, where we are likely to encounter coral reefs, we can expect a boundary layer shear flow due to friction at the bottom of the water column. This is the basic response to tides and geostrophic forcing, to which we can superpose the effects of wind stress at the surface, stratification and wave-induced mixing. Note that the only concept of laminar flow is in the viscous layer at the bottom, and eddy diffusion prevails throughout the water column.

Mixing due to currents is driven by the vertical shear in the horizontal velocity of the water in the column and is carried out by eddies in the vertical plane. A commonly assumed model for the vertical eddy viscosity, N_z, is the linear model given by

$$N_z = ku_*(h - z), \tag{5}$$

which leads to the logarithmic bottom-friction layer,

$$u(z) = \frac{u_*}{k} \ln\left(\frac{h-z}{z_0}\right), \tag{6}$$

where z is the distance from the surface (positive downwards), h is the water depth, z_0 is the thickness of the viscous layer, u_* is the friction velocity and k is the von Karman constant. The vertical gradient in the horizontal flow has a shearing tendency which induces mixing.

Mixing of the vertical column due to bottom friction is strongest near the bottom where velocity shears are greatest. However with strong currents and shallow water this can impact on the mixing of the upper solar-heated layer. One important thing about this simple theory of mixing in the logarithmic boundary layer is that it gives us a conceptual reference frame for turbulent mixing in the water column when velocity shears are caused by other phenomena.

One such phenomenon is the formation of a horizontal eddy on the lee side of vertical obstructions (reefs, islands, etc.) [see Wolanski et al., 1984]. Such eddies are formed by large horizontal shears between the main flow and the shadow of the structure, where the horizontal turbulence scales are favorable (i.e., appropriate Reynolds number). As this mechanism requires significant currents, it is likely that the bottom-friction mixing associated with the currents will be substantial.

Mixing due to wind stress

Wind at the surface of the sea produces momentum transfer to the water, and hence a wind stress velocity at the surface. The velocity at the surface is transferred down through the column by eddy diffusion. If we assume that the vertical eddy viscosity, controlled by the stress at the surface, grows linearly with depth then we have a mathematical form similar to the bottom friction layer with

$$N'_z = k u'_* z, \tag{7}$$

where z is the distance from the water surface (positive downwards), and u'_* is the stress velocity at the surface.

The actual eddy viscosity in the water column is a combination of N'_z and N_z, and the velocity profile is a combination of the bottom boundary layer and the surface boundary layer. This leads to complications in numerical modeling of the currents and various schemes have been suggested for combining the eddy viscosity terms.

It is clear that the velocity shears induced by wind at the surface of the water have a significant role in the vertical mixing of the surface solar heated layer. Bleaching weather suggests little to no wind; in practice, low-speed winds generally exist and need to be considered.

Stratification

Stratification imposes an impediment to mixing due to the potential energy of the stratification. The Richardson number compares the potential energy (PE) of stratification and the turbulent kinetic energy (KE) and provides us with an index to measure the severity of the stratification. Under these conditions the turbulent kinetic energy works to erode the stratified layer.

Following the approach of Simpson and Hunter [1974], de Silva Samarasinghe [1989], and others, we consider the rate of loss of potential energy to be equal to some small fraction per second, σ, of the turbulent kinetic energy as

$$\frac{\partial}{\partial t}(PE) = -\sigma(KE), \tag{9}$$

which can be written as

$$\frac{\partial}{\partial t}\left\{\int_0^h gz(\rho-\bar{\rho})dz\right\} + \sigma\bar{\rho}\int_0^h N_z\left(\frac{\partial u}{\partial z}\right)^2 dz = 0 \tag{10}$$

per unit area of the water column, where h is the depth of the water column, g is gravitational acceleration, ρ is the density, $\bar{\rho}$ is the mean water density in the column, u is the horizontal velocity, and N_z is the vertical eddy viscosity. Simpson and Hunter [1974] found $\sigma = 0.0037\ s^{-1}$ in the Irish Sea; Hearn [1985] derived a similar value.

Equations 9 and 10 give us important insights into the stability of the upper layer in the water. When the sun heats the surface layer, the density of that layer is reduced, the potential energy (in Equation 9) is increased, and the water column is likely to be inherently stable. When PE dominates there is a high risk of coral bleaching. If the KE term in Equations 9 and 10 is large then mixing is enhanced and the density gradient is eroded. When KE dominates the solar heated water is mixed with deeper water and the risk of coral bleaching is reduced.

An elementary model for predicting the conditions for coral bleaching can use observed or estimated currents (u) and local water depth (h) for gauging the relative importance of the energy terms. This approach is to carry out a calibration (using previous data) of current speed against vertical mixing (or even against the mitigation of bleaching).

If there is wind stress driving currents in the warm surface layer then there is likely to be an enhanced current shear at the bottom boundary of the stratification which can assist mixing.

Another form of stratification is the thin surface layer which is evaporatively cooled by water vapor (latent heat) flux from the ocean to the atmosphere. This is a thin layer of the order of millimeters, with a regeneration time constant of several seconds if it is destroyed, for example, by a micro-breaker [Mobasheri, 1995]. This layer is unstable in the water column and promotes mixing. When we put this micro-layer mixing in the context of solar insolation on the order of a meter depth below the surface it is quickly lost in the scales of energy transfer and penetration depth. A more significant effect of the "skin layer" is that it is this layer which provides the infrared radiation used by satellite radiometers to measure the surface temperature. The skin layer reduces the brightness temperature by up to half a degree-Celsius (and perhaps more in tropical waters). This is not a random error, but is a variable offset in the measured temperature which depends on the nature of the skin layer.

The skin layer has little impact on coral bleaching because it is so thin that it does not contain much heat energy.

Mixing due to wave breaking

In the open ocean, most of the wave energy is conserved and not lost to mixing processes. It is only when the waves become non-linear that they lose energy to turbulence. It is the process of wave breaking that dominates the transfer of wave energy to mixing. At reef fronts the transfer is almost complete with only a remnant of wave energy being reflected back to the ocean, some of it transferring to a forward bore in the breaking wave, and a significant fraction going into turbulence at the breaker location. For a propagating surface gravity wave most of the energy is in the upper part of the water column. This is illustrated in Figure 3 where we show the depth profile of the horizontal surge velocity for a wave with 1 m amplitude (mean to crest) and 6 s period. This is a typical oceanic wind wave and the graph shows how the velocity decreases rapidly with depth. If there is any non-linearity or breaking then the associated energy becomes available for mixing.

The kinetic energy density for a wave with amplitude, a, angular frequency, ω and wavenumber, k is given by

$$KE(z) = \frac{1}{2}\rho(u^2 + w^2), \tag{11}$$

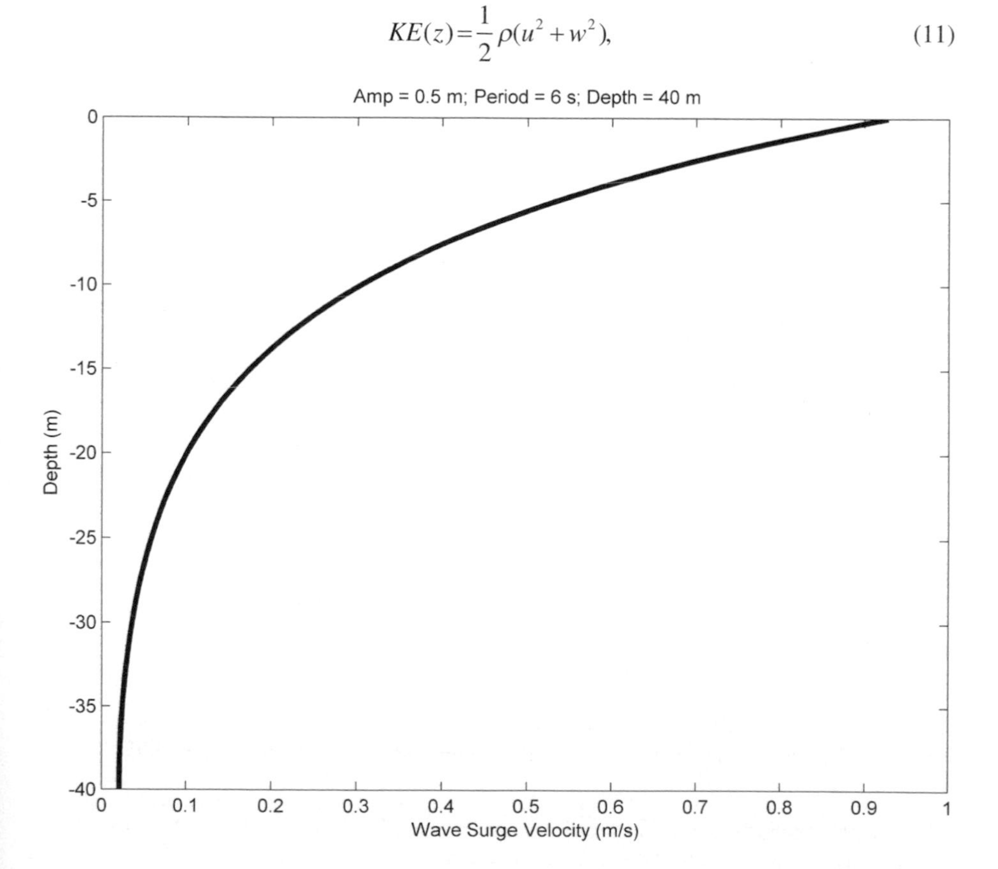

Figure 3. Horizontal surge velocity versus depth for a typical wind wave. Most of the wave energy is in the top few meters. The vertical surge velocity profile follows the same curve near the surface but departs and goes to zero at the bottom of the water column, set to 40 m here.

where u, and w are the instantaneous horizontal and vertical depth-dependent particle velocities:

$$u^2 = \frac{\pi a^2 g^2 k^2}{\omega^2} \frac{\cosh^2(k(h-z))}{\cosh^2(kh)} \tag{12}$$

$$w^2 = \frac{\pi a^2 g^2 k^2}{\omega^2} \frac{\sinh^2(k(h-z))}{\cosh^2(kh)} \tag{13}$$

where ρ is the density, g is gravitational acceleration, h is the depth of the water column and z is the (positive-downward) distance below the sea surface.

This wave energy is generally not available for mixing on shelf waters. However, when a wave encounters a reef front it loses most of its energy and provides a dominant mixing effect for the solar heated layer near the surface. This effect is so dominant that it is difficult to think there would be coral bleaching on the weather side of a reef except in very flat-calm conditions. Waves breaking on the reef front also send pulses of water forward across the reef flat. This pulsing bore is also well-mixed and we would expect mitigation of bleaching on the parts of the reef flat that are flushed with this water.

The physical processes of wave breaking on the reef front and the subsequent pulsing of water across the reef flat have a strong mitigating effect on coral bleaching.

Hydrodynamic Modeling for a Bleaching Event

The physical mechanisms that can influence bleaching are deterministic and, as such, can be modeled to predict the spatial variations in thermal stress. Bleaching weather conditions suggest maximum insolation and no significant wind-induced mixing. The inclusion of a parameter for swell is yet to be done. The effect of swell on mixing depends on the swell direction and the bathymetry of the reef and its surrounds. Swell waves are very effective mixers where they exist and can mix the water when they impinge on a reef. However, they are not capable of cooling an entire reef and will not be available for every reef. Swell may therefore contribute to the variability of coral bleaching on local scales; however, it is yet to be included in the hydrodynamic model presented here.

This leaves currents as the only mechanism which the model considers for altering spatial patterns of SST. The vertical temperature profile is determined by surface heat flux, dominated by solar radiation. Currents then mix this vertical profile via bottom friction and turbulent kinetic energy. The spatial pattern of mixing then modulates the vertical temperature profiles to create patterns of low to high SST during a bleaching event. To date, coral bleaching has occurred in regions of high SST, with the regions of cool water remaining relatively stress free. The case study described in the next section illustrates the success of linking hydrodynamic modeling of currents to the modulation of coral bleaching.

Case study - Palau heat stress model

During the latter half of 1998, Palau experienced unprecedented bleaching that resulted in significant mortality and the loss of significant proportions of one of the few remaining pristine coral reefs in the world [Wilkinson, 2002]. Prior to and since 1998, little to no coral bleaching has been observed. Figure 4 is a plot of accumulated heat stress at Palau as measured by the NOAA Coral Reef Watch DHW satellite product. A DHW value of 4 or more indicates significant bleaching [Liu et al., 2003; Skirving et al., 2006]. Note the DHW = 4 line in Figure 4; the 1998 accumulated stress easily surpassed that mark and is the only year to have done so since 1985. While the DHW product provides a large-scale description of coral bleaching events, it does not describe the smaller-scale variations of thermal stress. An understanding of these variations will lead to improved management.

The Nature Conservancy and the Palau Government joined forces to design and implement a Protected Areas Network (PAN) for Palau's coral reef ecosystem. They recognized bleaching as one of the major future threats to the Palau coral reef ecosystem. However, with only one poorly documented bleaching event, it is hard to understand the response of this ecosystem to coral bleaching and then build resilience to such events into the PAN.

At the same time, NOAA and the Australian Institute of Marine Science (AIMS) were collaborating on the development of hydrodynamic models to predict heat stress during a bleaching event. In 2003, it was decided to combine these efforts and for NOAA and AIMS to produce a heat stress model for Palau for use in the PAN.

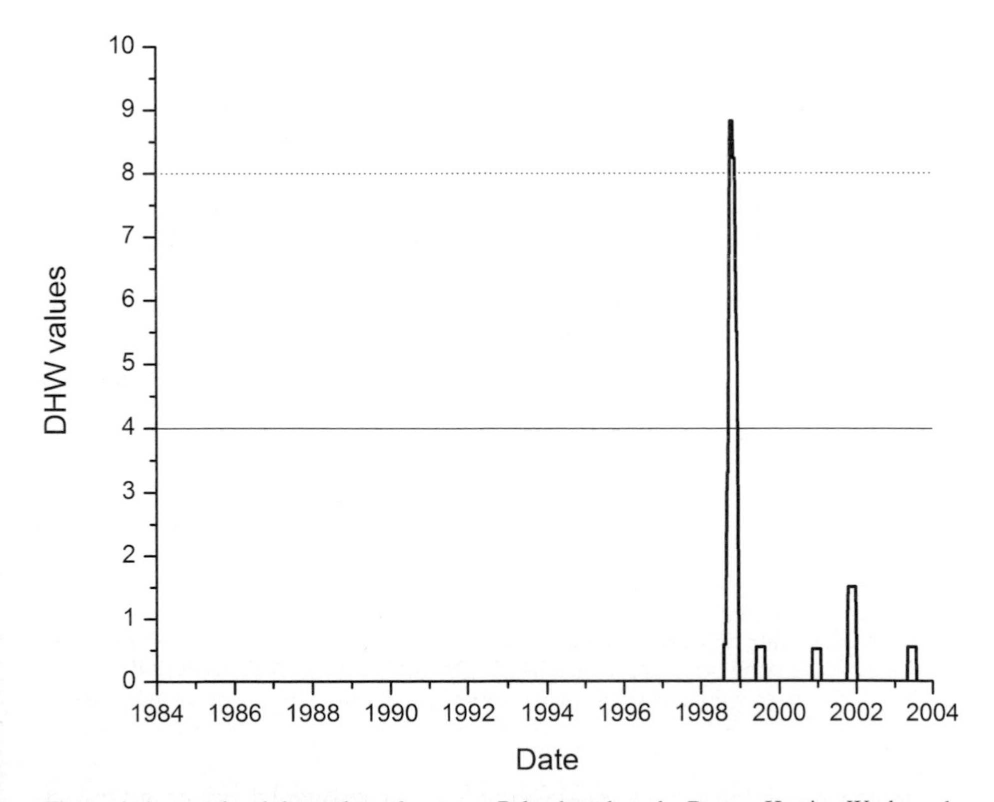

Figure 4. Accumulated thermal coral stress at Palau based on the Degree Heating Week product for the period 1985 to 2004.

For the model to be constructed, NOAA and AIMS needed [Skirving et al., 2005]:

1) The Palau bathymetry: Due to a lack of available data, NOAA derived the bathymetry from a combination of Landsat data and bathymetric transects taken with a depth sounder from a small boat. This produced a chart with 256.5 meter horizontal resolution and an rms error of approximately 1 meter vertically.
2) Low frequency currents: The Naval Research Laboratory [NRL] Layered Ocean Model (NLOM) and NOAA's Ocean Surface Current Analyses – Real time (OSCAR) were used to derive the seasonal low frequency currents around Palau.
3) High frequency currents: A combination of tide gauge data collected in and around Palau and a global tidal model was used to derive a model that could accurately predict the tides. The bathymetry and the tidal model were then used with the Princeton Ocean Model (POM) to build a model of tidally induced currents in and around Palau. Field data collected over a period of 5 months were used to calibrate and validate the output of this model.
4) Vertical temperature profile: This was derived by modeling a patch of water with a homogeneous temperature and applying a diurnal cycle of solar radiation and a constant, low wind of 2.6 m/s for a period of two weeks.

Simpson and Hunter [1974] provide the parameterization that was used to distinguish between stratified and well-mixed water by combining the currents with the bathymetric data. This information was then used in conjunction with the vertical temperature profile to determine the likely spatial distribution of sea surface temperature. To account for advection and tidal variation, while providing a static image for use in the PAN, the surface cooling due to mixing was accumulated over one tidal cycle (two spring-neap cycles). This accumulation parallels the DHW approach used by NOAA/CRW, with units of temperature-time.

This model is better described as a measure of thermal capacitance than as accumulated cooling. Thermal capacitance is the ratio of heat absorbed to the resultant temperature rise. Areas of low thermal capacitance will exhibit a larger increase in temperature for a given amount of heat input than areas of high thermal capacitance. Regions with complete vertical mixing (greater accumulated cooling) have high thermal capacitance, while stratified regions (lesser or no accumulated cooling) have low thermal capacitance.

The result of this is that the well-mixed regions in the Palau model represent regions of mild thermal climatology (i.e., less temperature variation and hence less thermal stress), whereas the stratified regions represent those areas that will experience the most extreme temperature range (i.e., greater thermal stress). Plate 1b is a thermal capacitance map derived from the Palau model; the blue regions have high capacitance (well-mixed, mild climate), while the red regions have low capacitance (stratified, variable climate).

A chart of this type can be extremely useful when designing a PAN. In general, most PANs are currently designed so as to provide protection to "representative bioregions". This means that as much as possible, every type of bioregion within the ecosystem of interest should be equally represented within the PAN. However, an ecosystem is not only made up of different species. It is also important to recognize that an ecosystem is made up of organisms that have unique physiological characteristics within each species. It is adaptation to the local climate (mild or variable) that will define these physiological characteristics.

When designing a PAN, it is relatively straight forward to map bioregions on the basis of species composition; however, the unique physiological properties within each species are not represented within these techniques. These physiological characteristics are likely grouped into areas that mimic the relative thermal capacitance through the region. As such, even without knowledge of the individual characteristics, it is possible to incorporate the spatial variation of physiology by means of the thermal capacitance.

When considering the ecosystem response to an individual thermal stress (bleaching) event, it is the regions of high thermal capacitance which will moderate the temperature rise, thereby experiencing lesser thermal stress. For this reason, high-capacitance areas should be selected for protection. It is important to note, however, that during extreme thermal stress events (such as was seen in Palau in 1998) even these areas may experience bleaching conditions.

When considering the sustainability of the ecosystem with respect to long-term climate change, it is the low thermal capacitance regions which should be protected. Low-capacitance regions experience variable climates exposing organisms to extreme temperatures and, thus, frequent periods of thermal stress. This exposure may lead to increased resilience to rising temperatures, via physiological adaptation, and aid their survival as climate change occurs. Representing equal proportions of high and low thermal capacitance areas is necessary and ideal for the short- and long-term protection of coral reefs.

Conclusion

Hydrodynamic modeling can provide us with a relatively accurate glimpse of what future stress events may hold for corals. However, for the full potential of the model to be realized and employed in management, additional work is required. Advancements in coral physiology to determine the response of organisms to thermal stress, including issues of acclimation and adaptation, will improve management during the existing climate regime. In addition to these, the inclusion of improved climate models (e.g., in spatial resolution and accuracy) would allow more accurate predictions of future bleaching events.

A careful examination of the facts surrounding the physical conditions during thermally-induced mass coral bleaching events leads to a few surprising conclusions:

1) Mass coral bleaching is a weather event and is not necessarily linked with climate.
2) In general, the vast majority of coral reefs around the world are not predisposed to bleaching during an ENSO.
3) Twice as many of the world's reefs bleached during the 1998-99 La Niña than during the 1997-98 El Niño.
4) Bleaching weather is characterized by cloudless (sunny) skies, low to no wind and low currents.
5) 90% of the sun's energy is absorbed in about the top 2 meters of the water column.
6) The SST patterns during a bleaching event are dominated by spatial variations in vertical mixing.
7) The hydrodynamic processes that cause mixing and hence create most of the SST patterns during a bleaching event are largely predictable.

In general, this means that although the timing of a coral bleaching event is unknown and cannot be predicted with current technology, the relative patterns of SST during the next bleaching event can be predicted using current techniques for hydrodynamic modeling. Hydrodynamic modeling, when combined with an improved knowledge of coral physiology, can go a long way to helping us understand the exact nature of mass coral bleaching, allowing for improved monitoring and predictions.

Implications for Management

Although the vertical profile of temperature can change from event to event, the mixing parameters change very little as the tidal and low frequency currents are cyclic and thus effectively predictable for any specific location. These parameters can be used to identify, using hydrodynamic modeling techniques, the variation of thermal capacitance across the region of interest. The result is that the SST pattern during a severe bleaching event is effectively static from one bleaching event to another, with the magnitude of temperature related to the input heat. Identification of these patterns allows a higher degree of management of coral reefs prior to and during the onset of thermal stress events.

References

Arzayus, L. F. and W. J. Skirving, The correlation between ENSO and coral bleaching events. *10th International Coral Reef Symposium*, Okinawa, Japan, 2004.

Berkelmans, R., Time-integrated thermal bleaching thresholds of reefs and their variation on the Great Barrier Reef. *Mar. Ecol. Prog. Ser.*, 229, 73-82, 2002.

Berkelmans, R., G. De'ath, S. Kininmonth, and W. Skirving, Coral bleaching on the Great Barrier Reef: Correlation with sea surface temperature, a handle on 'patchiness' and comparison of the 1998 and 2002 events. *Coral Reefs*, 23, 74-83, 2004.

Bernstein, L. S., A. Berk, and P. K. Acharya, Very narrow band model calculations of atmospheric fluxes and cooling rates using the MODTRAN code, *J. Atmos. Sci.*, 53, 2887-2904, 1996.

Bird, J. C., C. R. Steinberg, T. A. Hardy, L. B. Mason, R. M. Brinkman, and L. Bode, Modeling Sub-Reef Scale Thermodynamics at Scott Reef, Western Australia to Predict Coral Bleaching. *10th International Coral Reef Symposium*, Okinawa, Japan, 2004.

de Silva Samarasinghe, J. R., Transient salt-wedges in a tidal gulf: A criterion for their formation. *Estuarine, Coastal and Shelf Science*, 28, 129-148, 1989.

Dennis, G. D. and R. I. Wicklund, The relationship between environmental factors and coral bleaching at Lee Stocking Island, Bahamas in 1990. *In: Case Histories for the Colloquium and Forum on Global Aspects of Coral Reefs: Health, Hazards and History*, F15-F21, 1993.

Downs, C. A., J. E. Fauth, J. C. Halas, P. Dustan, J. Bemiss, and C. M. Woodley, Oxidative stress and seasonal coral bleaching. *Free Radical Biology and Medicine,* 33, 533-543, 2002.

Drollet, J. H., M. Faucon, S. Maritorena, and P. M. V. Martin, A survey of environmental physico-chemical parameters during a minor coral mass bleaching event in Tahiti in 1993. *Aust. J. Mar. Freshw Res.*, 45, 1149-1156, 1994.

Glynn, P. W. Coral reef bleaching: ecological perspectives. *Coral Reefs*, 12, 1-17, 1993.

Hearn, C. J. On the value of the mixing efficiency in the Simpson-Hunter h/u^3 criterion. *Deutsche Hydrographisches Zeitschrift*, 38(H.3), 133-145, 1985.

Heron, S. F. and W. J. Skirving, Satellite bathymetry use in numerical models of ocean thermal stress. *La Revista Gayana*, 68(2), 284-288, 2004.

Hoegh-Guldberg, O., Climate change, coral bleaching and the future of the world's coral reefs. *Mar. Freshwater Res.*, 50, 839-866, 1999.

Jerlov, N. G., Marine Optics, *Elsevier, Amsterdam*, pp. 232, 1976.

Jones, R. J., O. Hoegh-Guldberg, A. W. D. Larcum, and U. Schreiber, Temperature-induced beaching of corals begins with impairment of the CO_2 fixation mechanism in zooxanthellae. *Plant, Cell and Env.*, 21, 1219-1230, 1998.

Liu, G., A. E. Strong, and W. J. Skirving, Remote sensing of sea surface temperatures during 2002 Great Barrier Reef coral bleaching. *EOS*, 84, 137-144, 2003.

McField, M. D., Coral response during and after mass bleaching in Belize. *Bull. Mar. Sci.*, 64, 155-172, 1999.

Mobasheri, M. R., Heat transfer in the upper layer of the ocean with application to the correction of satellite sea surface temperature. *Ph.D. thesis, James Cook University, Townsville, Australia*, pp. 182, 1995.

Simpson, J. H. and J. R. Hunter, Fronts in the Irish Sea. *Nature*, 250, 404-406, 1974.

Skirving, W. J., The Hydrodynamics of a Coral Bleaching Event: The role of satellite and CREWS measurements. In: Hendee, J. C. (Ed.) *The effects of combined sea temperature, light, and carbon dioxide on coral bleaching, settlement, and growth and NOAA Research Special Report*, pp. 33-34, 2004.

Skirving, W. J., and J. Guinotte, The sea surface temperature story on the Great Barrier Reef during the coral bleaching event of 1998. In: Wolanski, E. (Ed.) *Oceanographic process of coral Reefs: Physical and Biological Links in the Great Barrier Reef. CRC Press, Boca Raton, Florida*, pp. 376, 2001.

Skirving, W. J., S. F. Heron, C. R. Steinberg, A. E. Strong, C. McLean, M. L. Heron, S. M. Choukroun, L. F. Arzayus, and A. G. Bauman, "Palau Modeling Final Report" *National Oceanic and Atmospheric Administration and Australian Institute of Marine Science*, pp. 46, 2005.

Skirving, W. J., C. R. Steinberg, and S. F. Heron, The hydrodynamics of a coral bleaching event. *ASLO/TOS Ocean Research 2004 Conference*, Honolulu, Hawaii, 2004.

Skirving, W. J., A. E. Strong, G. Liu, C. Liu, F. Arzayus, J. Sapper, and E. Bayler, Extreme events and perturbations of coastal ecosystems: Sea surface temperature change and coral bleaching. *Chapter 2 in Remote Sensing of Aquatic Coastal Ecosystem Processes, Richardson, L. L. and E. F. LeDrew (Co-Eds), Kluwer publishers*, 2006.

Stewart, R. H., Introduction to Physical Oceanography, *Texas A & M University, College Station, Texas U.S.A.*, pp. 354, 2005.

Thekaekara, M. P., A. J. Drummond, Standard values for the solar constant and its spectral components. *Nature Phys. Sci.*, 229, 6-9, 1971.

Wilkinson, C. R., Status of Coral Reefs of the World: 1998. *Global Coral Reef Monitoring Network and Australian Institute of Marine Science, Townsville, Australia*, pp. 184, 1998.

Wilkinson, C. R., Status of Coral Reefs of the World: 2000. *Global Coral Reef Monitoring Network and Australian Institute of Marine Science, Townsville, Australia*, pp. 363, 2000.

Wilkinson, C. R., Status of Coral Reefs of the World: 2002. *Global Coral Reef Monitoring Network and Australian Institute of Marine Science, Townsville, Australia*, pp. 378, 2002.

Winter, A., R. S. Appeldoorn, A. Bruckner, E. H. Williams, and C. Goenaga, Sea surface temperatures and coral reef bleaching off La Parguera, Puerto Rico (northeastern Caribbean Sea). *Coral Reefs*, 17, 377-382, 1998.

Wolanski, E., J. Imberger, and M. L. Heron, Island wakes in shallow coastal waters. *J. Geophys. Res.*, 89, 10555-10569, 1984.

9

Identifying Coral Bleaching Remotely via Coral Reef Watch – Improved Integration and Implications for Changing Climate

A. E. Strong, F. Arzayus, W. Skirving, and S. F. Heron

1. Overview of Coral Reef Watch

The world's first near real-time satellite global bleaching monitoring and early warning system was developed in 1996 at the National Oceanic and Atmospheric Administration's (NOAA) National Environmental Satellite, Data and Information Service (NESDIS), where Coral Reef Watch (CRW) is located [Strong et al., 1997]. This followed several years of student independent research projects in collaborations with midshipmen from the U.S. Naval Academy [e.g., Montgomery and Strong, 1994; Gleeson and Strong, 1995]. Bleaching "HotSpots" were inaugurated in 1997 as web-accessible "pilot" products to both the U.S. and the global coral reef communities. In 1997, the program included only one "nowcast" monitoring tool: bleaching HotSpots (Plate 1 (A)). This timing proved to be somewhat fortuitous given that we now know that 1998 turned out to be one of the most significant years for mass coral bleaching events on record [Wilkinson et al., 1999]. NOAA's satellite bleaching monitoring product demonstrated its utility as a tool to managers and researchers alike and to this day, it remains the only system of its kind available in the world. Until recently, the CRW products represented the only global suite of operational satellite products being used for the management of any marine ecosystem.

In 2000, NOAA's Coral Reef Watch Program (http://coralreefwatch.noaa.gov) was established as NESDIS' satellite bleaching monitoring system. A suite of new monitoring tools, including bleaching Degree Heating Weeks (DHW) (Plate 1 (B)), and Bleaching Alerts (Figure 1) was incorporated as a core component of this new program. NOAA's CRW is mainly a monitoring program established to provide early warnings and long-term monitoring for both U.S. and global coral reef ecosystems. It includes both satellite and *in situ* monitoring components.

In early 2004, as a result of a revised research structure within NOAA, a new integrated observing system was proposed to accomplish the observational requirements for corals set forth by President Clinton's Executive Order #13089 mandate. The aim of this new system was to incorporate coral reef efforts from most of NOAA's line offices into a coordinated Coral Reef Ecosystem Integrated Observing System (CREIOS). While all this was happening on the domestic front, CRW efforts were expanding into the international realm

Coral Reefs and Climate Change: Science and Management
Coastal and Estuarine Studies 61
Copyright 2006 by the American Geophysical Union.
10.1029/61CE10

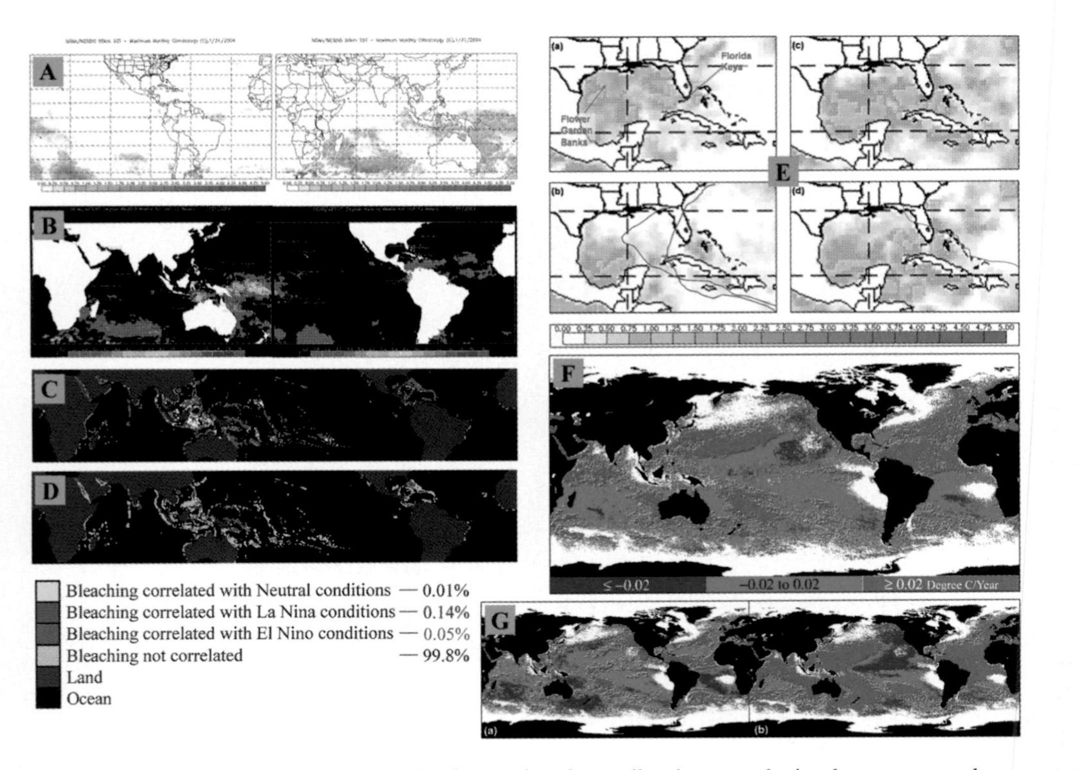

Plate 1. (A) Bleaching HotSpot Charts (E and W Hemisphere). Regions colored as yellow/ orange depict those waters where present sea surface water temperatures are 1 degree (C) or more higher than maximum climatologies have revealed. Blue/purple waters are currently above climatological maximums but still less than critical to be designated as a "yellow/orange" HotSpots. (B) Degree Heating Weeks Charts [E & W Hemisphere] depict regions where "HotSpots" have accumulated thermal stress over 12 weeks that are likely to have noticeable effects to most corals species – e.g., where levels exceed DHW = 4 (green color – see color bar) bleaching is highly probable. (C) Coral reefs: Bleaching observed 1985-2004 – red; no bleaching – white. (D) Correlations of bleaching events with ENSO – see scale. (E) Composite of HotSpots images for the Gulf of Mexico (a) 3 August 2004 indicating locations of National Marine Sanctuaries; (b) 17 August 2004 showing the tracks of Hurricanes "Bonnie" (red) and "Charley" (green); (c) 4 September 2004; (d) 7 September showing the track of "Frances" (brown); (bottom) HotSpot anomalies scale in degrees Celsius. (F) SST Trends: 1986-2002 (G) (a). SST Trends 1985-1996; (b) SST Trends 1991-2002 [Same scale as Plate 1-F].

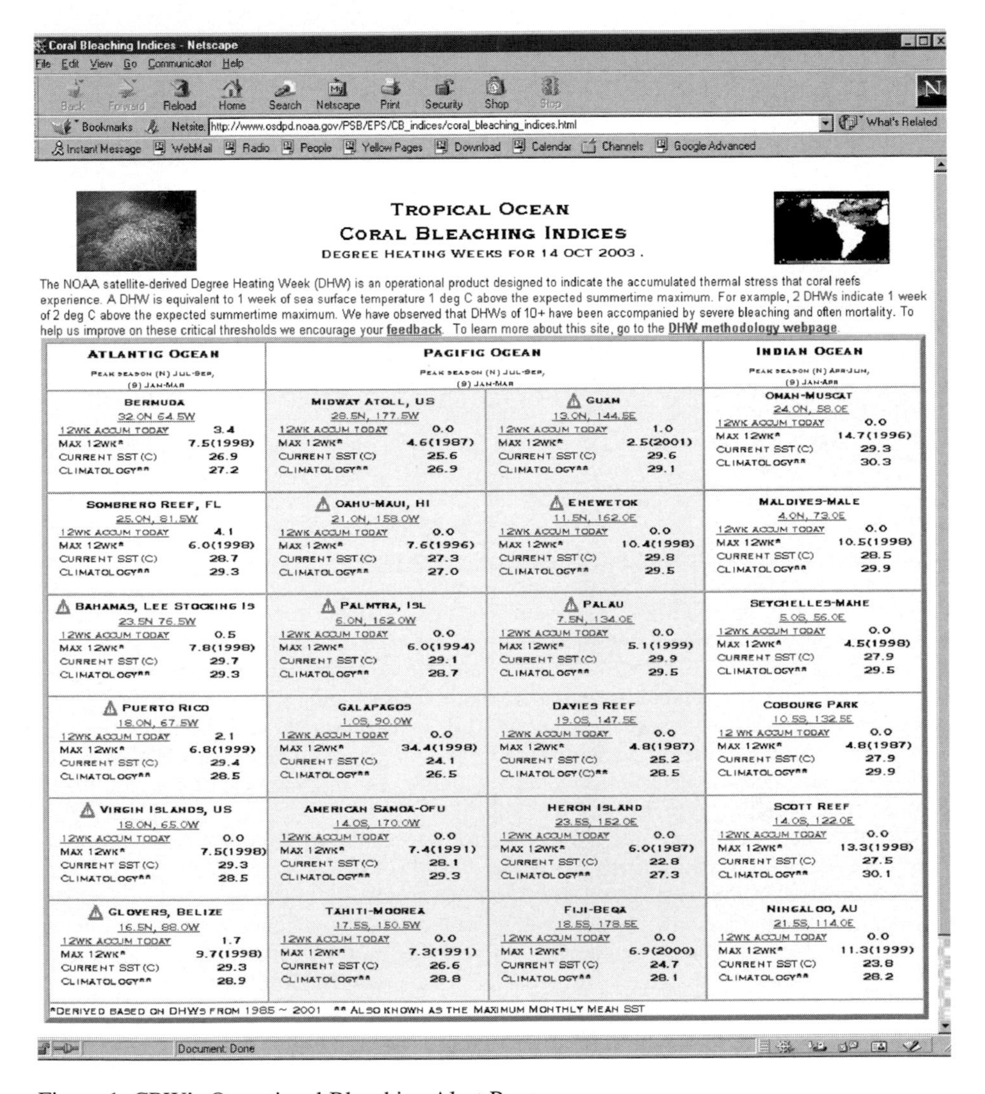
Coral Bleaching Indices - Netscape

File Edit View Go Communicator Help

Back Forward Reload Home Search Netscape Print Security Shop Stop

Bookmarks Netsite: http://www.osdpd.noaa.gov/PSB/EPS/CB_indices/coral_bleaching_indices.html What's Related

Instant Message WebMail Radio People Yellow Pages Download Calendar Channels Google Advanced

TROPICAL OCEAN
CORAL BLEACHING INDICES
DEGREE HEATING WEEKS FOR 14 OCT 2003 .

The NOAA satellite-derived Degree Heating Week (DHW) is an operational product designed to indicate the accumulated thermal stress that coral reefs experience. A DHW is equivalent to 1 week of sea surface temperature 1 deg C above the expected summertime maximum. For example, 2 DHWs indicate 1 week of 2 deg C above the expected summertime maximum. We have observed that DHWs of 10+ have been accompanied by severe bleaching and often mortality. To help us improve on these critical thresholds we encourage your feedback. To learn more about this site, go to the DHW methodology webpage.

ATLANTIC OCEAN PEAK SEASON (N) JUL-SEP, (S) JAN-MAR	PACIFIC OCEAN PEAK SEASON (N) JUL-SEP, (S) JAN-MAR		INDIAN OCEAN PEAK SEASON (N) APR-JUN, (S) JAN-APR
BERMUDA 32.0N, 64.5W 12WK ACCUM TODAY 3.4 MAX 12WK* 7.5(1998) CURRENT SST (C) 26.9 CLIMATOLOGY** 27.2	MIDWAY ATOLL, US 28.5N, 177.5W 12WK ACCUM TODAY 0.0 MAX 12WK* 4.6(1987) CURRENT SST (C) 25.6 CLIMATOLOGY** 26.9	GUAM 13.0N, 144.5E 12WK ACCUM TODAY 1.0 MAX 12WK* 2.5(2001) CURRENT SST (C) 29.6 CLIMATOLOGY** 29.1	OMAN-MUSCAT 24.0N, 58.0E 12WK ACCUM TODAY 0.0 MAX 12WK* 14.7(1996) CURRENT SST (C) 29.3 CLIMATOLOGY** 30.3
SOMBRERO REEF, FL 25.0N, 81.5W 12WK ACCUM TODAY 4.1 MAX 12WK* 6.0(1998) CURRENT SST (C) 28.7 CLIMATOLOGY** 29.3	OAHU-MAUI, HI 21.0N, 158.0W 12WK ACCUM TODAY 0.0 MAX 12WK* 7.6(1996) CURRENT SST (C) 27.3 CLIMATOLOGY** 27.0	ENEWETOK 11.5N, 162.0E 12WK ACCUM TODAY 0.0 MAX 12WK* 10.4(1998) CURRENT SST (C) 29.8 CLIMATOLOGY** 29.5	MALDIVES-MALE 4.0N, 73.0E 12WK ACCUM TODAY 0.0 MAX 12WK* 10.5(1998) CURRENT SST (C) 28.5 CLIMATOLOGY** 29.9
BAHAMAS, LEE STOCKING IS 23.5N, 76.5W 12WK ACCUM TODAY 0.5 MAX 12WK* 7.8(1998) CURRENT SST (C) 29.7 CLIMATOLOGY** 29.3	PALMYRA, ISL 6.0N, 162.0W 12WK ACCUM TODAY 0.0 MAX 12WK* 6.0(1994) CURRENT SST (C) 29.1 CLIMATOLOGY** 28.7	PALAU 7.5N, 134.0E 12WK ACCUM TODAY 0.0 MAX 12WK* 5.1(1999) CURRENT SST (C) 29.9 CLIMATOLOGY** 29.5	SEYCHELLES-MAHE 5.0S, 56.0E 12WK ACCUM TODAY 0.0 MAX 12WK* 4.5(1998) CURRENT SST (C) 27.9 CLIMATOLOGY** 29.5
PUERTO RICO 18.0N, 67.5W 12WK ACCUM TODAY 2.1 MAX 12WK* 6.8(1999) CURRENT SST (C) 29.4 CLIMATOLOGY** 28.5	GALAPAGOS 1.0S, 90.0W 12WK ACCUM TODAY 0.0 MAX 12WK* 34.4(1998) CURRENT SST (C) 24.1 CLIMATOLOGY** 26.5	DAVIES REEF 19.0S, 147.5E 12WK ACCUM TODAY 0.0 MAX 12WK* 4.8(1987) CURRENT SST (C) 25.2 CLIMATOLOGY(C)** 28.5	COBOURG PARK 10.5S, 132.5E 12 WK ACCUM TODAY 0.0 MAX 12WK* 4.8(1987) CURRENT SST (C) 27.9 CLIMATOLOGY** 29.9
VIRGIN ISLANDS, US 18.0N, 65.0W 12WK ACCUM TODAY 0.0 MAX 12WK* 7.5(1998) CURRENT SST (C) 29.3 CLIMATOLOGY** 28.5	AMERICAN SAMOA-OFU 14.0S, 170.0W 12WK ACCUM TODAY 0.0 MAX 12WK* 7.4(1991) CURRENT SST (C) 28.1 CLIMATOLOGY** 29.3	HERON ISLAND 23.5S, 152.0E 12WK ACCUM TODAY 0.0 MAX 12WK* 6.0(1987) CURRENT SST (C) 22.8 CLIMATOLOGY** 27.3	SCOTT REEF 14.0S, 122.0E 12WK ACCUM TODAY 0.0 MAX 12WK* 13.3(1998) CURRENT SST (C) 27.5 CLIMATOLOGY** 30.1
GLOVERS, BELIZE 16.5N, 88.0W 12WK ACCUM TODAY 1.7 MAX 12WK* 9.7(1998) CURRENT SST (C) 29.3 CLIMATOLOGY** 28.9	TAHITI-MOOREA 17.5S, 150.5W 12WK ACCUM TODAY 0.0 MAX 12WK* 7.3(1991) CURRENT SST (C) 26.6 CLIMATOLOGY** 28.8	FIJI-BEQA 18.5S, 178.5E 12WK ACCUM TODAY 0.0 MAX 12WK* 6.9(2000) CURRENT SST (C) 24.7 CLIMATOLOGY** 28.1	NINGALOO, AU 21.5S, 114.0E 12WK ACCUM TODAY 0.0 MAX 12WK* 11.3(1999) CURRENT SST (C) 23.8 CLIMATOLOGY** 28.2

*DERIVED BASED ON DHWS FROM 1985 ~ 2001 ** ALSO KNOWN AS THE MAXIMUM MONTHLY MEAN SST

Document: Done

Figure 1. CRW's Operational Bleaching Alert Page.

with renewed Memoranda of Understanding between NOAA and Australian organizations [the Australian Institute of Marine Science (AIMS), the Great Barrier Reef Marine Park Authority (GBRMPA) and the University of Queensland (UQ)]. Both CREIOS and this expanded international cooperation promise to build upon the scientific activities centered on coral bleaching, including ecosystem issues pertaining to climate change. Furthermore, in late 2004 Australian and U.S. scientists became involved with the new 15-year World Bank/Global Environment Facility (GEF) for Targeted Research in Corals that seeks to build capacity through reef managers in developing countries.

2. Satellite and *in situ* Contributions

During this period, NOAA's Oceanic and Atmospheric Research (OAR) Atlantic Oceanographic and Meteorological Laboratory (AOML) had been developing the Coral Reef Early Warning System (CREWS), an integration of meteorological and *in situ* oceanographic instrumented arrays (buoys and dynamic pylons) employing artificial intelligence software. These CREWS stations are being deployed as coral reef environmental monitoring stations to monitor for conditions theoretically conducive to coral reef bleaching [see Hendee et al., 2001], and provide continuing long-term data sets for other coral reef ecosystem modeling and for Marine Protected Areas (MPA) decision support. The CREWS marine expert system concept grew out of prototyping and experimentation collaborations with the Florida Institute of Oceanography's and NOAA's similarly instrumented-array SEAKEYS program, developed in the early 1990s for the Florida Keys National Marine Sanctuary [Ogden et al., 1994].

In an effort to expand NOAA's coral reef monitoring and bleaching alert capabilities NESDIS and OAR joined their complementary remote sensing and *in situ* coral activities under the nascent Coral Reef Watch program in 2000. CREWS temporally intensive multi-sensor data served to validate NESDIS satellite-derived spatially extensive sea surface temperature (SST) monitoring products, with the benefit of NESDIS satellite products extending coral reef bleaching monitoring to larger spatial scales and remote locations. Over the last few years CREWS software became available for several Pacific Ocean buoys operated by NOAA's National Marine Fisheries Service (NMFS) Coral Reef Ecosystem Division (CRED) located in Honolulu, Hawaii. Through the Coral Reef Watch Program, NESDIS in conjunction with OAR and NMFS attempts to maximize coral reef resources by joining existing coral reef monitoring strengths under a coordinated program. CRW seeks to fully utilize space-based SST observations combined with CREWS and CRED in-water derived data to continually monitor for early indications of thermally induced coral bleaching worldwide.

NESDIS' fleet of polar orbiting global monitoring satellites is equipped with Advanced Very High Resolution Radiometers (AVHRR) capable of detecting multi-channel sea surface temperatures. Coral Reef Watch HotSpots are a measure of the heat stress on a one-half degree latitude/longitude (approximately 50-km square) parcel of water. The amount of stress is deduced by comparing the SST with the Maximum Mean Monthly Climatology (MMM) for that parcel of water. Thus, a parcel of water with a HotSpot value of 1, would represent an anomaly that is 1 degree C above the MMM. Degree Heating Weeks or DHW, are the accumulation of thermal stress, where the HotSpot value is 1 or more, over a concurrent 12-week period. Therefore, a DHW value of 1 is equivalent to one week of MMM + 1; DHW = 2 is equivalent to two weeks of MMM + 1 *or* one week of MMM + 2. These algorithms are based on the fact that mass coral bleaching is the result of coral reefs being exposed to SSTs above a non-specific survival threshold [Goreau and Hayes, 1994]. Massive bleaching occurs when temperatures above this threshold are sustained for extensive periods. Empirical evidence has shown that a DHW of 4 or greater (e.g., 4 weeks with 1 degree C above threshold) usually returns mild to significant coral bleaching and DHW of 8 or greater indicates widespread bleaching and some mortality [Skirving et al., 2006]. As such, managers and stakeholders utilize these products to gain an awareness of the potential threat of coral bleaching.

The operational "Tropical Ocean Coral Bleaching Indices" page product, shown in Figure 1, was added to the HotSpot Suite in 2003. This allows managers easy access to environmental indices from 24 world wide coral reef sites – e.g., SST, wind, HotSpots, DHWs. In late-2004, CRW began issuing automated Satellite Bleaching Alerts (SBA) for these 24 Bleaching Index sites. Our automated e-mail SBA system informs registered subscribers of the most up-to-date status on thermal stress at their reefs as soon as new satellite SST data are collected and processed – presently this is twice-weekly. When the thermal stress at a selected reef site becomes detected by satellite as reaching pre-defined warning levels, in terms of bleaching HotSpot and DHW values, a corresponding e-mail warning is issued automatically to CRW subscribers. Currently, the warning levels are defined as bleaching "**watch**," "**warning**," and "**alert**" (*level 1* and *level 2*). The bleaching **watch** is the lowest reported stress level that is issued when HotSpot levels initially develop. An **alert** for *level 2* is the highest level that is issued, being activated when accumulated active thermal stress reaches 8 DHWs or above. Although a free service, users are requested to subscribe to receive the SBA. Recipients are encouraged to provide reports on bleaching status and feedback to the new CRW alert system. At present, 24 pre-selected individual reef sites are available for SBA services. It is anticipated that additional sites will be added, based on feedback from our users.

Currently at 50 km resolution, CRW's satellite bleaching monitoring is capable of monitoring coral bleaching only in association with large-scale thermal stress having spatial coverage over both inshore water and offshore water. Over the next few years higher resolution products are envisioned, providing increased SST detail to extend the monitoring capability from basin and regional scale thermal-bleaching events to local and reef scale events.

A NOAA project recently completed in the Republic of Palau developed a hydrodynamic model to gain an improved understanding of the role of tidal mixing on a bleaching event (presented elsewhere in this monograph). Satellite data, with *in situ* validation, were used to produce a high-resolution bathymetry. A vast array of oceanographic instruments was deployed to provide validation data for the model [Steinberg et al., 2004]. The predictability of the tides and monitoring of major current systems by satellite and models allows an understanding of why some reefs are more likely to bleach than others during a bleaching event. For those areas that do bleach the model results should prove helpful for understanding reef recovery and/or resilience, and aid in locating areas primed for MPA development. The first hydrodynamic model for understanding "bleaching risk" was developed for the Great Barrier Reef [Skirving et al., 2004].

These high-resolution (250m) hydrodynamic models can also be used to derive connectivity and water quality (as a result of rainfall/runoff and re-suspension) products, once the relevant validation fieldwork has been carried out. The only difference between the bleaching hydrodynamic models and the connectivity/water quality models is that wind input is assumed to be zero in the former (a characteristic of bleaching events) and included in the latter. Thus the investment in a regional hydrodynamic model for bleaching could then be feasibly used for other purposes important for MPA design and reef management.

3. More than a Temperature Event?

Coral bleaching can be caused by a number of stressors [Hoegh-Guldberg, 1999]. Among them, sea temperature has been demonstrated to be the most reliable predictor of large-scale coral bleaching events, often referred to as mass coral bleaching [Glynn, 1996; Brown, 1997]. Reported above-average sea temperatures can be brought on by extreme El Niño conditions [Wilkinson, 1999], however, as will be shown later in the chapter, a cursory assessment of ENSO events does not support a clear relationship between coral bleaching and all intensities (weak, moderate, and strong) of El Niño or La Niña events [Arzayus and Skirving, 2004].

On a regional scale, the development of conditions that can lead to coral bleaching is likely to be influenced by regional weather patterns, especially conditions that lead to high air temperatures and extended periods of low-wind and low-cloud cover [Liu et al., 2003]. As previously stated, in well monitored areas, such as the Great Barrier Reef, the occurrence of ENSO events has not been a reliable predictor of hot water anomalies over the last twenty years. The extent, maximum, and persistence of warm or dry weather conditions are the main indicator of bleaching risk. In either case, monitoring weather patterns can provide an early warning of the development of conditions that may lead to bleaching over advance timescales of weeks to months. Local weather forecasts are readily accessible and can be an adequate basis for an indication of whether bleaching risk is increasing or decreasing in the short-term.

Table 1 lists the major climate variables that are known to influence bleaching risk by increasing sea temperatures. Longer term (i.e., seasonal) forecasts can be used to assess the probability of weather conditions occurring that are conducive to increasing sea temperatures, while shorter term (i.e., weekly weather) forecasts indicate whether sea temperatures will increase or decrease in coming days and weeks. For example, seasonal outlooks for summer that predict above-average air temperatures and decreased monsoonal activity would indicate that there is an increased probability of conditions that may lead to stressful sea temperatures. Once summer has begun, forecasts of hot weather with clear skies indicate that surface waters are likely to warm significantly, especially if these conditions are accompanied by low wind and neap tides (which reduce the potential for mixing of hot surface waters with cooler deep water). In contrast, forecasts of cooler monsoonal conditions, increased cloud cover, and strong winds indicate that sea temperatures may stabilize or decrease over the coming week.

TABLE 1. Climate variables and their influence on bleaching risk.

Climate variable	Implications for bleaching risk
ENSO	El Niño conditions increase sea temperatures in the eastern equatorial Pacific Ocean often promoting far-reaching affects through teleconnections in the atmosphere, and may in some instances increase the chances of stable hot conditions in the atmosphere overlaying key reef regions.
Air temperature	Hotter air temperatures due to increased solar radiation are an indication of higher sea surface temperatures.
Cloud cover	Cloud cover provides shade to lessen direct heating effects from the sun.
Wind and wind waves	Low winds during times of more intense insolation promote increased heating of surface waters, and no wind allows for greater depth penetration of solar radiation. Higher winds and breaking waves have the ability to mix heat to greater depths, while reducing the accumulations of heat at the surface.
Tidal currents	Tidal currents (especially the stronger spring tides) increase mixing and provide the potential to reduce temperatures of surface waters by mixing cooler bottom waters into the upper water column.
Low frequency currents	Synoptic-scale currents (e.g., equatorial and western-boundary currents) can also influence the mixing of the water column, particularly for reefs at the continental shelf edge.

4. Role of ENSO and PDO in Bleaching Events

Thanks to the satellite era, ENSO events are now known to directly influence a large expanse of the tropical Pacific Ocean that teleconnect with the atmosphere bringing characteristic El Niño "weather"-associated anomalies around the globe. It is appropriate to improve our understanding of the impacts of these 3-7 year ENSO oscillations on coral reef bleaching events. All ENSO events, both El Niño and La Niña, when examined statistically for any role in mass bleaching events globally, are apparently rather **insignificant** [Arzayus and Skirving, 2004]. It may be shown subsequently that the more intense ENSO events have higher correlations, but at this time the major factor influencing mass bleaching is found to be extended periods of calm, clear weather when the sun is highest in the sky, during and immediately surrounding what is climatologically the warmest month of the year. Plate 1 (C) gives an overview of reported bleaching events since 1985. Those reefs that experienced no bleaching are shown as white. In Plate 1 (D) those locations that show a correlation for occurrence during the same quarter of the year with El Niño, Neutral and La Niña conditions are identified. Reefs that have bleached at least 70% of the time during one specific ENSO regime represent less than 1% of all observations.

One ENSO relationship that has been established is a reduction of hurricane/tropical cyclone activity in the Atlantic Basin during El Niño conditions [Pielke and Landsea, 1999; Camargo and Sobel, 2005]. During these periods SST levels are higher over much of the eastern tropical Pacific.

While severe weather events such as cyclones and hurricanes are often associated with deleterious effects on coral reefs, owing to significant turbulence from the pounding of high seas and swell, these tropical systems sometimes offer a much needed respite from extended periods of excessive thermal stress and bleaching. In 2004 (a non-El Niño year), regions around Florida were showing severe levels of thermal stress throughout August. By late September, following a record number of visits by hurricanes to that state, vertical mixing caused by strong winds and pounding waves, had been so complete that no high temperatures remained (Plate 1 (E)), potentially preventing severe bleaching events on the Florida coastlines [Heron et al., 2004].

Another ocean oscillation operating over longer timescales, approximately 30 years, has direct and far reaching impacts spanning all latitudes of the Pacific Ocean – the Pacific Decadal Oscillation (PDO) [or Variability (PDV)]. The PDO is still poorly defined and mostly thought of as a North Pacific oscillation – though presumed to have a counterpart in the South Pacific as well. To be able to predict, with any measure of accuracy, future anomalous SST events leading to bleaching certainly requires a more complete understanding of both PDO and ENSO, as well as their interactions. As will be shown in the next section it appears that a PDO reversal during the late 1990s may have established quite a different weather regime over the next 20 to 30 years until PDO again switches to the characteristics shown throughout the Pacific from the 1970s ["The Great Pacific Climate Shift" – Bratcher and Giese, 2002] - until the turn of the century.

5. 20 Year Trends

Satellite data sets used for monitoring environmental conditions over coral reefs have been available for over 20 years. The most robust data set comes from NOAA's Advanced Very High Resolution Radiometers (AVHRR). Satellite SSTs have maintained a continuous record of stable statistics when compared to *in situ* SSTs from drifting and moored buoys, a validation process that began in 1982. Most importantly these statistics show no statistically significant trend in any biases observed between these two measurement types. Only two interruptions have been observed in this 23-year satellite SST time series. These came from volcanic eruptions – El Chichón in 1982; Mt. Pinatubo in 1991. SST corrections have been made for most of the 2 year stratospheric aerosol contamination (negative offset in SST derivations) period attributed to Pinatubo, while the earlier El Chichón event came too early in NOAA's AVHRR SST program to be corrected. This means we have a well-maintained time series of SSTs from 1985 to present.

At CRW we have been looking at SST trends (more properly called short-term trends or tendencies) over the tropical oceans with the desire to understand any climate tendencies or shifts in tendency that might effect coral reefs. Over the 18 year period 1985-2002, tropical SSTs are showing nearly the same average rate of increase that has been identified over land for the past 100 years. As Figure 2 shows, latitudinal rates of increase are higher in the northern hemisphere's oceans and can be seen to grow in their tendency with increasing latitude.

Global SST trends in this same interval (1985-2002) have shown substantial regions sporting decreasing SSTs, though even more regions, particularly in the tropics, are showing rising trends. These regions of positive and negative tendencies are shown in Plate 1 (F).

Plate 1 (G) shows 12-year subsets of the trend charts; (a) 1985-1996 and (b) 1991-2002. From these shorter periods, there appears to be a distinct shift (reversal?) that took place during the last half of the 1990s.

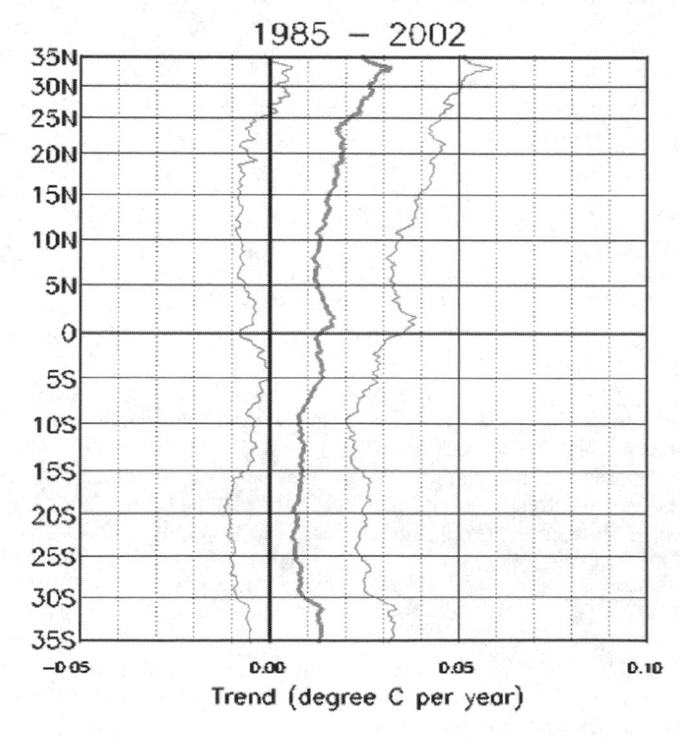

Figure 2. Latitudinal SST trends and their associated variances from 35S to 35N over the period 1985-2002.

A remarkable shift is obvious: from cooling to warming over the tropical Indian Ocean and western Pacific; from moderate warming to cooling over the eastern Pacific; and enhanced warming over vast portions of the Atlantic. What will be instructive is whether the present shift continues to evolve into the first and second decade of the 21st Century as the new phase of PDO or is merely a transition phase of our changing climate.

6. Extrapolations/Expectations for 2050

As has been stated above, to properly understand what will happen with baseline SSTs under the likelihood of a changing climate over the tropics between the present and the year 2025 may well require coming to a better understanding of PDO's role. This knowledge would seem critical for putting past dramatic SST rises and bleaching, and for predicting future events, into perspective. By the middle of the century, with PDO most probably completing any cycling that will be experienced between now and then; some higher SSTs seem unavoidable as present overall trends could easily sustain 0.5°C increases in the mean with many places feeling significantly higher rates. The major question is how much higher can tropical SSTs rise from a physical point of view. Air-sea interactions/processes would undoubtedly produce more cloud cover, hopefully, alleviating these worrisome increases in many locations.

NOAA has embarked upon some work with Princeton University and NOAA's Geophysical Fluid Dynamics Laboratory to extend the existing General Circulation Models (GCM) into the coastal environment. At present, GCMs only include our oceans up to the continental shelves [e.g., Donner et al., 2005]. We are planning to inquire of these improved models questions such as: what happens in this important region where ocean meets land?; are temperatures in these regions likely to be enhanced as our global land mass heats more rapidly than our oceans under a changing climate scenario based on increased greenhouse gases?; will enhanced sea breezes (and possibly weaker nighttime land breezes) provide some additional relief during the daytime when SSTs might otherwise be higher; will cloud regimes be altered significantly to decrease the incoming heating from insolation? We are hopeful that these enhanced models may shed some new light on these important questions with regard to our coral reef ecosystems and their expected evolution into the 21st Century in a changing climate.

7. Coral Bleaching: The Use of CRW DHW Product to Place 1998 into Context

Coral bleaching in 1998 effectively destroyed 16% of the world's coral reefs, with losses in the Indian Ocean being almost 50% [Wilkinson, 2004]. During that year, 1000-year old corals died and some areas reported their first bleaching, e.g., the Republic of Palau. Many suspected that this was a sign of reef destruction for the future, but there has been no repeat of bleaching like 1998 in the following 6 years (with the unprecedented 2005 Caribbean mass bleaching event possibly now becoming a more recent notable, albeit regional, exception). The NOAA Coral Reef Watch DHW product is able to be used to assess the effects of accumulated heat stress on coral reefs from 1985, and hence provide a context within which to place the 1998 bleaching event.

Whenever CRW issues bleaching warnings via the Internet, coincident *in situ* observations are consistently showing a correlation between bleached corals and Degree Heating Week values of 4 and greater. More than 200 bleaching warnings have been issued since 2000, and on more than 100 occasions field reports were received. All of these reports confirmed that coral bleaching had occurred. Thus, Degree Heating Weeks provide an effective indicator of the onset of coral bleaching (if not a conservative one).

Figure 3 is a plot of the percentage of the global coral reef area experiencing DHW values greater than 4 from 1985 to 2002. Note that 1998 is clearly a standout value and is not part of the background trend. This may indicate that 1998 was not part of the climate change story, but rather a statistical anomaly. It will be interesting to see if this remains the case when paleo-climate records are used to significantly lengthen the satellite data set. Nevertheless, this plot is extremely interesting because even without the 1998 spike, there is a positive "background" trend (the line in Figure 3). If this trend were to continue at the rate shown here, the percentage of global bleaching each year would reach 1998 levels in around 50 years. Although this is considerably less than some current perceptions of the increase in bleaching, it is nevertheless climatologically significant.

Nevertheless it is important to view the 1998 event in the correct perspective; it is an unusual event in the 20-year satellite record. The satellite record is far too short to derive accurate climate trends that are not overwhelmed by potential natural variability. However, if the upward trend of the baseline persists, events like 1998 could become commonplace in the latter half of this century. This is consistent with the predicted climate trends as reported by the IPCC [Houghton et al., 2001].

8. Integrating Coral Reef Observing Systems: Challenges

The time is now ripe to implement a coordinated effort that will synergistically integrate, seamlessly complement and leverage *in situ*, satellite remote sensing, paleoclimatic and

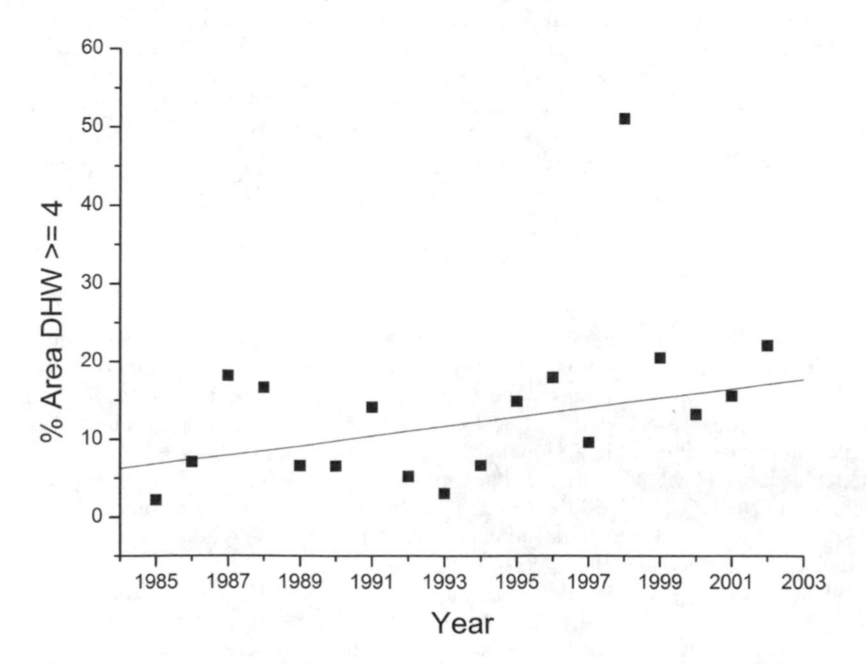

Figure 3. Annual percentages of global coral reef area where DHWs >4 (1985-2002).

institutional assets to achieve common observational goals. This should make it possible to place field sites in their larger context and thus to extend the results from expensive and labor-intensive field work over much larger areas, supporting more effective management action to respond to the stresses threatening coral reef ecosystems. This requires improved access to remote sensing products, technical capacity building in reef monitoring programs, and targeted financial support, possibly assisted by regional image processing and coordinating centers.

Given the new emphasis by the National Oceanic and Atmospheric Administration (NOAA) to coordinate operational and research efforts between their line office organizations, the thrust for an integrated approach to observing coral reef ecosystems was given a high priority by NOAA's coral reef program managers. The Coral Reef Ecosystem Integrated Observing System (CREIOS) was formed on the premise that NOAA's line offices should provide the stepping stones to the larger Integrated and Sustained Ocean Observing System (IOOS). The CREIOS vision is to provide a diverse suite of long-term ecological and environmental observations and information products over a broad range of spatial and temporal scales. The CREIOS goal is to understand the condition and health of and the processes influencing coral reef ecosystems, to assist stakeholders in making improved and timely ecosystem-based management decisions to conserve coral reefs.

Currently, CREIOS is composed of members encompassing scientists, managers, engineers and local experts driven by the NOAA coral program requirement to provide information for effective management of coral reef ecosystems. Because CREIOS is a small community, it is highly effective at responding to user needs and requirements by providing specific local-products and responses and delivering unique data to specialized users.

NOAA, as a member of the Integrated Global Observing Strategy (IGOS), also fulfills the requirements and responds to the issues set forth in the recently accepted Coastal Theme, and in specific, the Coral Sub-theme. These issues include: 1. To develop a strategy identifying user's present and future needs and optimizing the reef information obtainable from buoys or fixed instrument platforms, scientific monitoring and amateur monitoring to fit those needs. 2. To complement and reinforce *in situ* coral reef monitoring and assessment programs by integrating a remote sensing component. 3. To determine the capacity of integrated *in situ* and remote sensing monitoring to provide data that can be reliably generalized or extrapolated over large reef areas. 4. To establish or reinforce regional centers able to provide remote sensing products in support of coral reef monitoring programs.

Bridging the Observational Gap

For coral reefs, there is a significant gap between observations possible from satellite and airborne platforms with a maximum resolution of 1-5 meters, and *in situ* observations such as video transects or manta-towed divers. To detect the extent and impact of a coral bleaching event, for instance, or to observe shifts in surface cover from live corals to algae, a resolution of about 10-30 cm would be ideal, requiring new *in situ* or proximate sensing techniques. For satellite sensing, much can potentially be learned with better analytical methods (e.g., spectral unmixing) that can retrieve sub-pixel components of coral/algal cover from larger pixels. Research on the latter and on radiative transfer modeling will assist in optical satellite instrument design. Detecting the presence and significance of coral bleaching events will require information also on the percent of living coral cover, the amount of recovery of pigmentation after the bleaching event, and the amount of recently dead coral shortly after the event. These require good time series imagery from a hyperspectral sensor with 4-10 m pixel resolution, unmixing methods and good atmospheric correction. A nested set of observing techniques including satellite remote sensing, intermediate sampling with high-resolution techniques, and traditional *in situ* methods would be ideal for intensively studied areas.

Additionally, the most significant zone for reef framework development on many coral reefs is the seaward reef face. This zone slopes steeply and thus appears from the air as a narrow band of widely varying depths, making current airborne and spaceborne remote sensing techniques largely redundant. This gap in capacity to observe vertical reef faces and deeper reefs over large areas at appropriate scales will require the development of new instruments capable, for instance, of underwater remote sensing perpendicular to the reef slope.

Another significant need is to be able to detect reefs that are not obvious from the surface. There are probably large areas of coral reefs growing on submerged limestone banks 10 to 50 meters below the surface, which may prove to be important reservoirs of coral and fish larvae for shallow reefs damaged by coral bleaching or overfishing.

Underwater remote sensing instruments should be assembled to be used for independent close range remote sensing, and also for fast ground-truthing of satellite data. For diver use or deployment from a small boat or ROV, it is feasible to connect an underwater spectrometer with an underwater laptop equipped with a depth sensor, GPS (through a small buoy attached to the PC), and a digital still or video camera, providing a fully autonomous instrument capable of recording video transects together with reflectance spectra of substrates, coordinates, and depth. An even more useful instrument would have an imaging underwater spectrometer (based on CASI or AISA for example) in the package to cover wider reef areas and to get information about reef structure. Such an instrument should be capable of covering a 10 meter wide band with a resolution of 10 cm, as well as closer video transect work. This would avoid problems of observing across the air-water interface, and, if artificial illumination were incorporated, the problems of light attenuation in the water column.

Such an instrument could provide more detailed information about benthic habitat than just video transects, once algorithms are developed to resolve different substrates on the basis of their reflectance spectra. For more extensive surveys, a similar instrument package mounted on or towed behind a boat, with sonar to measure distance to the reef, could survey extended reef faces at higher speed.

The issue is to fill the observational gap between 10 cm and 4 m resolution, to overcome underwater light attenuation effects, and to survey sloping or vertical reef surfaces. New instrument packages for underwater remote sensing are required to fulfill this.

Integration of Remote Sensing and In Situ Information

To integrate *in situ* and remotely sensed data, the best tool is a Geographic Information System (GIS). The existing global coral reef database, ReefBase, could evolve into an interactive environmental management information system. Such an on-line GIS should have the ability for "data mining" all available *in situ* data. Users, scientists and participants in monitoring programs should be able to click on a map and zoom in on their reef from separate layers of satellite images to even include close-up underwater photographs. They should be able to query the health of the reef, input recent data and survey information, and obtain a response in chart format, with management advice on what to do about problems observed. Where possible, these GIS systems augmented with remotely sensed data should be integrated into decision support systems designed to identify the potential consequences of management decisions for reefs and reef-dependent people.

As observing systems improve, there is a growing need for specialist teams of scientists to provide advice on the analysis and interpretation of the observations and to translate the results into management recommendations. In particular, the capacity of scientists in developing countries responsible for reef systems requires strengthening, through education and training programs, and through joint research programs.

As noted above, NOAA is installing *in situ* monitoring stations at strategic coral reef areas for purposes of establishing long-term data sets, providing near real-time information products, and surface-truthing NOAA satellite "HotSpot" products which are used for coral bleaching predictions and early warnings. The sea temperature sensor data are automatically compared with satellite-monitored temperatures and thus provide near real-time feedback on the accuracy of the satellite-monitored temperatures. The CREWS stations also measure wind speed and direction, air temperature, barometric pressure, sea temperature, and salinity, as well as photosynthetically active radiation and ultraviolet-B above and below the water. Data are presented daily on the Web as well as saved for access via an online database at http://www.coral.noaa.gov/crw.

Improved Coordination among Existing Programs and Networks

Developing early warning systems for coral bleaching and other signs of damage as required by the Convention on Biological Diversity (CBD) will require extensive coordination among international partners and a significant effort to organize monitoring at the community level. However obtaining funding for international coordination of such operational networks is very difficult. The Global Coral Reef Monitoring Network (GCRMN), closely associated with IOC/GOOS and UNEP, is the existing framework for *in situ* scientific monitoring, while Reef Check provides a similar standard network for surveys by teams of volunteers under scientific supervision. Most coral reef remote sensing has been part of research programs rather than operational monitoring. The NOAA/NESDIS operational remote sensing work on sea surface temperature and coral reef HotSpots, and the Coral Reef Watch stations for a Coral Reef Early Warning System are the beginnings of a near real-time operational system. ICRI and ICRAN are the principal global mechanisms for coordinating management action on coral reefs. One essential step in a coherent coral reef observing strategy will be to establish mechanisms for regular coordination between these components, both globally and in specific countries and regions.

Operational Products and Services

There are several user groups requiring specific coral reef observational products and services:

- **commercial and private entities** that use coral reefs as a source of revenue, such as commercial fishermen and the tourist industry. Tourism is a major economic activity for many developing countries, but too often, for lack of adequate information, it damages the very coral reef resources that help to attract tourists. Fishermen generally become supportive of coral reef management when they see that it ensures the sustainability of their catch.
- **parks, marine reserves** and other areas dedicated to the conservation of coral reefs. Managers and planners need information to maintain and enhance the state of coral reefs under their care, to zone for various uses, and to monitor the effectiveness of management measures.

- **environmental and resource managers** in government and non-profit entities. Managers require information to create and enforce legislation and regulations in order to maintain environmental quality and resource productivity.
- **research and international conservation organizations**. Scientific organizations explore the complex interactions and feedback systems (among others) relevant to ecosystem functions and anthropogenic impacts, as well as reef biodiversity, ecology, pharmaceutical potential, etc.
- **planners**. Government planners (national and local), non-governmental organizations and stakeholder groups need to integrate many kinds of information at various geographic and temporal scales, often in near real-time, with multiple database platforms and Geographic Information Systems.

These groups have different capacities to pay for observing data depending on their informational needs, requiring a careful consideration of the economics of delivering coral reef observations, and particularly of a hyperspectral space mission appropriate for coral reefs. Both public and commercial providers may develop capacities for this, with different costs and advantages. Given the public benefit from better reef management, it would be desirable for image copyrights to be owned by a public or international entity, and for costs to be subsidized or covered systematically for scientists, managers and end-users in developing countries.

For bleaching prone reef areas, a single product combining *in situ* and remotely sensed data into "risk maps" for coral bleaching is needed (recent results in the Great Barrier Reef are promising). These maps would incorporate general hydrodynamic models of the 'risk' area coupled with satellite-derived HotSpot information as well as other *in situ* or remote sensed products needed for the particular area.

Operational data processing and management will be another requirement to be considered. Assuring repetitive global coverage is one thing; providing a mechanism to transfer large amounts of raw data and products to users is another. The characteristics to be included in an effective information delivery system are:

- free scientific use of the data;
- web or hard distribution;
- documentation of algorithms and performance (cal/val experiments);
- free availability of maintained graphic user interface (GUI) software for processing images from raw data into products;
- distribution of beta-products on CDs to users with poor Internet facilities;
- assistance in acquiring necessary computer infrastructure and training.

Given the diversity of coral reefs around the world, regional and thematic optimization of the products will require scientific inputs and controls at all stages. Even for the most routine applications such as geomorphological mapping, classification schemes have to be adapted and optimized to obtain thematically-relevant products from one region to another. Many of the end users who can most benefit from reef observing products may not have the skills and facilities to produce them on their own at the present time. This will require establishing regional facilities for the development, production, distribution and archiving of information products optimized for the region. Four such centers might be appropriate for the Caribbean/Atlantic, the Red Sea/Indian Ocean, South-East Asia, and the Pacific. Remote sensing algorithms can also be developed and automated to enable non-specialists to process images quickly to provide useful information for management use.

Consideration will have to be given to the daily data volume generated by the system, and the minimum and maximum extent of coverage, as related to the long-term acquisition plan. Existing regional remote sensing facilities will need to be evaluated for their capacity to handle the amount of data required for change detection analysis on a regional scale, and their ability to produce a line of information products relevant for their regions.

Given the sheer volume of potential observing systems in coral reef research and management communities, a web-based inventory of observational activities, including sources of both basic and advanced coral reef information, should be established and maintained. This should include the current fleet of operational satellites and their available products, research programs developing new types of observations, and archives, data sources, and public and private entities providing processed information products. What is envisioned must be a truly integrated "research to operations" effort. Some data for this are being assembled under NOAA-funded projects.

9. Management Considerations

On June 11, 1998, President Clinton issued Executive Order 13089 on Coral Reef Protection. The order directs Federal agencies to identify their actions that might affect U.S. coral reef ecosystems, to use their authorities and programs to protect and enhance these ecosystems, and, to the extent permitted by law, ensure that any actions they authorize, fund or carry out, will not degrade the conditions of these coral reef ecosystems. This Executive Order carried sweeping implications for managers of coral reef ecosystems located in U.S. waters, setting off an impetus for management protocols, the use of standardized instrumentation, metrics and sampling definitions, and a clearer understanding of how tools and technology could complement a sound management and implementation plan.

The coupling of remote sensing tools and *in situ* measurements to detect and model potential coral bleaching events can be a powerful management tool to observe the relationships between climate episodes, human and natural events and bleaching risks. "A Reef Manager's Guide to Coral Bleaching" (Marshall and Schuttenberg, 2006 and summarized elsewhere in this monograph) was designed to aid reef managers in the use of proven methods and technologies often used in the management plan of reef parks and sanctuaries. Many managers, stakeholders, scientists and engineers participated in the writing of this publication.

Based on this publication, a step-wise process follows the development of a management plan. At every step in this process, there is a need for real-time or near real-time information from the reef sites being managed. Managers' tasks identified in this publication involve assessing the seasonal water column temperatures and associated bleaching risks to then establish an enhanced "early warning" system for coral bleaching. Remote sensing data coupled with sporadic *in situ* measurement cross-checks, empirical observations, and computerized models, provide the manager with an outlook of the overall health of the reef on a day-to-day basis. If or when a bleaching event was to occur, managers would classify the affected areas of the reef and identify the least affected reefs. This information could then be used to manage reefs based on maximum resilience or sustainability (long term planning), or to manage stressors to increase survival (short term planning).

The standard suite of CRW tools provides a good starting point for any reef manager to get an idea of the stress and accumulated heat in a particular body of water. Moreover, an in-depth knowledge of these tools can provide a comprehensive assessment of the area, including changes in sea surface temperature (SST) over time, time of peak SST stress, and a rough visualization of ocean currents and speed. CRW, in liaison with The Nature Conservancy and the World Bank, has provided capacity building workshops for reef managers and stakeholders in the Philippines, Mexico, Palau and the U.S. Virgin Islands. These workshops provide information on new sensor and tool developments, the theory behind identifying coral reef bleaching through the use of remote sensing, and a full set of hands-on exercises with real-life scenarios.

Managers and stakeholders are increasingly being placed in positions where they need to report management plans, implement actions, and convey potential results to governing authorities within a short turn-around timeframe. The advent of advanced remote sensing techniques has placed a wealth of readily available information in the hands of knowledgeable managers and stakeholders. CRW has built its suite of satellite tools to provide both frequent and comprehensive environmental information that is accessible with minimal user interface and local processing time. This is necessary since most users are located in remote parts of the world and often need to rely on low bandwidth internet connections. The current emphasis is on providing low resolution data with high management impact so that managers can use their knowledge of local stressors, such as fishing, tourism and water quality, and compare these against the several SST products, to produce a long-term resilience plan. In the not-so-distant future, applications of remotely sensed data will include high resolution visible data to identify waters being cooled by upwelling and mixing; detecting ocean currents that can flush toxins; detecting turbid waters that may reduce the exposure of corals to harmful UV radiation; and finding conditions that are conducive to coral recolonization.

Current technologies, in both remote sensing characterizations and *in situ* assessments, primarily allow managers to detect ecosystem changes, and perhaps pinpoint the onset of coral bleaching events. Further technological advances will increase the monitoring capacity with these tools. While this knowledge is extremely useful, it is the use of this information through a sound management plan that can assist in the sustainability of these magnificent ecosystems, and perhaps reduce or halt their chronic disappearance from our planet.

10. Conclusions

Much can be accomplished by fully implementing a widely available suite of integrated satellite and *in situ* derived products based on sensors presently available. At the same time, space agencies should consider developing new sensors to increase the resolution and accuracy needed to do coral reef monitoring.

Integration of remote and *in situ* sensors is essential to coral reef observations. There are very few *in situ* sensing stations providing near real-time data, and their number and the coverage of critical reef areas need to be increased. Data assimilation techniques need to be developed that can combine satellite data and traditional field data for coral reef monitoring across the different scales necessary for management action.

The ability to obtain high-resolution imagery of coral reef areas is critical. The present ability to perform multiple scale mapping is limited, expensive and difficult to produce in a timely fashion. The ability to map coral reefs at high resolution in multiple geographical scales would be an asset to reef managers and researchers because it would allow for three-dimensional analysis of coral reefs and for visual confirmation of the health of the reef. A high spatial and spectral resolution instrument is required to assess changes in the community structure of affected reefs.

An integrated remote sensing program that links physical oceanographic processes (e.g., causes of disturbance on reefs or transport of reproductive larvae) with biological measurements of the consequence (e.g., coral cover) would revolutionize understanding of climate-induced phenomena on coral reefs. Furthermore, by identifying physical or biological regions that seem to resist particular disturbances, remote sensing will help governments and conservation agencies prioritize efforts to protect such ecosystems.

References

Arzayus, L. F., and W. J. Skirving, Correlations between ENSO and Coral Reef Bleaching. 10th International Coral Reef Symposium. Okinawa, Japan, 2004.

Brown, B., Coral bleaching: causes and consequences. *Coral Reefs*, 16: S129-S138, 1997.

Bratcher, A., and B. Giese, Tropical Pacific decadal variability and global warming, *Geophysical Research Letters*, 29: 24-1 to 24-4, 2002.

Camargo, S. J., and A. H. Sobel, Western North Tropical Cyclone Intensity and ENSO, *Journal of Climate*, 18: 2996-3006, 2005.

Donner, S. D., W. J. Skirving, C. M. Little, M. Oppenheimer, and O. Hoegh-Guldberg, Global assessment of coral bleaching and required rates of adaptation under climate change, *Global Change Biology*, 11: 2251-2265, 2005.

Gleeson, M. W., and A. E. Strong, Applying MCSST to coral reef bleaching, *Adv. Space. Res.*, 16(10): 151-154, 1995.

Glynn, P. W., Coral Reef Bleaching: facts, hypotheses and implications, *Global Change Biology*, 2: 495-509, 1996.

Goreau, T. J., and R. L. Hayes. Coral bleaching and "ocean hot spots". *AMBIO*, 23: 176-180, 1994.

Hendee, J. C., E. Mueller, C. Humphrey, and T. Moore, A data-driven expert system for producing coral bleaching alerts at Sombrero Reef in the Florida Keys, *Bull. Marine Science*, 69 (2): 673-684, 2001.

Heron, S. F., G. Liu, L. F. Arzayus, W. J. Skirving, and A. E. Strong, A benefit from hurricanes, PORSEC2004, Concepcion, Chile, 2004.

Hoegh-Guldberg, O., Coral bleaching, Climate Change and the future of the world's Coral Reefs, *Marine and Freshwater Res.*, 50: 839-866, 1999.

Houghton, J. T., Y. Ding, D. J. Griggs, M. Noguer, P. J. van der Linden, and D. Xiason (Eds), Climate Change 2001: *The Scientific Basis*. Cambridge University Press, UK, pp. 944, 2001.

Liu, G., W. Skirving, and A. E. Strong, Remote Sensing of Sea Surface Temperatures During the 2002 (Great) Barrier Reef Coral Bleaching Event, *EOS*, 84(15): 137, 141, 2003.

Marshall, P., and H. Schuttenberg, A Reef Manager's Guide to Coral Bleaching, Great Barrier Reef Marine Park Authority, Townsville, Australia, 2006.

Montgomery, R. S., and A. E. Strong, Coral Bleaching threatens oceans, life. *Eos, Transactions, American Geophysical Union*, 75: 145-147, 1994.

Ogden, J., J. Porter, N. Smith, A. Szmant, W. Jaap, and D. Forcucci, A long-term interdisciplinary study of the Florida Keys seascape, *Bull. of Marine Science*, 54(3): 1059-1071, 1994.

Pielke, R. A. Jr., and W. Landsea, La Niña, and Atlantic Hurricane damages in the United States, *Bull. Amer. Meteor. Soc.*, 80: 2027-2033, 1999.

Skirving, W. J., C. R. Steinberg, and S. F. Heron, The hydrodynamics of a coral bleaching event. *ASLO/TOS Ocean Research 2004 Conference*, Honolulu, USA, 2004.

Skirving, W. J., A. E. Strong, G. Liu, L. F. Arzayus, C. Liu, and J. Sapper, Extreme events and perturbations of coastal ecosystems, pp.11-25, Richardson, L. L., and E. F. LeDrew (eds), 2006, In *Remote Sensing of Aquatic Coastal Ecosystem Processes, Springer Remote Sensing and Digital Processing series*, Vol. 9, pp. 324, 2006.

Steinberg C. R., S. F. Heron, W. J. Skirving, C. McLean, and S. M. Choukroun, Palau Oceanographic Array Data Report, August 2003 – January 2004. Report to The Nature Conservancy, Australian Institute of Marine Science and National Oceanic and Atmospheric Administration, pp. 246, 2004.

Strong, A. E., C. S. Barrientos, C. Duda, and J. Sapper, Improved satellite Techniques for Monitoring Coral Reef Bleaching. Proc 8th International Coral Reef Symposium, Panama City, Panama, pp. 1495-1498, 1997.

Wilkinson, C. R., Global and local threats to coral reef functioning and existence: review and predictions, *Marine and Freshwater Res.*, 50: 867-78, 1999.

Wilkinson, C., O. Linden, H. Cesar, G. Hodgson, J. Rubens, and A. E. Strong, Ecological and socioeconomic impacts of 1998 coral mortality in the Ocean: An ENSO impact and a warning of future change? *AMBIO*, 28(2), 188-196, 1999.

Wilkinson, C. R., Status of Coral Reefs of the World: 2004, Global Coral Reef Monitoring Network and Australian Institute of Marine Science, Townsville, Australia, pp. 301, 2004.

10

Management Response to a Bleaching Event

David Obura, Billy Causey, and Julie Church

Abstract

Mass coral bleaching events have become increasingly frequent and severe since the 1980s, significantly impacting coral reefs around the world. Bleaching events threaten Marine Protected Areas (MPAs) as these have not been designed with the threat of bleaching in mind, and have been impacted as severely as unprotected areas. Thus MPA managers must develop tools to mitigate the impacts of bleaching to protect the values for which the MPAs were established. The primary trigger for mass bleaching events is the global increase in atmospheric greenhouse gasses such as carbon dioxide. This mismatch in scale against the limitation of management actions to local levels has resulted in despondency that management cannot hope to relieve the impact of coral bleaching on coral reefs. We present case studies and a framework that demonstrate that managers can take positive actions at the local level, and that these can in turn influence higher level policy processes at a similar scale to climate change drivers. While managers are limited in their ability to reduce the primary threat of bleaching they can take actions to limit other interacting threats and they can take actions to promote recovery. Further, mangers can develop early warning, monitoring and communications systems to inform all levels of management responses, engage the public and stakeholders and influence larger scale policy. These not only improve the effectiveness of individual management actions but also establish an enabling framework that fosters a stewardship approach amongst stakeholders that further empowers management. Lessons from the Florida Keys National Marine Sanctuary in the United States, which suffered repeated and increasingly severe bleaching events in the 1980s and 1990s, show all of the above possibilities for management action. Lessons from the Kiunga Marine National Reserve in Kenya, where bleaching coincided with the establishment of participatory monitoring programme, show how communications and dialogue between managers and stakeholders provide an essential framework for establishing an enabling environment for co-management. We present the case that progressive improvements in management actions will occur as science and management progress and learn from bleaching events and that increasing effectiveness in managing the impacts of bleaching events will occur.

Introduction

As mass coral bleaching and mortality events increase in frequency and intensity globally [Williams and Bunkley Williams, 1990], they threaten the survival of coral reefs as we

Coral Reefs and Climate Change: Science and Management
Coastal and Estuarine Studies 61
Published in 2006 by the American Geophysical Union.
10.1029/61CE11

know them today [Hoegh-Guldberg, 1999]. Significant investments have been made, and will increasingly be made to conserve the best coral reef areas, to manage heavily utilized ones to sustain their provision of goods and services, and to rehabilitate degraded reefs. However these efforts are all undermined by the pervasive influence of coral bleaching caused by climate change – irrespective of whether this is of natural or anthropogenic origin, or a combination of both. In the context of other chapters in this volume, this one sets out to justify the need to adapt management to the threat of coral bleaching, using the threat of bleaching as a tool to focus and prioritize actions, and develop new interventions. It will present an organizing framework for targeting management actions and summarize interventions that have been possible to date in response to bleaching events. The chapter will provide 2 case studies illustrating the importance of conventional management approaches for dealing with climate change impacts, and a discussion of key areas for future development.

Past Bleaching Events and Impacts on Management Areas

Coral bleaching that results in mass mortality of corals results in degradation to the entire coral reef ecosystem [see Glynn, 1993, Brown, 1997, Wilkinson et al., 1999]. This is due to losses in coral cover and diversity, and the consequent changes in ecosystem function (see other chapters, this volume), generally to a simpler less diverse and less productive alternate state [Nÿstrom et al., 2000]. As a result, the social and economic values of reefs based on biodiversity and productivity also decrease as the provision of goods and services declines [Möberg and Folke, 1999].

The impacts of coral bleaching have to date applied indiscriminately with respect to Marine Protected Area (MPA) boundaries, affecting protected and unprotected areas alike around the globe [Wilkinson, 2000, 2002, 2004]. In many cases, bleaching events have repeated at increasingly severe levels and shorter time intervals, such as in the Florida Keys, where massive coral bleaching events have been observed in 1983, 1987, 1990, 1997 and 1998. Here, each of these bleaching events reached a new threshold in the geographical extent, duration, or overall impact to the health of the living coral reefs. In many cases, associated with bleaching events other related impacts such as accompanying coral diseases and fish die-offs have been observed. Such events undermine the values underpinning the establishment of MPAs, threatening their very reason for existing. It is thus critical to be able to quantify the impacts of bleaching on MPAs, resources and management areas in general, not only in biological or ecological terms, but also in social and economic terms.

Coral reefs have been rated among the most valuable of marine ecosystems globally [Costanza et al., 1997] alongside mangrove forests, for the varied goods and services they supply to adjacent coastal populations, as well as their regional and global contributions. Thus while justifications for mitigating coral bleaching and climate change impacts to coral reefs can range from biodiversity conservation to economic, it is instructive to ground management responses (and their costs) in economic terms. Depending on the uses to which they are put, coral reefs have been valued at \$ 20,000 – 270,000 km^{-1} yr^{-1} [Southeast Asia, Burke et al., 2002] and \$ 100,000 – 600,000 km^{-1} yr^{-1} [Caribbean, Burke and Maidens, 2004]. The lower values relate to reefs used only for subsistence fishing and extraction, the upper values to those also supporting tourism industries and shoreline protection in built-up areas. Losses in value associated with coral bleaching have been documented in a number of case studies (Table 1). These demonstrate that even at small-island (El Nido, Philippines; Zanzibar, Tanzania) and municipality (Mombasa, Kenya) levels, estimated economic losses associated only with tourism exceed tens of millions of dollars

TABLE 1. Socio-economic assessments of impacts of coral bleaching on the value of coral reefs.

Location	Description	Findings	Source
El Nido, Philippines	Questionnaire survey of tourists in 2000, on impact of 1998 coral bleaching	Annual losses of U$ 1.5 M; 20 year losses of U$ 15-27 M[a]	Cesar, 2000
Bolinao, Philippines	Survey of fisheries post bleaching (1999) compared to pre-bleaching (1997)	No discernible impact of bleaching on catch rates	Pet-Soede, 2002
Zanzibar, Tanzania	Survey of tourist divers before (1997) and after (1998) bleaching	While tourists stated a preference for pristine areas, bleaching had no impact on measured activities and revenues	Andersson, 2003
Sri Lanka, Maldives	Survey of tourist arrivals in 1999 and analysis of secondary sources	Potential economic costs estimated at $ 19 M (Maldives) and $ 2.2 M (Sri Lanka) from tourist preferences	Westmacott et al., 2000b
Mombasa (Kenya), Zanzibar (Tanzania)	Questionnaire survey of tourists in 2000, on impact of 1998 coral bleaching, compared to baseline surveys in 1996	Potential economic costs estimated at $ 1.8-2.8 M (Zanzibar) and $ 10.1-15.1 M (Mombasa) from tourist preferences	Westmacott et al., 2000b

[a] M – million.

over the long term [Cesar, 2000, Westmacott et al., 2000b, Pet-Soede, 2002, Andersson, 2003]. By including other values of coral reefs, such as for fisheries and coastline protection, these estimates would increase considerably. At national and regional levels, the annual income value of coral reefs exceeds billions of dollars, estimated at $ 2.4 billion per year in Southeast Asia [Burke et al., 2002] and over $ 3.1-4.6 billion per year in the Caribbean [Burke and Maidens, 2004].

The economic incentive to develop management interventions to limit the damage caused by coral bleaching is therefore enormous, at all levels from local to national and global economies. The question being posed, however, is what can management do to reduce the impacts of coral bleaching? The problem is particularly complex as the triggers for coral bleaching – rising concentrations of atmospheric greenhouse gasses such as carbon dioxide, warming sea surface temperatures and increased UV radiation to shallow corals – are driven by global processes beyond the reach of managers, while the impacts are felt from the smallest local scales of individual reef diving and fishing sites to regional coral reef biomes.

A number of workshops and publications have addressed the question of local management responses to climate change [Westmacott et al., 2000a, Salm and Coles, 2001, Schuttenberg, 2001]. Larger scale actions to mitigate climate change impacts to coral reefs rise above the classical realm of management to policy at national and international levels, though can be influenced by management actions and outputs. This paper will focus on the local management actions that directly address coral bleaching, as well as discuss how effective local action and institutions can influence larger scale policy processes. Because of the growing importance of Marine Protected Areas (MPAs) as a primary management tool for achieving coral reef conservation, the paper focuses on implementation of actions within MPAs. However what happens to reefs outside of MPAs, in less intensive management regimes or under management regimes that are not area-specific, is just as important, and many of the concepts and interventions expressed here can apply equally to coral reefs outside of MPAs.

To be effective, management actions need to be highly specific and targeted, cost-effective and logistically and financially achievable. While management actions and planning processes for marine habitats in general and coral reefs in particular have been well documented over time [Kelleher and Kenchington, 1991, Salm et al., 2000], coral bleaching and climate change pose an unprecedented challenge. Thus a firm understanding of the dynamics and biological features of coral bleaching is essential. The next section provides a simplified model of coral bleaching that can aid the development and targeting of management actions, and that can incorporate improvements in the scientific understanding of coral bleaching through complementary changes in management processes.

Framework for Management Responses

Coral bleaching is a result of physical and biological processes that occur from the microscopic scale of cellular physiology to the large scale of oceanic and global processes. A simplified model of coral bleaching converts the physical-biological processes of bleaching (Figure 1a) to a management framework that specifies loci and scales for possible interventions to reduce the impacts and threat of bleaching (Figure 1b). In a management context, coral bleaching can be viewed in four conceptual boxes:

1) The physical stressors or triggers, primarily heat and light, influenced by global levels of climate change drivers, primarily anthropogenic emissions of the greenhouse gas carbon dioxide.
2) Bleaching of live corals, which is the whitening of corals signalling a mid to late stage of physiological stress from the triggers above.
3) Mortality of corals, which may occur following bleaching of corals, if cellular damage from heat and light stress exceeds the capacity of the bleaching response to minimize damage. Under heat and light stress conditions, it is possible that some corals may not bleach, but nevertheless suffer mortality from the stress.
4) Recovery, which may be by regrowth of bleached or unbleached corals that did not suffer 100% mortality and recruitment by coral larvae from the plankton. Multiple levels of recovery processes in other taxa must also occur for a site to recover full ecological functionality.

Processes affecting the transitions between these stages determine the nature of management interventions that can be undertaken in response to the bleaching event. At the level of the heat and light triggers of bleaching, which occur in response to large scale regional and global processes, there is little scope for management interventions that operate on smaller, local scales. However as the system progresses through to the bleaching, mortality and recovery stages, the spatial and temporal scale of events approaches that of management authorities, and an increasing variety of mitigating actions become possible. In general, actions available for a management response can be classified into four general areas (Figure 2):

1) *reducing threats and impacts.*

The primary triggers of coral bleaching – increased heat and light stress – are beyond the scope of management. However, it may become possible at the level of protected areas and of small scale sites within them to influence the bleaching process. The primary variables of heat

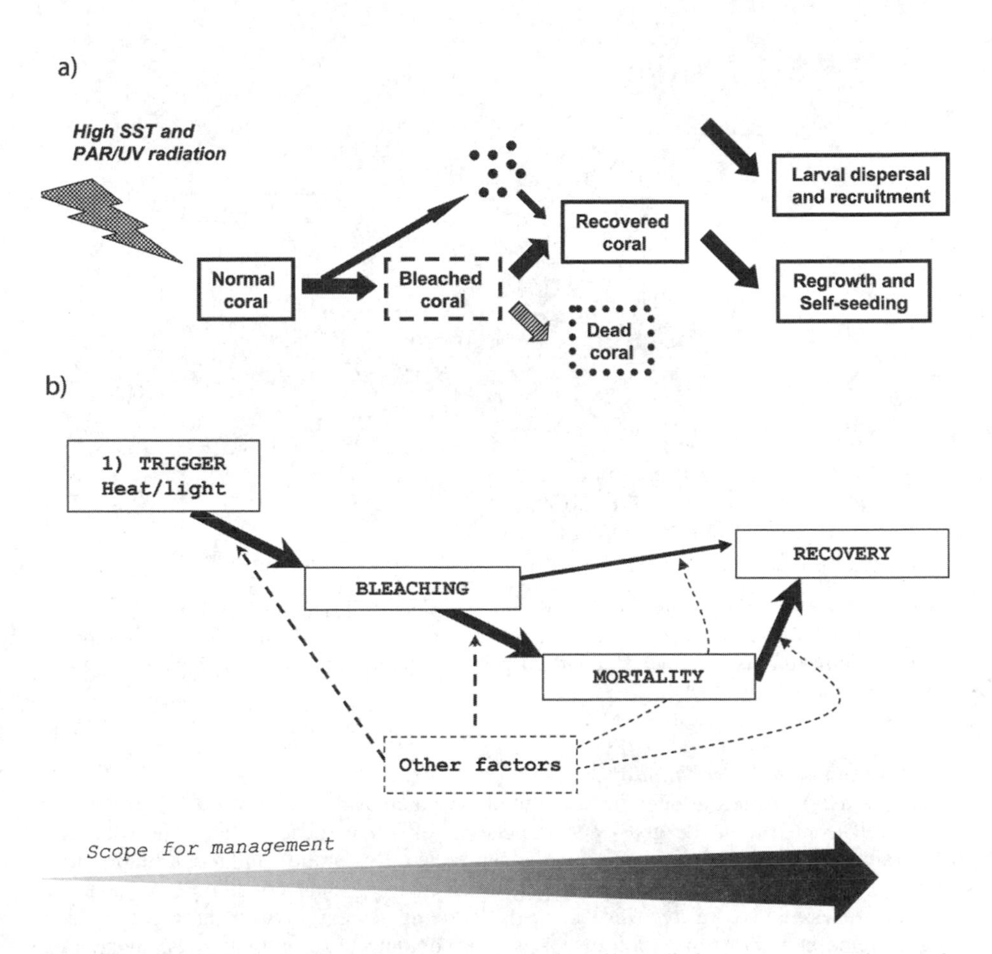

Figure 1. a) Bleaching of corals in response to thermal (and/or radiation) stress occurs by expulsion of zooxanthellae. Bleached corals may recover or die depending on the nature and severity of the stress and other mitigating factors. Recovery of individual colonies from bleaching occurs by repopulation of remnant zooxanthellae within the tissue, and/or by colonization of coral tissue by free-living zooxanthellae. Following mortality of corals, recovery of the coral population or community occurs by growth and self-seeding by reproduction of surviving colonies and recruitment of larvae dispersed from distant sources [Figure from Obura, 2005]. b) A management model for coral bleaching, identifying the triggers (heat/light), the response (bleaching) and outcomes (mortality and recovery). Multiple factors may influence all stages of the process. The scope for management interventions to influence the triggers, responses and outcomes increases from left to right, as the spatial and temporal scales of factors affecting the bleaching process (a) approach those of local management regimes.

and light are in general not accessible to management actions, however management already addresses other threats such as overfishing, water quality control, physical impacts to corals and other uses that may exacerbate the vulnerability of corals to excess heat and light. At this level, management may affect bleaching, mortality and recovery processes; for example, with an early warning system managers can alert stakeholders of a potential bleaching event and solicit their assistance to alleviate direct user pressures on coral reefs.

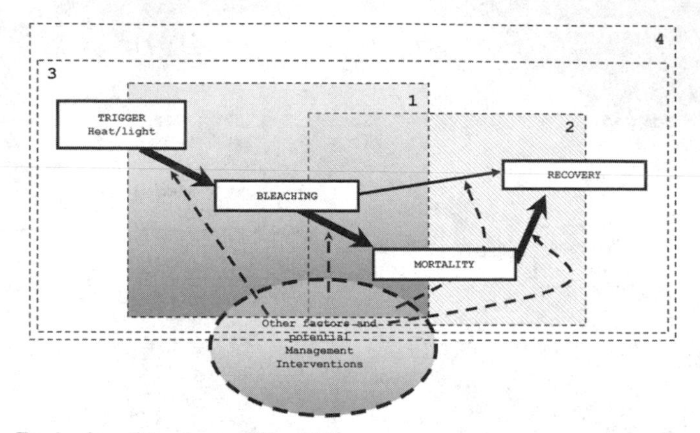

1 - Reducing threats and impacts
2 - Restoration and rehabilitation
3 - Monitoring and evaluation
4 - Communications, awareness, stakeholder involvement

Figure 2. Potential loci for management responses to coral bleaching. Numbers relate to the boxes in which they are nested, and are explained in the text: 1) reducing threats and impacts, 2) restoration and rehabilitation, 3) monitoring and evaluation and 4) communications, awareness, stakeholder involvement.

2) *restoration and rehabilitation.*
Efforts at rehabilitation of degraded coral reefs are increasingly common, though controversy still exists about the goals and targets for effective rehabilitation, and logistical constraints of techniques, cost, effort and area affected. Restoration and rehabilitation may relate to two areas of activity: restoring general environmental conditions to maximize natural processes of coral recovery (e.g., by reducing sewage inputs that degrade water quality), and b) direct restoration of corals to repopulate local communities, such as by transplantation from wild stocks, managed aquaria or other stocks, or enhancing larval settlement. However, it is as yet not feasible to deal with any scale above 100s of meters of impacted coral reef.
3) *monitoring and evaluation.*
Effective management responses rely on the availability of adequate information and its evaluation. The dedication of monitoring programmes to cover all aspects of the bleaching phenomenon will become essential components of future management responses to bleaching events. The broad extent of the bleaching phenomenon will require collaboration across many disciplines, including climate science, oceanography, marine biology and social sciences, and encompassing all aspects of management responses under the preceding items 1) and 2).
4) *communications – awareness and policy.*
Because bleaching threatens the values of coral reefs and protected areas that human populations rely on, it is a direct threat to the livelihoods and interests of all stakeholder groups. A comprehensive communications programme that alerts stakeholders and the public to the findings of the above responses is an essential tool in minimizing the risks and vulnerabilities of stakeholders exposed to bleaching impacts. Effective communication is even more necessary when possible management responses entail restricting or

curtailing pre-existing uses and access, either on temporary or permanent bases, to preserve ecological integrity of the affected ecosystems. Additionally, faced with unprecedented changes imposed by climate change, managers will have to experiment with interventions, and public goodwill is an essential component to facilitate management actions when all likely outcomes cannot be predicted.

Successful predictions or anticipation of coral bleaching events from observations made through an early warning system tends to build confidence in management on the part of stakeholder groups. This has several positive results, including the heightening of the stakeholders' awareness of the impact of coral bleaching, convincing stakeholders they can be a part of the solution, and communicating to stakeholders that this is a global phenomenon that requires a global solution that has to start locally. Enlisting stakeholders' assistance in taking local management actions will serve to make them aware that there are short-term local and regional management actions that can be implemented while long-term management strategies are planned and implemented. It is important that people not rationalize giving up on conserving coral reefs because coral bleaching is a global problem, but rather that they know they can make a difference in managing coral reefs at the local scale while the large scale and global components of coral bleaching are addressed through other means.

There have been a number of initiatives searching to develop recommendations for management responses. Initially, these have been based on the 'conventional wisdom' of minimizing other threats that managers already have experience with (stage 1 Figure 2, Westmacott et al., 2000a), and have proceeded rapidly to more formalized explorations of scientific principles around which to base recommendations for management responses and the planning of protected area systems [Done, 2001; Salm and Coles, 2001; West and Salm, 2003; Salm et al., this volume; Obura, 2005]. Already, packages condensing these principles into action strategies for implementation by coral reef managers have been developed [Hansen, 2003; TNC, 2004; Marshall and Schuttenberg, *in review*]. However capacity to deal with the triggers, threats and impacts is still low, as understanding of the bleaching process (Figure 1) is at early stages for advocating specific management actions, and the scale on which the triggers and threats occur are generally larger than the local scale of management capacity. These issues are dealt with in a later chapter [Salm et al., this volume] and explored in Obura [2005].

Restoration and rehabilitation attempts (stage 2, Figure 2) in response to mass coral bleaching and mortality events have been implemented in a number of locations. Direct actions have generally focused on promoting the recovery of impacted sites through transplantation of coral fragments from other locations, and in some cases to relieving pressure from threats to facilitate faster recovery. In the former case, efforts range from low-cost low-technology transplantation of individual coral fragments [e.g., Bowden Kirby, 1997; Church and Obura, 2004b; Lindahl, 1998], intensive farming of coral fragments [Heeger and Sotto, 2000], translocation of entire reef communities [Muñoz-Chagin, 1997] to high-cost technology-intensive efforts such as using electrical currents to increase deposition rates of calcium carbonate and growth of manually transplanted coral fragments [van Treek and Schumacher, 2002]. In all cases the spatial scale of efforts are negligible, less than or approximately 100 m^2 reef areas restored, compared to the surrounding impacted reef areas, and to the regional scale of impacted reefs. New experiments on facilitating settlement of concentrated slicks of coral larvae are underway, but currently too preliminary for management applications. In all cases, these restoration attempts will be as vulnerable as natural communities to future bleaching events, unless other actions to improve conditions for recovery and/or maximize 'resistance and resilience to bleaching' [West and Salm, 2003; Salm et al., this volume; Obura, 2005] are undertaken. In the latter case, relieving pressures from other threats, such as pollution, boat groundings, dive and snorkel damage and overfishing, is an option well within the scope of management, as these have been the focus of management recommendations for many years [e.g., Westmacott et al., 2000a; Salm and Coles, 2001]. Temporarily closing coral reefs that are stressed from coral diseases or coral bleaching is a further management option. Mass coral bleaching and mortality have strengthened commitments to improve the effectiveness of management in all its ramifications [Hockings et al., 2002; Pomeroy et al., 2004], and MPAs and management regimes are increasingly incorporating the threat of coral bleaching into their rationales for reducing other threats [e.g., WWF, 2003; Chadwick and Green, 2002].

The most common and immediately feasible response of management organizations to major coral bleaching and mortality events has been to start or improve ecological monitoring programmes (stage 3, Figure 2) to provide real-time and improved information on coral bleaching events and their aftermath (Table 2). Examples include the Bali Barat National Park, Indonesia, where in response to bleaching from the El Niño of 1997-98, the World Wide Fund for Nature (WWF) collaborated with the Great Barrier Reef Marine Park Authority (GBRMPA) and the International Centre for Living Aquatic Resource Management (ICLARM) to develop a Coral Bleaching Monitoring Program for Bali Barat and for broader application globally [WWF, 2003; Oliver et al., 2004]. In the Western Indian Ocean, the ocean-wide mortality of corals in 1998 [Wilkinson et al., 1999] led to the initiation of the regional CORDIO programme with funding from the Swedish Development Cooperation Agency, Sida [Coral Reef Degradation in the Indian Ocean; Linden and Sporrong, 1999; Souter et al., 2000; Linden et al., 2002], as well as refocusing of a number of site-specific and national monitoring programmes, such as the Kiunga Marine National Reserve, Kenya [Church and Obura, *in review*], Mafia Island Marine Park and other parts of Tanzania [Muhando, 1999; Mohammed et al., 2002] and the Mozambique National Coral Reef Monitoring Programme [Schleyer et al., 1999; Motta et al., 2000, 2002]. Pre-existing long term monitoring of MPAs and unprotected sites in Kenya since the 1980s enabled assessment of bleaching–related damage to coral reefs with and without protection [McClanahan et al., 2001].

In general, the above programmes had two primary objectives: a) to improve the provision of ecological data to better understand trends and prospects for the study areas, and b) to better communicate the threat of coral bleaching to stakeholders, the public and policy makers [stage 4, Figure 2]. In the case of Bali Barat and the Kiunga Marine National Reserve, an important element of the monitoring programmes was the involvement of local stakeholders in monitoring the onset and progress of bleaching so that they had first-hand knowledge of the phenomenon. They then acted as messengers to their peers facilitating broader acceptance of management interventions specific to bleaching and for other threats, and promoting participation in co-management structures. A strong communication programme based on good quality monitoring and observations of degradation can also lead to larger scale interventions in the policy arena. In case of the Florida Keys National Marine Sanctuary in the US, long term observations by the management agency benefited from detailed understanding of individual events provided by a diverse range of research scientists and organizations active in the Florida Keys. Together, these enabled the construction of a consistent picture of long term reef decline associated with a common set of oceanographic conditions related to major degradation events. The quality of the information available has played an important role in raising awareness of the threat of coral bleaching in government and policy circles within the US, and influenced national policy on coral reefs.

To date, management responses to major bleaching and mortality events have been widespread, but occurred on a relatively *ad hoc* basis due to the enormity and surprise of the events since the 1980s. They have focused on monitoring and research, and communications and education programmes (stages 3 and 4), while attempts to deal with the causes have necessarily focused on precautionary responses by altering manageable threats (stage 1)

TABLE 2. Examples of management responses to major coral bleaching events.

Site/Bleaching event	Event	Action	By whom	Source
Florida Keys National Marine Sanctuary (USA)	Long term decline 1980s to present. Multiple bleaching	Long term observations and monitoring of stressful reef con	FKNMS staff; FWC CRMP; NURC-UNCW; FIO; NOAA; EPA; MML; stakeholder groups, including divers; volunteers	Causey, 2001
Kiunga Marine National Reserve (Kenya)	Bleaching event: 1998	Dedicated monitoring programme building on rapid assessments at time of bleaching in 1998	WWF, KWS and CORDIO, with local fisher communities	Church and Obura, 2005
Mamanuca Islands (Fiji)	Bleaching event: 2000	Greater education and networking, expansion of voluntary reef surveys using ReefCheck methods.	Variety of stakeholders (communities, dive operators, etc.), science surveys led by Coral Cay Conservation	Harding et al., 2005
Bolinao, El Nido (Philippines), Mombasa (Kenya), Zanzibar (Tanzania), Siri Lanka, Maldives	Bleaching event: 1998	Studies on economic impacts of bleaching on fisheries and tourism industries	Consultant researchers often working through local/ national institutions, funded by donors.	Cesar, 2000, Westmacott et al., 2000b, Pet-Soede, 2002, Andersson, 2003
Bali Barat National Park (Indonesia)	Bleaching event: 1998	Development of a Bleaching monitoring protocol	Development by WWF, GBRMPA and ICLARM* for implementation by local monitoring teams of scientists and stakeholders	Oliver et al., 2004
Western Indian Ocean	Bleaching event: 1998	Improved support and networking of local and national coral reef monitoring teams, and of reporting and dissemination	Major regional programmes include CORDIO, WCS and COI based on foreign funding and implemented by local and national organizations	Linden and Sporrong, 1999, Souter et al., 2000, Linden et al., 2002, Goreau et al., 2000

Abbreviations used in tables: COI – Indian Ocean Commission; CORDIO – Coral Reef Degradation in the Indian Ocean; EPA – Environmental Protection Agency; FAO – Food and Agriculture Organization; FIO – Florida Institute of Oceanography; FWC CRMP – Fish and Wildlife Conservation Coral Reef Monitoring Program; GBRMPA – Great Barrier Reef Marine Park Authority; GCRMN – Global Coral Reef Monitoring Network; ICLARM – International Centre for Living Aquatic Resources Management (now WorldFish Centre); KMNR – Kiunga Marine National Reserve; KWS – Kenya Wildlife Service; MML – Mote Marine Laboratory; NOAA – National Oceanographic and Atmospheric Administration; NURC-UNCW – National Undersea Research Center, University of North Carolina at Wilmington; Sida/SAREC – Swedish International Development Cooperation Agency; UNEP – United Nations Environment Programme; WCS – Wildlife Conservation Society; WWF – World Wide Fund for Nature; WWF – Worldwide Fund for Nature; FKNMS – Florida Keys National Marine Sanctuary.

and trials at restoring damaged reefs but on a small spatial scale (stage 2). Rapid advances in the understanding of mass bleaching events and the responses of healthy versus degraded coral communities are already leading to more well defined recommendations for relief of stressful conditions during initial stages of bleaching and promoting processes that facilitate recovery (*see late chapters, this volume*). The next section of this paper details two of the case studies mentioned above, the Florida Keys National Marine Sanctuary (USA) and the Kiunga National Marine Reserve (Kenya), to illustrate in more detail the linkages between and importance of monitoring and communications programmes.

Case Studies

The Florida Keys National Marine Sanctuary, USA

The Florida Keys National Marine Sanctuary (FKNMS) is one of thirteen National Marine Sanctuaries that are administered by the National Oceanic and Atmospheric Administration (NOAA) in the United States Department of Commerce. It encompasses 9800 square kilometers (2900 square nautical miles) of coastal waters off the Florida Keys. The Sanctuary (Figure 3) extends approximately 404 km (220 miles) southwest from the southern tip of the Florida peninsula and includes nationally significant marine environments,

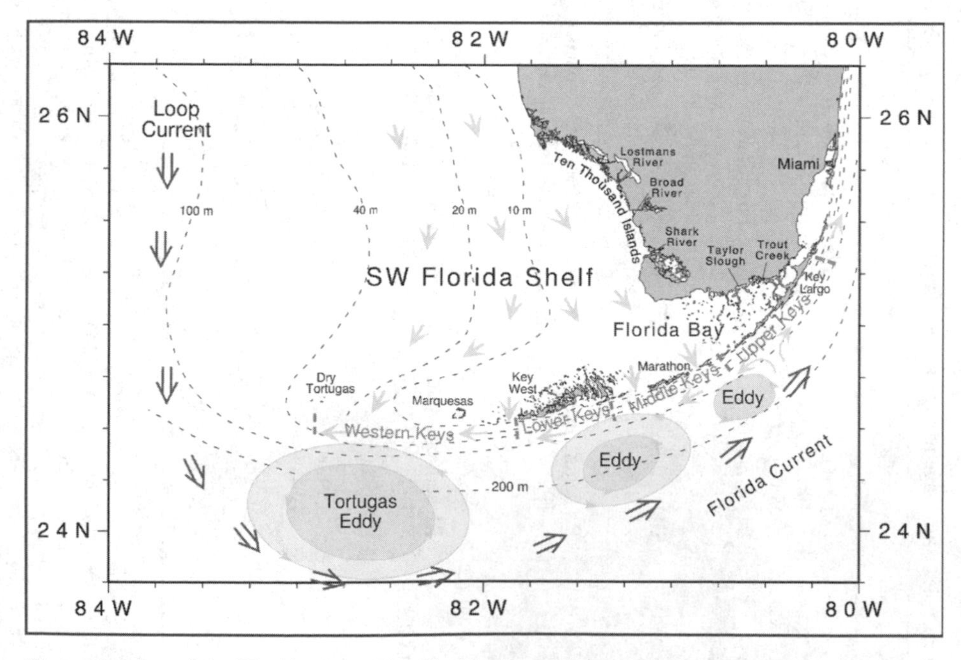

Figure 3. Map of the Florida Keys. The Florida Keys National Marine Sanctuary, encloses the entire chain of islands from Key Largo to the Dry Tortugas, covering 9,844 km^2. The red hexagon shows the location of Looe Key, in the flow of warm water from the Gulf of Mexico through the lower keys onto the Florida reef tract. Interactions of these waters with the Florida current are shown, which result in this warm water being entrained westwards within the reef tract.

including seagrass meadows, mangrove islands and extensive living coral reefs. These marine environments support rich biological communities possessing extensive conservation, recreational, commercial, ecological, historical, research, educational, and aesthetic values which give this area special national significance. The clear tropical waters, bountiful resources, and appealing natural environment are among the many fine qualities that have attracted visitors to the Keys for decades. Over three million visitors spend 13.3 million visitor-days and spend approximately US$ 1.2 billion each year in the Florida Keys. The majority of the visitors either snorkel or SCUBA dive on the coral reefs of the Keys.

A number of factors have contributed to the FKNMS providing a signature case study of the benefits of long term monitoring and observation, and of the critical role this and communications strategies play in managing bleaching impacts. First is the wide coverage of the FKNMS across the entire Florida Keys, forcing a system-wide approach to observation and management. In addition, the long term stability of its management regime, key personnel, and in particular the manager or Superintendent, have allowed the development of a long timeline of experience, knowledge and institutional memory. Third, the concentration of many research scientists and institutions in South Florida and their focus on local reefs in the Florida Keys has enabled deeper interpretations of events revealed by long term monitoring and observation.

Table 3 summarizes a chronology of significant impacts to coral reefs of the FKNMS. The table juxtaposes observations of major events indicative of ecological stress to the coral reefs with responses by management, scientists and stakeholders. The string of events reported in the first column show the importance of a long institutional memory – even without detailed monitoring programmes, clear observations and good record-keeping allowed connections to be made retrospectively of past events in the light of new knowledge and information. The observations show a clear trend of intensifying stress to various components of the coral reef system – primarily hard corals, but including fish and other anthozoans – in terms of increasing frequency (with back-to-back bleaching of corals in 1997-98], severity (increasing severity and spatial scale of bleaching, increasing mortality following bleaching) and geographical scope. All of the impacts are associated with doldrum summer conditions in July – September, when normal hot, calm conditions are intensified by increasingly calm wind and sea conditions.

The observation record also points to a strong influence of local geomorphoglocial-oceanographic conditions. Among the first reefs to be impacted each year is Looe Key reef in the mid-Keys area, which is located where warm waters from Florida Bay/Gulf of Mexico flow southwards towards the reefs on the outer reef tract. Here local oceanographic conditions force this water westwards within the reef system before it mixes with the east-flowing cooler Florida Current offshore of the Keys and thence into the Gulf Stream. Observations have shown that corals in the Looe Key area are among the most susceptible to doldrum conditions that intensify their already-heated state. Depth and reef structure were also significant factors, with shallow offshore reefs most susceptible to bleaching, followed by deeper reefs, then lastly inshore reefs.

Warm doldrum conditions also impact other components of the reef ecosystem, documented in the Florida Keys for microbial mats on extensive sandy areas in 1987. This also suggests that increases in coral and fish diseases during doldrum conditions may be due to enhanced growth of disease microbes. Also likely to enhance microbial activity is nutrification of shallow reef waters and sediments by terrestrial outflows into Florida Bay, and storm- and waste-water flows from the Florida Keys – both of which are increasing with increasing population pressure in South Florida. Water from Florida Bay flows westerly over the shallow banks where storm- and waste-water flows from the islands are concentrated, before turning south and east after mixing with the Florida Current. This circulation spreads nutrients over the outer reef tract and may also enhance stress from doldrum conditions.

TABLE 3. Long term observations of coral reef decline in the Florida Keys National Marine Sanctuary, and responses by local managers, scientists and the public [adapted from Causey, 2001].

Timeline and monitoring activity	Stakeholders, networking and implications
June-July 1979	
Sponge die-off, Lower Florida Keys. Possibly in relation to water outflow from Florida Bay.	No water quality monitoring during event, relation to Florida Bay inferred from later physical oceanographic studies.
July 1980	
Fish die-off throughout Florida Keys, during doldrum conditions over 6 weeks. Minor coral bleaching also observed in offshore corals.	Scientists called to sample fish, but were too late.
July 1983	
Sea urchin (*Diadema*) die-off in the Florida Keys but no pathogen isolated; Caribbean-wide phenomenon. Coral bleaching (but low mortality) of offshore reef shallow communities during doldrum conditions	Scientists on hand to study sea urchin die-off and coral bleaching. Studies linked bleaching to flow of warmer Florida Bay waters and potentially to stormwater and wastewater from the Keys. Coral later reported from other parts of the Caribbean and other parts of the world.
May 1986 and summers to 1990s	
Black band disease on corals, with mortality of 200 year-old colonies. Most intense during warm summer months, spread throughout offshore corals. Coincided geographically with previously bleached coral reefs	Dedicated monitoring programme established to track disease outbreak
June-August 1987	
Coral bleaching during doldrum conditions, including offshore and deep reefs and throughout Florida Keys and later the Caribbean. Low mortality observed. Increased microbial growth in mats over extensivesandy areas	After 1st week of doldrum conditions MPA staff started watching for coral bleaching,and reports came in from divers. Small scale monitoring programmes documented loss in coral cover, but no area-wide monitoring programme in place.
August 1989	
Bleaching of single genus, *Agaricia* in limited area. No mortality.	Reports from wider network of similar bleaching in Puerto Rico and the Bahamas.
July-September 1990	
Bleaching in corals and the zooanthid *Palythoa caribaeorum*, and for the first time in inshore coral reefs. Lost 65% of fire coral (*Millepora*) on Looe Key reef crest.	MPA received calls from concerned public. MPA staff established monitoring transects and recorded for first time substantial loss of coral cover.
1993-1994	
Coral and fish diseases and die-offs widespread, affecting many genera and species and with high mortality.	Increased interest and monitoring of diseases and discovery of new undescribed syndromes.

TABLE 3. (*Continued*)

Timeline and monitoring activity	Stakeholders, networking and implications
July 1997-January 1998	
Mass coral bleaching during doldrum conditions with widespread mortality. Bleaching lasted 3 months longer than previously documented. Later reports from the Caribbean and globally	Reports from long-term residents that this was the first episode of this kind they had ever seen or heard of.
1998	
Widespread coral bleaching and mortality. Warm conditions persisted through the winter of 1997/98 and to summer of 1998 and repeat doldrums conditions	First back-to-back bleaching of corals reported. Five-year monitoring programme was in place, but multiple threats (bleaching, hurricane, water quality degradation) made it hard to identify clear causality.
1999	
A long-term decline of 30% of living corals documented at some reefs by scientists	Increased awareness within management and science circles of the alarming levels of degradation, the common thread of doldrum conditions and linkages to climate change, and interactions with local threats and activities.
1999-2004	
No massive coral bleaching events since 1998. The long-term coral monitoring programme has recorded no substantial or increase in living coral cover since 1999	Increased collaboration among coral monitoring programmes (FWC-CRMP/ NURC-UNCW/FIO/EPA/ Mote).
The occurrence of large scale coral diseases has slowed down since 2002, with primarily outbreaks of diseases on *Acropora* species since 2002	The lull in bleaching events has allowed the development of systematic management responses, including an early warning system (with GBRMPA and MML) and a comprehensive monitoring system incorporating rapid response assessments and long term monitoring

Abbreviations used in tables: COI – Indian Ocean Commission; CORDIO – Coral Reef Degradation in the Indian Ocean; EPA – Environmental Protection Agency; FAO – Food and Agriculture Organization; FIO – Florida Institute of Oceanography; FWC CRMP – Fish and Wildlife Conservation Coral Reef Monitoring Program; GBRMPA – Great Barrier Reef Marine Park Authority; GCRMN – Global Coral Reef Monitoring Network; ICLARM – International Centre for Living Aquatic Resources Management (now WorldFish Centre); KMNR – Kiunga Marine National Reserve; KWS – Kenya Wildlife Service; MML – Mote Marine Laboratory; NOAA – National Oceanographic and Atmospheric Administration; NURC-UNCW – National Undersea Research Center, University of North Carolina at Wilmington; Sida/SAREC – Swedish International Development Cooperation Agency; UNEP – United Nations Environment Programme; WCS – Wildlife Conservation Society; WWF – World Wide Fund for Nature; WWF – Worldwide Fund for Nature; FKNMS – Florida Keys National Marine Sanctuary.

The responses to the string of stress events in the FKNMS illustrate various aspects of the scope and variety needed in developing communications to deal with major bleaching events. Initial responses were typical of the surprise occurrence of unexpected conditions – low levels of ongoing monitoring, few relevant research programmes (or at least not adequately linked into the management system) and after-the-fact responses that may be too late to be of use. However, as the realization grew within the management system first of increasing likelihood of predictable adverse weather conditions, and secondly of the correlation between severe degradation events and these weather conditions, responsiveness in all sectors increased. By 1983 scientists were alert to the incidence of the *Diadema* die-off and coral bleaching. By the late 1980s dedicated monitoring programmes were in place at some locations to track overall reef health and disease outbreaks. Also in the late 1980s MPA staff had established an informal response protocol to be on the lookout for coral bleaching during doldrum conditions, and the widening scope of coral bleaching regionally and globally resulted in improved networking to report similar bleaching of coral reefs in geographically distinct places under other management authorities.

The intensification of coral bleaching in 1987, 1990 and 1997 resulted in the first widespread public concern among residents in the Florida Keys. Increasing incidences of coral diseases between these years, and back-to-back coral bleaching in 1997-98 convinced the scientific community and the FKNMS that the problem was significant and worsening. The realization by 2000 that a large proportion (30%) of the Florida Keys reef tract had been lost within a span of 20 years focused attention on the need to reverse the trend, and if possible, to use coral bleaching to learn how to improve reef management. The greatest level of decline in the coral cover was documented between 1996 and 1999 (FWC-CRMP 2004]. A decline of approximately 30% of the living coral cover occurred as a result of the unprecedented back-to-back 1997-1998 coral bleaching events in the Florida Keys (Causey, personal observations), consistent with that reported for other coral reefs in the Wider Caribbean.

There has not been a massive coral bleaching event in the Florida Keys since 1998 and the Sanctuary's long-term coral monitoring program has documented little variation in the amount of living coral cover over the past five years. Unfortunately, unlike some areas in the Pacific that have rebounded from the 1997-98 bleaching events, the amount of living coral cover in the Florida Keys has not increased. Sanctuary managers attribute this lack of recovery to a number of factors including water quality problems, algal overgrowth, over-fishing and low recruitment of new corals. Importantly, the time between the 1997-98 coral bleaching events and the present has given Sanctuary managers the necessary time to work with the Great Barrier Reef Marine Park Authority and Mote Marine Laboratory to prepare an early warning program, as well as a rapid assessment and monitoring program for future coral bleaching events.

The recognition of the vulnerability of one of the foremost coral areas in the USA has played a powerful role in increasing the support for coral reef conservation, management and science at local, state and national levels. Along with coral reef losses in other U.S. waters and globally, lessons from the FKNMS helped drive a political process that has supported a series of consultations from local to national levels, workshops, policy statements and decisions on threats to coral reefs. Importantly, these have included resolutions about climate change passed by the U.S. Coral Reef Task Force that have: 1) identified climate change as one of six key threats to U.S. Coral Reefs (Resolution 1, 8th USCRTF Meeting, Puerto Rico); 2) called for a public-private partnership to strengthen understanding and management of reefs for climate change (Resolution 5, 8th USCRTF Meeting, Puerto Rico); and 3) catalyzed the development of management strategies to respond to the mass coral bleaching (Resolution 6, 10th USCRTF Meeting, Guam/CNMI). These resolutions initiated an interagency managers meeting hosted by the U.S. Environmental Protection

Agency (EPA), the National Oceanic and Atmospheric Administration (NOAA), and the Department of the Interior (DOI) in 2003 that focused exclusively on coral reefs, climate change and bleaching. Within the State of Florida, lessons from the FKNMS played a key role creating support for a resilience and climate change initiative, where complementary interest from management agencies (in particular the FKNMS), prominent NGOs and scientists gave credibility to trialing the new concept of resilience in management and giving it strong policy support. Among the reasons the FKNMS was instrumental in this regard is its crossing of local to national boundaries: the FKNMS is a federal government institution that acts at the local level and shows strong leadership among Florida Keys communities, and at the State level in Florida. This has enabled local lessons to rise to national prominence in influencing national policy. This leadership role played by the FKNMS in local, state and national coral reef affairs was recognized by the USCRTF in its December 2004 meeting by awarding a "Coral Reef Champion" award to the Superintendent of the FKNMS for the leadership role played by he and the FKNMS team, locally and nationally, in promoting the protection of coral reefs.

Kiunga Marine National Reserve, Kenya

The Kiunga Marine National Reserve (KMNR) (approx. 1°45'S 41°20'E) lies in the Bajun Archipelago at the northern extreme of the Kenya coast, at the border with Somalia (Figure 4), in the Indian Ocean. The reserve extends over some 60 km in length by 3-5 km in width and is characterized by a linear series of barrier islands sheltering extensive mangrove stands in protected lagoons, large areas of seagrass with intermittent reefs on rocky substrates on the shallow outer fringing reefs, and a barrier rock reef some 3-4 km offshore. Coral reefs of the KMNR, and the Bajun Archipelago in general, are marginal and broken up, unlike the continuous fringing reef of southern Kenya or the extensive reef systems of Tanzania and northern Mozambique. Before the El Niño of 1997-98, coral cover of healthy reefs averaged from 10% on the outer barrier reef to 20-25% on shallow inner and fringing reefs, and the reefs were dominated by algae (turf, fleshy and *Halimeda*) at levels of 40-80% [Church and Obura, 2005]. Coral bleaching and mortality during the El Niño were severe, at levels of >80% at all depths. From 1998 to 2003 recovery progressed slowly in shallow inner and outer reefs to levels varying from 20-80% of pre-bleaching levels. Fish populations in the KMNR are the most abundant and robust reported for coral reefs in Kenya, supporting an artisanal and small-scale commercial reef fishery. Fish densities are low and depleted close to fishing villages, increasing towards northern and offshore sites were pressure is lower. Local and migrant fisher communities are the primary users of the marine reserve, and the primary stakeholder group with which comanagement is being negotiated [Church and Obura, 2005].

Management actions related to coral bleaching in the KMNR have focused on monitoring, community participation and education (Table 4). In Kenya, the 'Reserve' designation allows for managed (traditional) use of resources, while 'Parks' are strict no-take zones. The Kenya Wildlife Service (KWS) manages the KMNR with the support of the Fisheries Department however difficult access, high management costs, insecurity, unclear and conflicting mandates, and lack of an enabling environment meant that management was not fully implemented. In 1996 a consensus-building approach was adopted with assistance from the World Wide Fund for Nature (WWF) and under this a coral reef monitoring programme was developed in 1998 for application by joint teams of KWS rangers, local staff and stakeholders, primarily artisanal fishers. Using local taxonomic names for fish and invertebrates (keyed to scientific names) enabled easy recruitment of personnel into the monitoring programme, and,

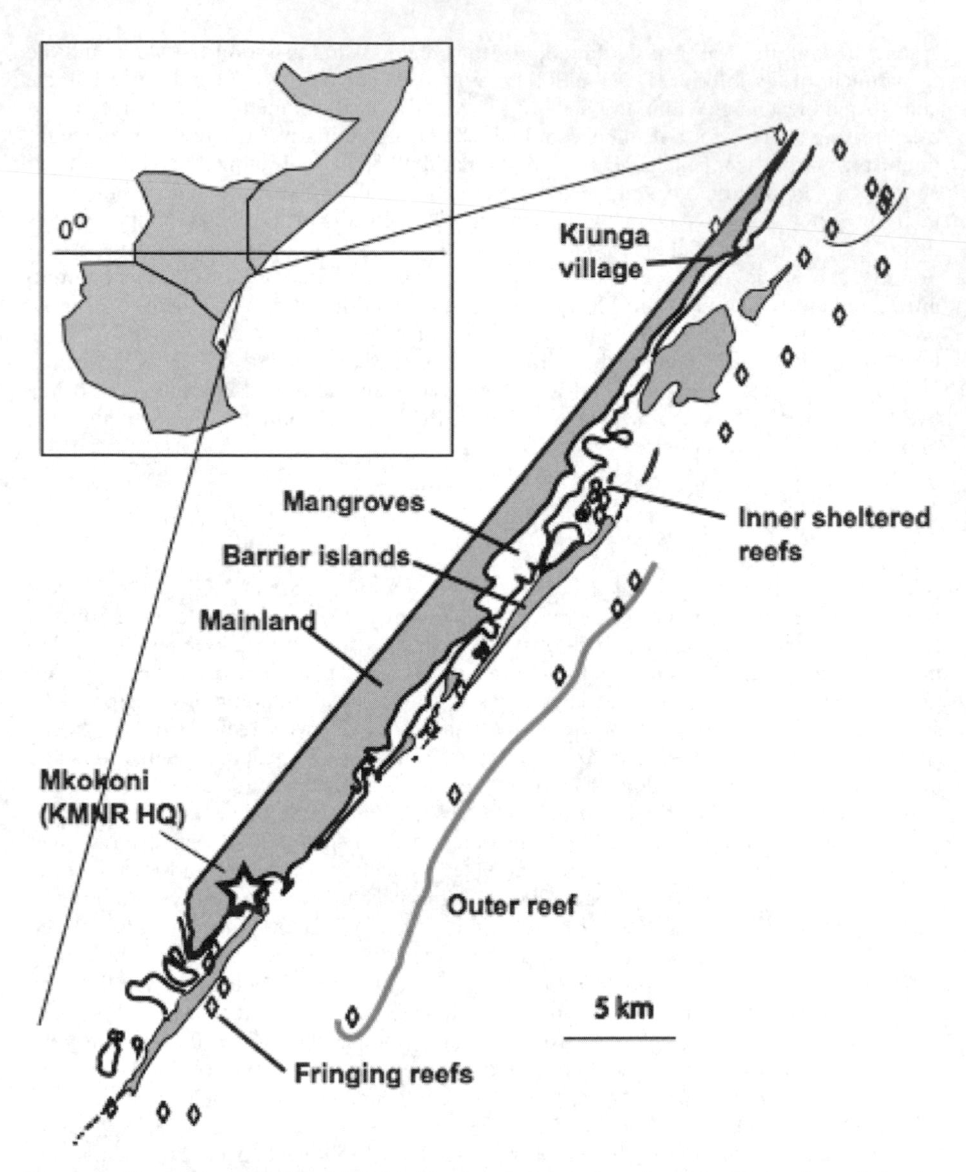

Figure 4. The Kiunga National Marine Reserve is located south of the Kenya-Somali border (2°S), in a transitional zone between the East African coral reef 'ecoregion' to the south and the upwelling Somali Current system to the north. The map shows the reserve headquarters, and spread of monitoring sites on inner, fringing and deep outer reef sites.

as importantly, it facilitated communication back to communities as well as to local government. Primary objectives of the KMNR Project, and of the coral reef monitoring programme, are to provide information on the vulnerability and status of coral reef resources in the reserve area, to make this information useful in management, and to communicate effectively with local stakeholders to build support for management, conservation and protection.

TABLE 4. Management responses in the Kiunga Marine National Reserve project to coral bleaching in 1998, focusing on the monitoring programme and communications results, and restoration trials. Quotes indicate statements made by fishers during monitoring, awareness and feedback sessions.

Timeline and monitoring activity	Community responses and involvement
March-April 1998	
Rapid assessment initiated in the KMNR byWWF and KWS, with funding from FAO/UNEP	Consultations on and selection of monitoring sites
	Quotes from fishers: "Why are the stones going white?" "Is this God's doing?" "Why are large areas of the stones going from white to brown?"
Coral bleaching first noticed around March 20 by monitoring teams	The term 'bleaching' was adopted by many fishers, who were later able to explain this phenomenon to fishers, leaders and children in the region
May-December 1998	
Repeated site visits to sample bleaching and mortality trends, first year of data analysed and information on coral bleaching effects fed back to the community	Fishers involved in the training were able to explain to their leaders and fellow fishers that underwater stones lived, and that external causes led to their bleaching and death. There was some discussion about the activities importance of managing their fishing to reduce direct pressure on the coral reefs
March/April 1999	
2nd year of KMNR coral reef monitoring. A blend of techniques recommended by the GCRMN were adapted to suit the needs of the KMNR	Participants noted the low coral cover and ubiquitous covering of brown turf algae. An inshore reef was closed to gill-netting activities in February 1999, and has since been successfully maintained by the local fishers
February-April 2000	
3rd year of KMNR coral reef monitoring, and SCUBA training provided to 10 local fishers, staff members and KWS	Further discussions were held about the need to manage local activities and the fishers asked for alternative, sustainable fishing methods. Answers were not easily provided given the complexities of marine resource management, access, high management costs and low literacy levels and management experience
March/April 2001	
4th Year of KMNR coral reef monitoring. Coral transplantation introduced in 3 sites in the KMNR	"Corals are like flowers. They flower and then take time to grow." "But you can plant them with cement and then they just grow – but very slowly"
March/April 2002	
5th year of KMNR coral reef monitoring; 2nd year of coral reef transplantation and	Fishers and local MPA staff showed sponsored secondary school children the technique

TABLE 4. (*Continued*)

Timeline and monitoring activity	Community responses and involvement
expansion of the transplantation process	required to develop a coral garden. The children were fascinated and for the first time realized the importance of their coral reefs
March/April 2003	
6th year of KMNR coral reef monitoring; further monitoring in out of reserve areas, coral transplantations expanded to 3 new reef sites	The need for an enabling environment to legally allow for joint management of the coral reef resources grew. Local fishers, their children and other villagers asked to have a legal role in management of the MPA
March/April 2004	
7th year of KMNR coral reef monitoring; on-site capacity expanded through a team leader. A zoning workshop with fishers and local government held to identify reserve regulations and zoning plan	KWS, WWF and fishers conducted the majority of the monitoring program without direct supervision for external experts. Workshop participants made many recommendations on fisheries restrictions, endorsed "growing corals" in the transplantation programme but stopped short of identifying restricted zones to limit fishing activity

Abbreviations used in tables: COI – Indian Ocean Commission; CORDIO – Coral Reef Degradation in the Indian Ocean; EPA – Environmental Protection Agency; FAO – Food and Agriculture Organization; FIO – Florida Institute of Oceanography; FWC CRMP – Fish and Wildlife Conservation Coral Reef Monitoring Program; GBRMPA – Great Barrier Reef Marine Park Authority; GCRMN – Global Coral Reef Monitoring Network; ICLARM – International Centre for Living Aquatic Resources Management (now WorldFish Centre); KMNR – Kiunga Marine National Reserve; KWS – Kenya Wildlife Service; MML – Mote Marine Laboratory; NOAA – National Oceanographic and Atmospheric Administration; NURC-UNCW – National Undersea Research Center, University of North Carolina at Wilmington; Sida/SAREC – Swedish International Development Cooperation Agency; UNEP – United Nations Environment Programme; WCS – Wildlife Conservation Society; WWF – World Wide Fund for Nature; WWF – Worldwide Fund for Nature; FKNMS – Florida Keys National Marine Sanctuary.

Significantly for the impact of the monitoring programme, the first year's baseline assessment coincided with the onset of mass coral bleaching in March 1998, and gave the first report of bleaching in Kenya during the event. The white corals were clearly observed by participants, but their significance was not apparent due to the common belief that corals are 'stones' and therefore dead (the local Swahili name for coral is 'stone' – '*mawe*'). Participation in the monitoring programme offered a direct avenue for dialogue between scientists, managers and stakeholders to explain what was happening and the significance of bleaching. As in many other cases fishers were not easily convinced by scientists and managers preaching on the negative impacts of fishing, however the understanding that their reefs and future fisheries were being threatened by a completely external threat was a powerful eye-opener. The monitoring programme used the opportunity to emphasize the need to manage local threats, primarily the continual increase in destructive fishing practices within the reserve. This resulted in one group of local fishers enforcing the closure of all netting activities on one susceptible inner coral reef in February 1999. Though not legally enforceable, this continues to be respected and managed by the local fishers with a little external support from KWS when needed.

In subsequent years, repeated sampling on the same reefs inculcated in the monitoring team a real appreciation for how slow and variable recovery of coral communities could be; this set the stage for the project to appeal to another traditional activity of the local communities - farming. The central tenet of farming as compared to fishing – nurturing versus gathering – was a value the project wanted to bring into the management system of the MPA. In 2001 the possibility of transplanting corals to reefs showing limited recovery was raised, and the idea eagerly taken up by fishers, project staff and MPA rangers. The monitoring teams were warned that true rehabilitation would not happen only with transplanted corals; that would take restoring fish communities and many years of slow ecological change. The transplantation exercise grew in 2002-2004 to become the most popular aspect of the annual monitoring programme, and it involved more regular visits to check on transplants (every 3 months in some cases). Through this activity the local teams have formed particular associations with the specific sites of transplantation, and even requested other fishers not to use those sites in case transplants were damaged by nets. Together with the project, they have organized visits to the sites taking local government officials, school children and outside visitors. Through recording the slow growth and survival of transplanted fragments, participants have a clearer appreciation of the organic nature of the reef ecosystem, and of how their fish depend on a living ecosystem, and that corals are not just 'rocks' on the bottom. Fishers and marine reserve staff that participated in the coral reef monitoring and transplantation programme have become key resource people in promoting the need for conservation management, and in facilitating the inclusion of their communities in planning activities.

This commitment and enthusiasm has also cemented the need for sound coral reef conservation management in the minds of the 35 WWF sponsored secondary school children from the MPA. Where once coral reefs and their resources were viewed as areas solely for uncontrolled use, the children of KMNR fishers are now dedicated conservationists who want to become marine scientists and marine resource managers, departing from the more common ambitions of becoming a fisher, politician, teacher or trader.

A fisheries stakeholders workshop was held in May 2004 to develop the beginnings of a consensus-based zoning and fisheries management plan for the KMNR. The leaders of the monitoring teams played a strong and active role in the workshop. The workshop attempted to cover three key areas for management: development of a zoning plan, development of fisheries regulations to foster sustainable fishing and development of a co-management mechanism for the reserve. While participants agreed on the concept of co-management and identified a number of fisheries regulations that included clear limitation of effort and cutbacks to promote sustainability, the workshop was unable to broach the topic of zoning, and in particular no-take zones within the marine reserve. Even with the communications successes of the monitoring and transplantation programmes this topic remained too sensitive to raise. This may be attributed to the top-down user-exclusion approach adopted by resource managers in Kenya in the past [Glaesel, 1997; King, 2000], with little explanation or dialogue with affected communities, Further sensitization work perhaps using the co-management group will be required for the project to address full protection.

Over the 7 years of operation to 2004, a varying mix of fishermen, WWF staff, KWS rangers, Fisheries Officers and researchers from the Kenya Marine and Fisheries Research Institute in Mombasa have participated in the coral reef monitoring and transplantation programme. It has been one of the largest joint annual activities carried out in the KMNR, and indeed in any protected area in Kenya, that focuses exclusively on marine conservation issues. This has given the programme a social and communications prominence that has served to raise awareness of environmental issues in general, and particularly of the primary threats to coral reefs, of both global and local origins. In the sensitive context of artisanal fisheries in Kenya, focusing on coral bleaching as a primary threat to coral reefs has facilitated the promotion of better management of reefs within the MPA.

Synthesis/Recommendations

The case studies emphasize how awareness of the threat of coral bleaching, real-time monitoring of impacts and obtaining support from stakeholders go hand in hand to facilitate management responses to coral bleaching. From acceptance and involvement by local fishermen to support and policy responses at different levels of government, clear monitoring and communications strategies are or paramount importance. Management actions that facilitate these, including the contributions of small-scale restoration efforts to awareness raising, are important tools for the manager. Across the board – from the case studies mentioned here to those referenced from Australia and Indonesia [Chadwick and Green, 2002; WWF, 2003] – improvements in early warning systems, real-time observation and long term monitoring provide managers with the information needed to convince stakeholders. In both case studies partnerships between management agencies and scientific institutions were necessary to develop mechanisms that combine practicality of implementation with high quality information for evaluation of threats and trends. Training needs and human and equipment resources are key elements needed to maintain this capacity within management agencies.

Given the primary importance of communication and participation strategies between management agencies and stakeholders, these need to be formalised. Clear roles, responsibilities and responses need to be identified and their value agreed beforehand to empower management to take necessary action during an emergency. Prior planning and the linking of decisions to specific monitoring outputs and thresholds will enable both managers and stakeholders to act with as little conflict as possible when a decision is needed. In many coral reef areas, this will entail the recognition and empowerment of groups representative of local resource user groups, such as artisanal or small-scale commercial fishers, and strengthening of their legal status. With increasing recognition of comanagement as the most feasible management model for many coral reef areas [Pomeroy, 1995; Johannes, 2002], a supportive environment needs to be developed that will facilitate management responses to coral bleaching. Thus the development of comprehensive policy instruments for local and national governments is a key prerequisite to enable local management responses in many countries with coral reefs.

The threat of coral bleaching poses a unique challenge to management because it applies across the board from local to global scales [Brown, 1997; Marshall and Schuttenberg, *in review*; Obura, 2005]. Local management responses need to be embedded in progressively larger scale planning, institutional, policy and legal contexts (*see later chapters*) that reflect the biological and geographical scaling up of bleaching from local to regional and global scales. The drivers of the causative agent of coral bleaching, i.e., increased demographic, development and globalization trends that lead to increased emissions of greenhouse gases from fossil fuel burning, likewise extend from local to global scales. Interventions to mitigate the threat of coral bleaching need to occur at all of these levels and in the linked spheres of environment, utilization, management and governance [Payet and Obura, 2004].

Positive experiences and learning need to be emphasized to help foster the motivation needed to facilitate local actions, and these need to be broadcast through a variety of media – public television, radio and print media, local plays, dances and events, as well as professional outlets such as scientific articles and management publications. For example, with the repetition and intensification of coral bleaching in the Florida Keys since 1983, Sanctuary Managers have been able to predict the onset of coral bleaching by monitoring local atmospheric and hydrographic conditions. When doldrum-like weather patterns replace the normal summer trade winds in the Florida Keys the marine environment begins to respond. If the skies remain clear and almost cloudless and the seas are slick calm for over a period of a couple of weeks the shallow waters begin to heat up. After only a couple of weeks of these weather patterns corals and reef fish begin showing signs of extreme stress. The development of credible early warning systems helps prepare stakeholders for difficult conditions ahead, and strengthens trust in the values of and need for management.

Local Management Actions are Possible – Promoting Stewardship over Despair

To answer the frequent question from the general public, as well as some coral reef managers, about what can be done about coral bleaching is an immediate priority; how can we protect coral reefs from coral bleaching when it is linked to such a long-term event as climate change? Herein lies the important question that needs to be explored by scientists, educators and managers alike.

There are several management mitigation tools that coral reef managers can implement immediately (Table 5) and increasing numbers of publications detailing management responses [Hansen, 2003; TNC, 2004; Marshall and Schuttenberg, *in review*]. Surely, as this important discussion takes place more tools will be identified and incorporated into management models. It is critical that at a time when coral reef managers are attempting to lessen direct and indirect human impacts on coral reefs by soliciting the public's support and assistance, that a sense of hopelessness is not portrayed in the place of stewardship. With increasing experience and linkages between management experience and science, increasingly specific management recommendations become possible.

Local Management Actions Influence Larger Scale Policy

Coral bleaching has been reliably documented as having impacted over 15% of the world's coral reefs in 1998 alone [Wilkinson, 2004]. Though there are clear signals that local and global lessons are informing national policies, there are still gaps in implementation at the most important national and international levels, where national priorities are decided. Even the threat to the largest coral reef protected area in the world, the Great Barrier Reef in Australia, is insufficiently large to influence national policy to reduce climate change drivers. At the international level, policy to facilitate mitigation of climate change drivers, is currently negotiated through the United Nations Framework Convention on Climate Change (UNFCCC, or the Kyoto Protocol), and world opinion is divided. On the one hand, the majority of governments, some of which are impacted by coral bleaching and all of which are impacted by other aspects of climate change, support the minimal reduction in greenhouse gas emissions prescribed by the IPCC and based on global scientific consensus [IPCC, 2001]. On the other hand, the few primary greenhouse gas-emitting governments, now and in the future, block acceptance of IPCC reductions for short term economic and political reasons.

Nevertheless, as with the need for a positive outlook on action at the local level, there are positive signals that lessons from local to global degradation of coral reefs are being heard. The ratification of the Kyoto Protocol by Russia in September 2004, and measured success of the UNFCCC and Kyoto Protocol Conference of Parties in December 2005 provide a strong global message. While its proscriptions for reducing emissions are not sufficient to

TABLE 5. Management actions can mitigate coral bleaching.

1. Use the ability to predict bleaching events to enhance coral reef monitoring programs – try to obtain pre and post bleaching data
2. Establish monitoring protocols to answer specific questions about the cause and effect of bleaching events
3. Use remote sensing tools to increase the level of predictability
4. Use the ability to predict the bleaching event to gain the attention of the public and to solicit their assistance in coral reef conservation
5. Use the severe impacts of coral bleaching as a way to leverage other conservation measures such as reducing point and non-point sources of pollution
6. Use coral bleaching events as a way to increase the public's awareness and peer pressure as to the need to cease destructive fishing practices
7. Contact coral reef users and encourage them to lessen their direct impact on the coral reefs during these stressful periods
8. Engage divers in education and outreach messages about coral reefs so they can take direct action to lessen their physical impacts on the corals during stressful periods
9. Communicate the long-term impacts of coral bleaching to reef users and solicit help in communicating to decision-makers the kinds of appropriate actions that need to be taken regarding climate change
10. Identify coral reefs that are resistant to bleaching and develop criteria that will aid in the design of Marine Protected Areas
11. Establish Fully Protected Reserves in areas resistant to coral bleaching
12. Enlist the scientific community to assist in communicating the long-term trends that we can expect if current trends of climate change continue
13. Integrate the geological and biological sciences in such a way as to hind cast our observations into geological times, so as to forecast the long-term expectations for coral reefs
14. Use current levels of research and monitoring data to communicate to decision-makers that coral reefs serve as "ecosystem indicators" of climate change and they are in need of support

effect significant reductions in warming trends, the commitment of willing countries to the UNFCCC, and of a broad cross section of industry and companies around the world are resulting in significant emissions reductions on a scale that would make a difference on a large scale if others were to follow [IUCN, 2004]. Key factors influencing the willingness of decision-makers to adopt strategies that would reduce climate change are messages of local impacts or successes, personal stories, personal experiences. It is thus imperative that local management actions to mitigate the risk of coral bleaching are pursued aggressively, their successes advertised, and a sense of hope and stewardship aroused. Aggregated to higher levels, this contributes to larger scale policies and initiatives to reduce climate change drivers, and to implement larger scale management recommendations covered in later chapters.

Conclusions

Coral reefs are being affected by a wide variety of anthropogenic and natural perturbations worldwide. Coral bleaching is one more of these impacts and immediate action needs to be taken to stem some of the negative consequences of coral bleaching events. The scientific debate will continue with regard to the causes and impacts of coral bleaching. Such debate is essential to the scientific process, and to developing new and better management interventions. However, as the debate takes place, management measures can already be taken to mitigate in various ways the impacts of coral bleaching.

As the intensification and spread of coral bleaching has been observed in the Caribbean and Atlantic over the past two decades, similar trends are now occurring in the Pacific and Indian Oceans. What an enormous benefit it would be to the science of coral bleaching if a coral reef research and monitoring program could be put in place, ahead of a massive bleaching event where one had not been previously reported. Marine Protected Areas can be among the primary contributors to this proactive approach to the threat of coral bleaching, with cascading benefits to larger national and regional levels as the value of management and the possibility of positive action are recognized by wider society.

Acknowledgments. We would like to thank the staff and stakeholders of the Florida Keys National Marine Sanctuary and the Kiunga Marine National Reserve, who provided the primary learning material for this paper. In addition, a number of colleagues have provided constructive comments and critiques, and we acknowledge Arthur Paterson, Al Strong, Jordan West, Heidi Schuttenberg and two anonymous reviewers.

References

Andersson, J., The Recreational Cost of Coral Bleaching- a Stated and Revealed Preference Study of International Tourists. Beijer International Institute of Ecological Economics, The Royal Swedish Academy of Sciences, Stockholm, pp. 26, 2003.

Bowden-Kirby, A., Coral transplantation in sheltered habitats using unattached fragments and cultured colonies. Proceedings of the 8th International Coral Reef Symposium, Panama. 23-27 October 2000 2: 2063-2068, 1997.

Brown, B., Coral bleaching: causes and consequences. *Coral Reefs*, 16: 129-138, 1997.

Burke, L., E. Selig, M. Spalding, Reefs at Risk in Southeast Asia. World Resources Institute, 2002.

Burke, L. A., Maidens, Reefs at Risk the Caribbean. World Resources Institute, 2004.

Causey, B. D., Lessons Learned from the Intensification of Coral Bleaching from 1980-2000 in the Florida Keys, USA. Proceedings of the Workshop on Mitigating Coral Bleaching Impact Through MPA Design, 2001.

Cesar, H., Impacts of the 1998 Coral Bleaching Event on Tourism in El Nido, Philippines. U.S. Department of State, East Asia and Pacific Environmental Initiative (US-EAP-EI)/Coastal Resources Center, University of Rhode Island, Narragansett, Rhode Island, pp. 21, 2000.

Chadwick, V., A. Green, Managing the Great Barrier Reef Marine Park and World Heritage Area through critical issues management: Science and management. Proceedings of the 9th International Coral Reef Symposium, Bali. 23-27 October 2000 2: 681-686, 2002.

Church, J., D. Obura, Sustaining coral reef ecosystems and their fisheries in the Kiunga Marine National Reserve, Lamu Kenya. Proceedings of the 10th International Coral Reef Symposium, 2005.

Costanza, R., R. d'Arge, R. de Groot, S. Farber, M. Grasso, B. Hannon, K. Limburg, S. Naeem, R. O'Neill, J. Paruelo, R. Raskin, P. Sutton, The value of the world's ecosystem services and natural capital. *Nature*, 387: 253-260, 1997.

Done, T. Scientific principles for establishing MPAs to alleviate coral bleaching and promote recovery. Salm, R., Coles, S. eds., In *Coral Bleaching and Marine Protected Areas.* Proceedings of the Workshop on Mitigating Coral Bleaching Impact Through MPA Design. Asia Pacific Coastal Marine Program Report #0102, The Nature Conservancy, Honolulu, Hawaii, USA., Bishop Museum, Honolulu, Hawaii, 29-31 May 2001. pp. 118, 2001.

FWC-CRMP. Fish and Wildlife Conservation – Coral Reef Monitoring Programme, 2004.

Glaesel, H., Fishers, parks, and power: the socio-environmental dimensions of marine resource decline and protection of the Kenya coast. University of Wisconsin-Madison D.Phil.Geog. Thesis. pp. 1997.

Glynn, P., Coral reef bleaching: ecological perspectives. *Coral Reefs*, 12: 1-17, 1993.

Goreau, T., T. McClanahan, R. Hayes, and A. Strong, Conservation of coral reefs after the 1998 global bleaching event. *Conservation Biology*, 14: 5-15, 2000.

Hansen, L., Increasing the resistance and resilience of tropical marine ecosystems to climate change. Hansen, L., Biringer, J., Hoffman, J. eds., *In Buying Time: a user's manual for building resistance and resilience to climate change in natural systems.* WWF. www.panda.org/climate/pa_manual, Washington DC, pp. 157-176, 2003.

Harding, S. P., J. Comley, J-L. Solandt, A. R. Harborne, and P. S. Raines, Coral Reef Status in the Mamanucas Islands, Fiji: an Assessment of Three Years of Reef Check data. Poster presentation. Proceedings of the 10th International Coral Reef Symposium, 2005.

Heeger, T., F. Sotto, Coral Farming Atool for reef Rehabilitation and Community Ecotourism. GTZ, pp. 94, 2000.

Hockings, M., S. Stolton, N. Dudley, Evaluating effectiveness; A summary for park managers and policy makers. WWF, IUCN, 2002.

Hoegh-Guldberg, O., Climate change, coral bleaching and the future of the world's coral reefs. *Marine and Freshwater Research*, 50: 839-866, 1999.

IPCC, Climate Change 2001: Synthesis Report. An assessment of the Intergovernmental Panel on Climate Change, 2001.

IUCN, Turning down the Heat: Managing risk and uncertainty. Managing Risk in a Changing World Theme. World Conservation Forum, 18 to 20 November 2004, Bangkok, Thailand. http://www.iucn.org/, 2004.

Johannes, R., The renaissance of community-based marine resource management in Oceania. Annual Reviews in Ecology and Systematics, 33: 317-340, 2002.

Kelleher, G., R. Kenchington, Guidelines for establishing marine protected areas. IUCN, 1991.

King, A., Managing without institutions: the role of communication networks in governing resource access and control. University of Warwick PhD Thesis. pp, 2000.

Lindahl, U., Low-tech rehabilitation of degraded coral reefs through transplantation of staghorn corals. Ambio 27: 645-650, 1998.

Linden, O., D. Souter, D. Wilhelmsson, and D. Obura, Coral Reef Degradation in the Indian Ocean. Status report 2002. CORDIO, pp. 108, 2002.

Linden, O., N. Sporrong, Coral Reef Degradation in the Indian Ocean. Status reports and project presentations, 1999. Sida/SAREC Marine Science Programme, Sweden, pp. 108, 1999.

Marshall, P., H. Schuttenberg, (in review) A Reef Manager's Guide to Coral Bleaching. Great Barrier Reef Marine Park Authority, Townnsville, Australia.

McClanahan, T., N. Muthiga, and S. Mangi, Coral reef and algal changes after the 1998 coral bleaching: interaction with reef management and herbivores on Kenyan reefs. *Coral Reefs*, 19: 380-391, 2001.

Möberg, F., C. Folke, Ecological goods and services of coral reef ecosystems. Ecological Economics, 29: 215-233, 1999.

Mohammed, S., C. Muhando, and H. Machano, Coral reef degradation in Tanzania: results of monitoring 1999-2002. In *Coral Reef Degradation In The India Ocean.* Linden, O., Souter, D., Wilhelmsson, D., Obura, D. eds., Status report 2002. CORDIO, Kalmar, Sweden pp. 21-30, 2002.

Motta, H., M. Pereira, and M. Schleyer, International Coral Reef Initiative/CORDIO Country Report: Mozambique. In *Coral Reef Degradation In The India Ocean.* Souter, D., Obura, D., Linden, O. eds., Status reports 2002. CORDIO/SAREC Marine Science Program, pp, 2002.

Motta, H., M. J. Rodrigues, and M. Schleyer, Coral reef monitoring and management in Mozambique. In *Coral Reef Degradation In The India Ocean.* Souter D, Obura D, Linden O eds., Status reports 2000. CORDIO/SAREC Marine Science Program, pp, 2000.

Muhando, C., Assessment of the extent of damage, socio-economic effects, mitigation and recovery in Tanzania. In *Coral Reef Degradation in the Indian Ocean.* Linden, O., Sporrong, N. eds., Status Reports and Project Presentations, 1999. Sida/SAREC Marine Science Programme, Sweden, pp. 43-47, 1999.

Muñoz-Chagin, R., Coral transplantation program in the Paraiso coral reef, Cozumel Island, Mexico. Proceedings of the 8th International Coral Reef Symposium, Panama. 23-27 October 2000 2: 2075-2078, 1997.

Nyström, M., C. Folke, and F. Möberg, Coral reef disturbance and resilience in a human-dominated environment. Trends in Ecology and Evolution, 15: 413-417, 2000.

Obura, D., Resilience and climate change - lessons from coral reefs and bleaching in the Western Indian Ocean. Estuarine Coastal and Shelf Science, 2005.

Oliver, J., P. Marshall, N. Setiasih, and L. Hansen, A global protocol for assessment and monitoring of coral bleaching. WWF, 2004.

Payet, R., D. Obura, The negative impacts of Human Activities in the Eastern African Region? An International Waters Perspective. Ambio 33: 24-34, 2004.

Pet-Soede, L., Effects of coral bleaching on the socio-economics of the fishery in Bolinao, Pangasinan, Philippines. Marine Science Institute, University of Philippines (UP-MSI), Manilla, pp. 46, 2002.

Pomeroy, R., Community based and co-management institutions for sustainable coastal fisheries management in Southeast Asia. Coastal Management 27: 143-162, 1995.

Pomeroy, R., J. Parks, and L. Watson, How is your MPA doing? A guide of natural and social Indicators for evaluating Marine protected Area Management effectiveness. IUCN, Switzerland and Cambridge, UK, 216, 2004.

Salm, R., J. Clark, and E. Siirila, Marine and Coastal Protected Areas: A guide for planners and managers. IUCN, Washington, DC, pp. 371, 2000.

Salm, R., Coles, S., eds., Coral Bleaching And Marine Protected Areas. Asia Pacific Coastal Marine Program Report #0102, The Nature Conservancy, Honolulu, Hawaii, USA., Bishop Museum, Honolulu, Hawaii, 29-31 May 2001. pp. 118, 2001.

Schleyer, M., D. Obura, H. Motta, and M-J. Rodrigues, A Preliminary Assessment of Bleaching in Mozambique. In *Coral Reef Degradation in the Indian Ocean.* Linden O, Sporrong N eds., Status Reports and Project Presentations, 1999. Sida/SAREC Marine Science Programme, Sweden, pp. 37-42, 1999.

Schuttenberg, H., Coral Bleaching: Causes, Consequences, and Response. Papers presented at the 9th International Coral Reef Symposium session on "Coral Bleaching: Assessing and Linking Ecological and Socioeconomic Impacts, Future Trends and Mitigation Planning." March 2001. Report No. Coastal Management Report #2230, Coastal Resources Center., Narragansett, RI, 2001.

Souter, D., D. Obura, and O. Linden, Coral Reef Degradation in the Indian Ocean. Status Report, 2000. Sida/SAREC Marine Science Programme, Sweden, pp, 2000.

TNC R2- Reef Resilience Toolkit, Version 2.0. The Nature Conservancy, 2004.

van Treek, P., H. Schumacher, Initial survival of coral nubbins transplanted by a new coral transplantation technology - options for reef rehabilitation. *Marine Ecology Progress Series*, 150: 287-292, 1997.

West, J., R. Salm, Environmental determinants of resistance and respilience to coral bleaching: implications for marine protected area management. *Conservation Biology*, 17: 956-967, 2003.

Westmacott, S., H. Cesar, L. Pet-Soede, and O. Linden, Coral bleaching in the Indian Ocean: Socio-Economic Assessment of Effects. In *Collected Essays on the Economics of Coral Reefs.* Cesar, H. ed., CORDIO, Coral Reef Degradation in the Indian Ocean, Kalmar, Sweden, pp. 244, 2000b.

Westmacott, S., K. Teleki, S. Wells, and J. West, Management of Bleached And Severely Damaged Coral Reefs. IUCN, Gland, Switzerland and Cambridge, UK, pp. 36. 2000a.

Wilkinson, C., Status of Coral Reefs of the World: 2000. Australian Institute of Marine Science, Townsville, Australia, pp. 363, 2000.

Wilkinson, C., Status Of Coral Reefs Of the World: 2002. Australian Institute of Marine Science, Townsville, Australia, pp. 378, 2002.

Wilkinson, C., Status Of Coral Reefs Of the World: 2004. Australian Institute of Marine Science, Townsville, Australia, pp, 2004.

Wilkinson, C., O. Linden, H. Cesar, G. Hodgson, J. Rubens, and A. Strong, Ecological and socioeconomic impacts of 1998 coral mortality in the Indian Ocean: An ENSO impact and a warning of future change? Ambio, 28: 188-196, 1999.

Williams, E., L. Bunckley-Williams, The world-wide coral reef bleaching cycle and related sources of coral mortality. *Atoll Research Bulletin*, 335: 1-71, 1990.

WWF, Case Study - Designing and implementing a coral bleaching response program, pp. 2, 2003.

11

Marine Protected Area Planning in a Changing Climate

Rodney V. Salm, Terry Done, and Elizabeth McLeod

Abstract

The establishment of Marine Protected Areas (MPAs) has become an important part of society's approach to conserving coral reefs. Protected area managers now must attempt to take account of climate change in the design and implementation of MPAs. A network of MPAs provides a logical way to distribute their benefits to the wider reef system of which they are part, and to minimize the risk that spatially unpredictable and unmanageable insults of any type may decimate all protected areas. This chapter reviews the types of environmental factors and settings that might be considered in the network design process. We focus mainly on environmental factors that appear to confer ecological resilience to bleaching, and also briefly touch on physiological resistance conferred to corals by some strains of zooxanthellae. Finally, our chapter provides a model outlining these four actions that MPA managers may take to build resilience into coral reef conservation programs: 1) spread risk by protecting multiple examples of a full range of reef types; 2) identify and protect coral communities that demonstrate bleaching resistance and may thus increased contribute genetically based bleaching resistance to recovering areas; 3) incorporate connectivity into MPA and network design; and 4) increase effectiveness and flexibility of reef management strategies.

Introduction

The increased frequency and severity of mass coral bleaching due to sea temperature rise has increased stresses on coral reefs and raised concerns about appropriate response strategies [Hughes et al., 2003]. Marine Protected Areas (MPAs) have been identified as one of the most effective tools for conserving reefs and related marine systems [Kelleher, 1999; Lubchenco et al., 2003; Palumbi, 2003]. However, protected area managers must incorporate climate change as well as increasing human pressures into their conservation strategies, or MPAs may not be able to safeguard biodiversity effectively. Networks of MPAs, including no-take areas totaling 30-50% of available reef area [Hoegh-Guldberg and Hoegh-Guldberg, 2004], have been suggested as a critical way to protect coral reefs from human stresses. No-take areas increase fish biomass and variability both inside and outside the boundaries of those areas [Halpern, 2003] and permit critical functional groups to persist, thus contributing to local ecosystem resilience Bellwood et al. [2004].

Coral Reefs and Climate Change: Science and Management
Coastal and Estuarine Studies 61
Copyright 2006 by the American Geophysical Union.
10.1029/61CE12

The Challenge: MPA Manager's Perspective

MPA managers in the 21st Century are faced with a number of synergistic environmental changes that can adversely affect coral survival, and over which they have no control [Buddemeier et al., 2004]: increasing temperature [Hoegh-Guldberg, 1999]; sea level rise [Done, 1999]; increased exposure to UV light [Catala-Stucki, 1959; Gleason and Wellington, 1993, 1995; Hoegh-Guldberg, 1999; Siebeck, 1988] and reduced alkalinity [Gattuso et al., 1998; Kleypas et al., 1999; Langdon et al., 2000, 2003; Leclercq et al., 2000; Marubini et al., 2001]. The evidence is incontrovertible that the Earth has already warmed 0.6-0.8° C since 1880. The projections that it will warm 2-6° C by 2100 are credible (IPCC 2001).

MPA managers, accustomed only to addressing direct and usual threats related to fishing and tourism, for example, find it difficult enough to address the impacts of coastal development and inland activities, and may consider climatic sources of environmental stress totally beyond their sphere of influence. This perception is reinforced when even well managed reefs in the remotest MPAs (e.g., Ngeruangel atoll in Palau and Aldabra atoll in the Seychelles) succumb to a climate related bleaching event. However, there are some direct actions that MPA managers can take immediately, even as the scientific understanding improves. Salm et al. [2001] proposed that it might be possible to mitigate the negative impacts of bleaching on coral reef biodiversity in two broad ways:

1. Identify and protect from direct anthropogenic impacts, specific patches of reef where local conditions are highly favorable for survival generally, and that also may be at reduced risk of temperature-related bleaching and mortality (i.e., coral assemblages with a high level of "resistance");
2. Locate such protected sites in places that maximize their potential contribution to their own resilience, and, through larval dispersal, to that of reefs that are interconnected with it.

In this chapter, we explore these ideas further.

Exploring Possible Solutions

Locations to be avoided are those where exposures to both anomalous high sea surface temperatures and solar radiation are likely. These factors interact to cause severe coral bleaching and mortality, and are responsible (either alone or in combination) for the majority of climate related disturbances [Fitt et al., 2001; Glynn, 2000; Wellington et al., 2001], West and Salm [2003] group factors that may protect areas into four broad categories: physical factors that reduce temperature stress; physical factors that enhance water movement and flush toxins; physical factors that decrease light stress; and factors that confer stress tolerance in the corals. Sites where these factors reliably occur would make good candidates for MPA selection and the investment of conservation effort and funds.

Reduction of Temperature Stress

There is strong evidence that vertical mixing of deeper cool waters up through the water column effectively reduces temperature-related stress to shallow water corals [Skirving, this volume]. Tidal-driven or long-wave currents can thus prevent temperature stress anomalies occurring within coral zones, during times when regional heat stress is generally widespread. Following the 1998 El Niño Southern Oscillation (ENSO), areas of local upwelling protected central Indonesian reefs from severe bleaching [Goreau et al., 2000; Salm et al., 2001]. Additionally, some reefs had lowered mortality due to cooling effects of strong upwelling: the outer reefs of Alfonse, St. Francois, Bijoutier atolls, Western Zanzibar, and certain areas in the Maldives [Goreau et al., 2000]; some outer reefs in the Great Barrier Reef following a 2002 heat wave [Berkelmans et al., 2004; Wooldridge and Done, 2004].

Strong Currents and Flushing

Flushing by strong currents, even if the water remains warm, may also protect corals to some degree, apparently working by removing free radicals that are a toxic byproduct of bleaching in corals [Nakamura and van Woesik, 2001]. In laboratory experiments, these authors demonstrated that *Acropora digitata* suffered high bleaching mortality under low-flow conditions, and none under high-flow conditions. While field evidence for this effect is weak, there have been observations in Palau and Indonesia that clearly demonstrate synergisms at work. At several sites in Palau[4] where corals had died on reef slopes below 2-4m, reef flats and shallow reef crests with strong currents showed much higher coral survival. At Komodo National Park and Nusa Penida Island in Indonesia, there was a clear vertical mixing effect of strong cool currents. At these locations, there was little or no bleaching or temperature related mortality in corals following the 1998 event.

Reduction of Light Stress

Under ideal conditions, corals thrive at high light levels [Jokiel and Coles, 1990]. However, in combination with a second factor (i.e., increased temperature), high light levels can become a stress. High north-south orientated islands can sometimes provide shading for corals during one or other half of the day, and undercut karst islands or coastlines often provide intensely shaded shelves on which corals can grow. Trees growing on the slopes of these coasts can further extend the shading effect seawards. The Rock Islands in Palau demonstrate well this effect of shading in reducing bleaching and related mortality. At Nikko Bay, in particular, one of the most diverse permanent monitoring sites of the Palau International Coral Reef Research Center, corals in shaded areas survived well during the 1998 bleaching event that, elsewhere in the vicinity, caused major coral mortality.

Suspended particulate matter in the water column may also protect corals by screening them from destructive high light levels. On silty and often turbid fringing reefs on the inner Great Barrier Reef, coral cover can approach 100% over large areas [Stafford-Smith and Ormond, 1992]. In parts of the Rock Islands and sheltered bays off Babeldaob, Palau, *Porites rus* and *Porites cylindrica* succumbed to bleaching and died in clear water areas, but survived in cloudy water. Turbidity may also have contributed to lower mortality following the 1998 bleaching event in the Gulf of Kutch, Southwestern Sri Lanka, Mahé [Goreau, 1998a, 1998b], along the 18-foot break inside the barrier reef in the Florida Keys National Marine Sanctuary (Causey, pers. comm.), and inside the lagoon of Alfonse atoll [Goreau, 1998c; Goreau et al., 2000].

---

[4]Lighthouse Reef flats, patch reef east of Ebiil Channel off NW Babeldaob Island, Fantasy Island area reef flats (these later severely impacted by the crown-of-thorns starfish *Acanthaster planci*)

Stress Tolerance

The algal symbionts of reef corals are extraordinarily diverse [e.g., Rowan and Powers 1991], and it is possible that reef corals may mitigate the effects of climate change by hosting unusually heat-tolerant symbiont communities as a response to changing temperatures [Baker et al., 2004]. Some corals have adapted to higher temperatures by hosting more heat tolerant zooxanthellae better able to resist coral bleaching [Rowan, 2004; Baker et al., 2004]. Baker et al. [2004] found that corals containing thermally tolerant *Symbiodinium* in clade D are more abundant on reefs after episodes of severe bleaching and mortality. Rowan [2004] also found that corals have adapted to higher temperatures by hosting thermally tolerant *Symbiodinium*. Rowan [2004] observed that *Pocillopora* spp. with *Symbiodinium D* resisted warm-water bleaching whereas corals with *Symbiodinium C* did not. He also noted that *Pocillopora* spp. living in water over 31.5°C hosted only *Symbiodinium D*; whereas those living in cooler habitats hosted predominantly [Rowan, 2004]. Baker et al. [2004] propose that adaptive shifts to heat resistant algal symbionts will increase coral reef resilience to future bleaching events [Baker et al., 2004]. However while the prospect of heat-tolerant strains of symbionts protecting entire habitats and multi-species coral assemblages is appealing, its realization is less certain.

In selection of sites for conservation management, quantitative information about the current level of such genetically based resistance in candidate areas should be taken into consideration, should it exist [e.g., Van Oppen et al., 2005). In practice, protection afforded to any given site will depend on factors currently unknown, notably the rates of propagation of heat-tolerant strains across the candidate area, and rates of propagation across all coral species and functional groups. Should such habitat and community-wide propagation occur, there would likely be a decrease in zooxanthellae diversity across the community. This may have consequences in relation to other phenotypic traits of the corals. For example, corals hosting the more heat-vulnerable *Symbiodinium C* grow more quickly than those hosting the less vulnerable *Symbiodinium D* [Little et al., 2004].

Acclimatization is a second means of increasing local site resistance. A history of regular exposure to severe conditions [Brown et al., 2000; Coles and Brown, 2003; Dunne and Brown, 2001] appears to acclimatize corals to cope with anomalous excursions in light and temperature. This includes corals in habitats exposed to intense solar radiation [Brown et al., 2000] and/or high temperatures [Craig et al., 2001; Jokiel and Coles, 1990; Marshall and Baird, 2000]. Corals on reef flats that emerge at low tide are exposed to conditions of heating, desiccation and rainfall. Such prior exposures have been posited as explanations for lower bleaching susceptibility recorded for corals in some inner reefs and lagoons relative to conspecifics from deeper waters [Hoeksema, 1991; Salm et al., 2001; West and Salm, 2003]. The central Indian Ocean in 1998 provides another example. Here, there was localized survival of corals in reef flat and lagoon areas [Spalding et al., 2000], probably reflecting the wide ambient variability in light and heat in these habitats.

Reef flat and lagoonal coral communities should thus not be overlooked in management plans or reef conservation. Indeed, reef flats will be the coral habitat most affected by sea level: given a strong supply of coral larvae, a rise in sea level could allow for successful recruitment and growth of a greater variety of corals than are currently found on some reef flats [Done, 1999].

Coral Community Type

It is possible that the presence of some of these factors described above is correlated with specific coral community types. Coral species with rapid growth rates, thinner tissue, and branching forms, (e.g., *Acropora* spp., *Stylophora* spp., *Pocillopora* spp.) tend to bleach sooner and more severely than slow-growing, massive corals with thicker-tissues (e.g., *Porites, Goniopora* spp.) [Gates and Edmunds, 1999; Loya et al., 2001]. However, even in a specific location, there can be great intra-specific variability in susceptibility to heat stress responses to temperature fluctuations [Marshall and Baird, 2000; Smith and Buddemeier, 1992].

In one study of three hard coral genera, *Acropora* spp. showed the most severe bleaching, *Pocillopora* spp. showed intermediate bleaching, and *Porites* spp. showed the least bleaching [Hoegh-Guldberg and Salvat, 1995]. It was the corals that showed the fastest growth and metabolic rates (*Acropora* spp.) that were the most susceptible. All *Porites* spp. colonies recovered from bleaching, while *Acropora* spp. did not recover well. Here, mass bleaching quickly changed the dominance relationships by decimation of a major component. A similar change in community structure was observed in Okinawa [Loya et al., 2001]. Community shifts away from branching corals might have negative impacts on these ecosystems, as many fish and invertebrate populations are obligate associates of intact branching corals [Goreau et al., 2000].

Strong and diverse coral recruitment and growth signify a site's resilience in respect of its corals. Strong recruitment is measured by both the number per unit area and the cover of small coral colonies established since a prior disturbance. The chronology of recovery is often reflected in the presence of well-defined size classes of corals. It can be argued that such recruitment is also a proxy measure of local reef "health"; i.e., suitability of the substratum and of water quality. The 'sources' of larvae that recruited may be some kilometers upstream, and themselves need to be identified and protected in MPAs to ensure strong and rapid recovery of the downstream 'sink' reef. It is likely that these source coral communities have survived because of one or more of these factors reduce the risk of bleaching at the site. Full protection of these "resistant" or low risk coral communities in MPAs is essential, whatever the underlying reasons for survival.

What Can MPA Managers do?

While there is little that MPA managers can do to control large-scale stresses at their sources, there are direct actions they can take to help reefs survive catastrophic bleaching events. The "Resilience Model" developed by The Nature Conservancy, is a simple tool to assist conservation planners and managers build resilience into coral reef MPAs (Plate 1).

The components of the model and their application to MPA network design are described below. The most effective configuration would clearly be a network of highly protected areas nested within a broader management framework. Such a framework might include a vast multiple-use reserve managed for sustainable fisheries as well as protection of biodiversity. The ideal MPA system would be integrated with coastal management regimes to enable effective control of threats originating upstream, and to maintain high water quality [e.g., Done and Reichelt, 1998].

1) Representation and Replication

Protect multiple examples of a full range of reef types, seeking to represent the area's total reef biodiversity. Replication within each type reduces the chance of any one type being completely compromised by an unmanageable impact such as a major bleaching event.

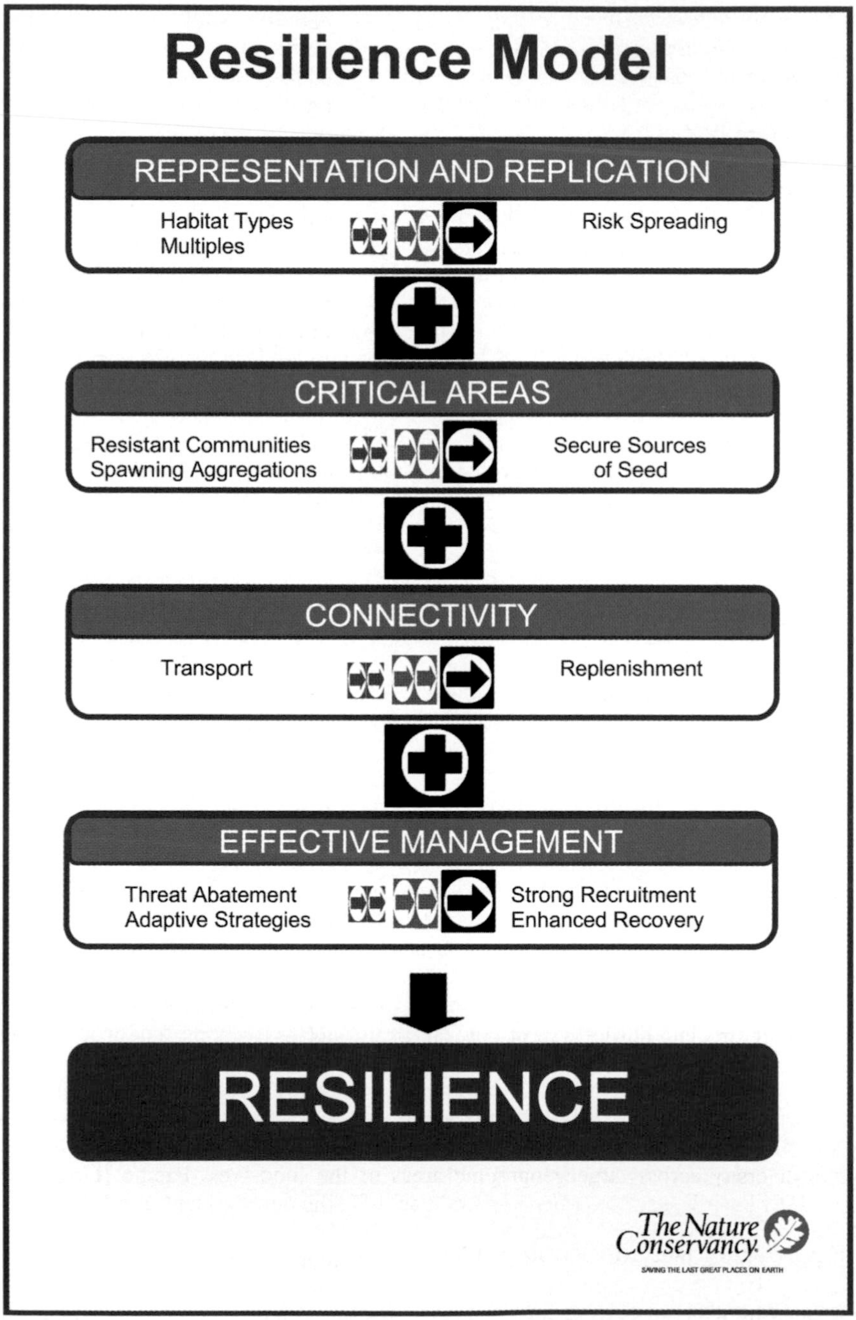

Resilience Model

REPRESENTATION AND REPLICATION

Habitat Types
Multiples
Risk Spreading

CRITICAL AREAS

Resistant Communities
Spawning Aggregations
Secure Sources of Seed

CONNECTIVITY

Transport
Replenishment

EFFECTIVE MANAGEMENT

Threat Abatement
Adaptive Strategies
Strong Recruitment
Enhanced Recovery

RESILIENCE

The Nature Conservancy
SAVING THE LAST GREAT PLACES ON EARTH

Plate 1. The Nature Conservancy's Resilience Model.

To fully represent regional biodiversity within fully protected areas, the range of reef types protected should include samples of offshore reefs (barriers, atolls) in areas with greater and lesser wave energy and exposure to trade wind, mid-shelf reefs (patch and fringing reefs) where these exist, and inshore fringing and patch reefs in sheltered locations. For long, linear coastlines, samples of all these reef types should be selected at regular intervals along the coast and reef tract. Wherever possible, multiple samples of each reef type should be included in MPA networks or larger management frameworks, such as multiple-use MPAs or areas under rigorous integrated management regimes. This overall approach has recently been put into practice with designation of around 30% of the Great Barrier Reef in 'no-take' areas in July 2004 [Day et al., 2002].

Selecting and securing multiple samples of reef types may also have the advantage of protecting essential habitat for a wide variety of commercially valuable fish and macroinvertebrates. Studies have confirmed that maintaining coral cover is important to maintaining fish biodiversity. According to an 8-year study in Papua New Guinea, a decline in coral cover caused by a combination of factors (e.g., coral bleaching, increase in sedimentation from terrestrial run-off, and outbreaks of crown-of-thorns) caused a parallel decline in fish biodiversity within marine reserves [Jones et al., 2004]. Therefore, it is essential to include replicates of representative habitats for two reasons: first, to spread the risk of losing any one of the representative habitat types through chronic or catastrophic events; and, second, to maintain a diverse range of habitat types and their associated communities.

2) Refugia

Identify and fully protect coral communities that can serve as refugia and thereby reseed other areas that are seriously damaged by bleaching.

Through analyzing local environmental factors that contribute to coral community resistance and resilience, managers can identify areas of cooling, shading, screening, stress tolerance, and strong currents, as described in the preceding section [West and Salm, 2003]. For example, in an area at the south end of Sulawesi, and along the Makassar Strait, a 17-year analysis (1985-2001) showed no thermal anomalies greater than 1°C in many parts of the region [WWF, 2003]. Kassem et al. [2002] suggests that this is due to a combination of oceanographic features including high current flow [WWF, 2003].

The Raja Ampat Islands in the heart of the Indo-West Pacific provide an example of a place that has escaped climate change impacts. This group of islands has over 75% of the world's known reef building corals, including 35 possible new species awaiting identification [Donnelly et al., 2003]. Raja Ampat is located in a major convergence of marine currents that circulate nutrients, transport larvae through the islands, and cool its waters, protecting the reefs from bleaching effects. The high biodiversity and apparent lack of coral bleaching, suggest that Raja Ampat is a key area, both as a major reservoir of biodiversity, and a secure source of propagules for reefs connected to it by currents. Historically, this region has been important in generating, maintaining and dispersing genetic diversity across large geographic areas of the Indo-West Pacific [Grigg, 1988; Veron, 1995] and is a strategically important area for the development of a MPA network with long term conservation objectives.

Finally, refugia must be large enough to support high species richness. Bellwood and Hughes [2001] suggest that high species richness in the Indo-Australian archipelago is maintained by large areas of suitable habitat. Large areas are more likely to support high genetic diversity because they may support larger populations which produce more offspring [Palumbi, 1997].

3) Connectivity

Identify patterns of connectivity among source and sink reefs, so that these can be used to inform reef selection in the design of MPA networks and provide rich stepping-stones for reefs, over longer time frames.

'Connectivity' describes the natural linkages among reefs that result from the dispersal and migration of organisms by ocean currents. The strength of connectivity depends on the abundance and fecundity of source populations, the longevity and pre-competency periods of their larvae, and the spawning sites and movement patterns of adults. Connectivity is thus a key driver of the strength and reliability of the replenishment of biodiversity on reefs damaged by natural or human-related agents. Ideally, to maximize a damaged site's chances of recovering from a bleaching event, it should have a bleaching resistant site upstream of it to supply its larvae.

Where protected areas are surrounded by intensively used lands and water, buffer zones are commonly established to provide a transition zone of partial protection. Such buffer zones will become increasingly important for coral reefs as sea level rises, potentially expanding the extent of some shallow water habitats for reefs and mangroves. As warm tropical waters extend more polewards, it may be timely to begin modeling future connectivity patterns. This would help guide planning of MPA configurations in light of possible expansions in latitudinal and longitudinal distributions of coral communities that presently are restricted by existing temperature ranges.

4) Effective Management

Manage reefs for both health and resilience, and monitor multiple indicators of the effectiveness of current actions as the basis for adaptive management.

Expanding the area of tropical seas managed for biodiversity conservation increases the impact on people's activities, access, and resource uses. Effective management must therefore address the socioeconomic impacts of both coral bleaching itself, and of the conservation measures introduced to counteract its ecological effects. Poverty reduction and sustainable development strategies become the cornerstones of effective MPA management. Partnerships are the key to both conservation and sustainable development.

For example, only through partnerships with local communities could managers hope to rehabilitate damaged sites by such actions as removal of crown-of-thorns starfishes or other coral predators, transplanting corals, restriction or reduction of fishing of herbivores, prevention of destructive practices, control of tourism impacts, improvement of water quality, and physical removal of macro-algal mats that are inhibiting coral settlement, survival, or growth. Only through the existence of influential partnerships among government agencies and the private sector could protection of coral reef resources be used as the leverage to bring about control of land-based sources of pollution. To engage communities in such ways, there is a clear need for managers to develop strategies that are attuned to local priorities and needs, with useful productivity and resilience of the reef resource system among those needs.

Socioeconomic Impacts of Bleaching

In many cases, MPA managers can best hope to engage local communities and mitigate any socioeconomic impacts of coral bleaching on them by managing MPAs as part of a sustainable fisheries program. Halpern [2003] reviewed 80 MPAs and discovered that on average reserves doubled abundance, tripled biomass, and increased both size and diversity of fish by a third. Additionally, the same data showed that increases usually became apparent within five years of protection [Halpern and Warner, 2002]. Importantly, fishing communities in some areas have noticed an increase in fish catch outside the protected area [Gell and Roberts, 2003; Russ, 2002]. Because it has yet to be conclusively demonstrated that increasing reef fish biomass will aid in reef resilience, it would be far better to emphasize the benefits that a resilient and productive coral reef system would provide for local communities.

Reef Restoration

Local population support for marine conservation has been gained by grass roots involvement of people in reef restoration projects. In Tanzania, for example, where dynamite fishing, wave action, and coral bleaching had seriously damaged the fringing coral reef around Mbudya Island [Wagner et al., 2001], local fishermen were key players in restoring damaged areas. They helped transplant hundreds of fragments of *Galaxea*, *Acropora*, *Porites*, and *Montipora* in seven dynamited sites, using cement filled, and disposable plastic plates. After three months, *Galaxea* showed 100% survival and *Porites* showed 55.7% complete survival and 13.9% partial survival. Over five months, *Galaxea* and *Porites* increased significantly in height, but not *Acropora*. Through such experiments, managers spread awareness of threats to the marine environment, determine the success of certain restorative options for degraded reefs, and gain an understanding of the ability of key coral species to reestablish on a damaged reef.

Predators, Herbivores and Nutrients

One of the greatest impediments to recovery of damaged coral areas is the pre-emption of the reef surface by carpets of algae: rates of algal production are too high relative to rates of export and consumption [e.g., Hughes, 1987; Littler and Littler, 1997; Williams et al., 2001]. This type of negative effect can be a result of sequential fishing down the fish food web, including herbivorous fish species [Bellwood et al., 2004; Dulvy et al., 2004; Pandolfi et al., 2003; Pauly et al., 1998]. For resilience in coral cover, there needs to be enough grazing by reef organisms to keep algal biomass on damaged reefs sufficiently low that corals can establish and flourish [Hatcher and Larkum, 1983; Sammarco, 1980; Steneck and Dethier, 1994]. In some circumstances, grazing rates can also be too high for corals to reestablish themselves. This can occur when top-level fish predators are fished out, causing populations of lower level grazers – notably sea urchins – to explode, eroding reef surfaces, and in the process, destroying small coral recruits [McClanahan, 1997]. Some intermediate level of grazing is therefore an important operational goal for coral reef management. Of equal importance to grazing, especially when coral reefs occur in enclosed waters, may be the reduction of runoff of nutrients that enhance the growth of seaweeds to the detriment of corals [Smith et al., 1981; McCook, 1999].

Monitoring and Adaptation

Coral reef management needs to be viewed as an adaptive, 'learning by doing' exercise; and baseline data and monitoring are essential bases for adaptive decision making. Not only is there a need to monitor the ecological well-being of the reef, but also to define and monitor indicators of the efficacy of reef-related governance measures and of reef-related socioeconomic trends [Pomeroy et al., 2004].

Retrospective surveys and follow-up studies of the spatial patterns of individual coral reef bleaching events in relation to the pattern of heat stress as recorded from satellites are a logical and informative means of assessing the vulnerability of particular reef systems at that time [Arceo et al., 2001; Berkelmans et al., 2004; Wooldridge and Done, 2004]. However, retrospective studies done long after the event are not always easy, as other factors (hurricane damage, disease, crown-of-thorns predation) may obscure the causes of mortality. There is no substitute for real-time tracking of an event with field work. This helps managers understand the vulnerabilities of particular species and communities in different locations and identify those places where corals merely bleach, as opposed to those where the corals die [Marshall and Baird, 2000].

In the long term, there is concern that the efficacy of MPAs and other management measures may also be compromised by several other manifestations of climate change [Pittock, 1999]: changes in ocean chemistry [brought about by increasing atmospheric CO_2 levels – Kleypas et al., 1999]; changes in salinity due to changed rainfall and runoff regimes [Pittock, 1999]; and changes in hydrography (sea level, currents, vertical mixing, storms and waves). Knowledge of their importance is building through both the accumulation of long-term monitoring data-sets of physico-chemical parameters at fixed sites (e.g., Hendee et al., 2001; Lough, 2000] and the understanding of processes based on intensive short term oceanographic studies. These studies may increasingly guide decisions about the configuration and management of MPAs in future, as we learn to anticipate changes better in the environment and reef communities. The results of such research and monitoring programs will provide an improved basis for recognizing truly bleaching-resistant sites for protection and for informing decisions relating to realization of the conservation goal.

Conclusion

The development of MPAs that can mitigate the impacts of global climate change needs a multidisciplinary approach based on the expertise of policy makers in government, MPA managers, a range of scientific disciplines, government agencies, conservation organizations, and local communities. One key message is that, with global changes affecting all countries irrespective of political boundaries, the design and management of MPAs will need to be adaptable, based on both improved scientific knowledge and the socio-political context at coral reef places around the world. The approach to MPAs is evolving away from a focus on individual sites to networks of mutually replenishing MPAs [Soto, 2002]. As iconic ecosystems, the plight of coral reefs should be one catalyst for increased efforts to mitigate the extent and effects of climate change worldwide. The only hope for coral reefs lies in both a slowing of the rate of global climate change, and actions that build survivability into reef systems. Strategically placed and well-managed MPAs seem to offer the most viable means of protecting and conserving key species and habitats in perpetuity.

References

Arceo, H. O., M. C. C. Quibilan, P. M. Aliño, W. Y. Licuanan, and G. Lim, Coral bleaching in the Philippines: Coincident Evidences of Mesoscale Thermal Anomalies. *Bulletin Marine Science,* 69(2): 579-594, 2001.

Baker, A. C., C. J. Starger, T. R. McClanahan, and P. W. Glynn, Corals' adaptive response to climate change. *Nature,* 430: 741, 2004.
Bellwood, D. R., T. P. Hughes, C. Folke, M. Nystrom, Confronting the coral reef crisis. *Nature* (London), 429: 827-833, 2004.
Bellwood, D. R. and T. P. Hughes, Regional-Scale Assembly Rules and Biodiversity of Coral Reefs. *Science,* 292: 1532-1534, 2001.
Bellwood, D. R., T. P. Hughes, S. R. Connolly, and J. Tanner, Environmental and geometric constraints on Indo-Pacific coral reef biodiversity. *Ecology Letters,* 8: 643-651, 2005.
Berkelmans, R., G. De'ath, S. Kininmonth, and W. J. Skirving, A comparison of the 1998 and 2002 bleaching events on the Great Barrier Reef: spatial correlation, patterns and predictions. *Coral Reefs,* 23: 74-83, 2004.
Brown, B. E., R. P. Dunne, M. S. Goodson, and A. E. Douglas, Bleaching patterns in reef corals. *Nature,* 404: 142-143, 2000.
Buddemeier, R. W., J. A. Kleypas, and R. B. Aronson, Coral Reefs and Global Climate Change: Potential Contributions of Climate Change to Stresses on Coral Reef Ecosystems. The Pew Center on Global Climate Change, Arlington, VA, USA, 2004.
Catala-Stucki, R, Fluorescence effects from corals irradiated with ultra-violet rays. *Nature,* 183: 949, 1959.
Coles, S. L. and B. E. Brown, Coral bleaching – capacity for acclimatization and adaptation. Advances in Marine Biology, 46: 183-223, 2003.
Craig, P., C. Birkeland, and S. Belliveau, High temperatures tolerated by a diverse assemblage of shallow-water corals in American Samoa. *Coral Reefs,* 20: 185-189, 2001.
Day, J., L. Fernandes, A. Lewis, G. De'ath, S. Slegers, B. Barnett, B. Kerrigan, D. Breen, J. Innes, J. Oliver, T. Ward and D. Lowe. The representative areas program for protecting the biodiversity of the Great Barrier Reef World Heritage Area. pp. 687-696. In M. K. Kasim Moosa, S. Soemodihardjo, A. Nontji, A. Soegiarto, K. Romimohtarto, Sukarno and Suharsono. 2002 (Editors) Proceedings of the Ninth International Coral Reef Symposium, Bali, Indonesia, October 23-27 2000. Published by the Ministry of Environment, the Indonesian Institute of Sciences and the International Society for Reef Studies. 1279 pp. ISBN 979-8105-97-4, 2002.
Done T. J. and R. E. Reichelt, Integrated coastal zone and fisheries ecosystem management: generic goals and performance indices. Ecological Applications 8 (Supplement): S110-118, 1998.
Done, T. J., Coral community adaptability to environmental changes at scales of regions, reefs and reef zones. *American Zoologist.* 39: 66-79, 1999.
Donnelly, R., D. Neville, and P. Mous, Report on the Rapid Ecological Assessment of the Raja Ampat Islands, Papua, Eastern Indonesia. Held on October 30-November 22, 2002. The Nature Conservancy, World Wildlife Fund, and Pemda Kabupaten Raja Ampat. 246 pp, 2003.
Dulvy, N. K., R. P. Freckleton, and N. V. C. Polunin, Coral reef cascades and the indirect effects of predator removal by exploitation. *Ecology Letters,* 7: 5, 410-416, 2004.
Dunne, R. and B. Brown, The influence of solar radiation on bleaching of shallow water reef corals in the Andaman Sea, 1993-1998. *Coral Reefs*, 20: 201-210, 2001.
Fitt, W., B. Brown, M. Warner, and R. Dunne, Coral bleaching: interpretation of thermal tolerance limits and thermal thresholds in tropical corals. *Coral Reefs*, 20: 51-65, 2001.
Gates, R. D., and P. J. Edmunds, The physiological mechanisms of acclimatization in tropical reef corals. *American Zoologist*, 39: 30-43, 1999.
Gattuso, J. P., M. Frankignoulle, I. Bourge, S. Romaine, and R. W. Buddemeier, Effect of calcium carbonate saturation of seawater on coral calcification. *Global and Planetary Change,* 18: 37-46, 1998.
Gell, F. R. and C. Roberts, The Fishery Effects of Marine Reserves and Fishery Closures. WWF U.S. and the University of York, Washington D.C. and York, 2003.
Gleason, D. F. and G. M. Wellington. Ultraviolet radiation and coral bleaching. *Nature*, 365: 836-838, 1993.

Gleason, D. F. and G. M. Wellington, Variation in UVB sensitivity of planula larvae of the coral *Agaricia agaricites* along a depth gradient. *Marine Biology* 123: 693-703, 1995.

Glynn, P. W. El Niño-Southern Oscillation mass mortalities of reef corals: a model of high temperature marine extinctions? pp. 117-133 in Insalaco, E., P. W. Skelton, and T. J. Palmer (eds.). Organism-Environment Feedbacks in Carbonate Platforms and Reefs. Geological Society, London, *Special Publications,* 178(151), 2000.

Goreau, T., T. McClanahan, R. Hayes, and A. Strong, Conservation of coral reefs after the 1998 global bleaching event. *Conservation Biology*, 14: 5-15, 2000.

Goreau, T. J., Coral Bleaching in Seychelles: impacts in the South Central Pacific during 1994. Report to the U.S. Department of State, Washington, D.C. Available through Global Coral Reef Alliance web site at http://www.fas.harvard.edu/~goreau, 1998a.

Goreau, T. J., Coral recovery from bleaching in Seychelles, December, 1998. Report to the Seychelles Marine Park Authority. Available through Global Coral Reef Alliance web site at http://www.fas.harvard.edu/~goreau, 1998b.

Goreau, T. J., Coral recovery from bleaching in Alphonse and Bijoutier. Report to the Seychelles Marine Park Authority. Available through Global Coral Reef Alliance web site at http://www.fas.harvard.edu/~goreau, 1998c.

Grigg, R. W., Paleoceanography of coral reefs in the Hawaiian-Emperor Chain. *Science,* 240(4860): 1737-1743, 1988.

Halpern, B. S, The impact of marine reserves: Do reserves work and does reserve size matter? *Ecological Applications*, 13: S117-S137, 2003.

Halpern, B. S. and R. R. Warner, Marine reserves have rapid and lasting effects. *Ecology Letters*, 5: 361-366, 2002.

Hatcher, B. G. and A. W. D. Larkum, An experimental analysis of factors controlling the standing crop of the epilithic algal community on a coral reef. *Journal of Experimental Marine Biology and Ecology*, 69: 61-84, 1983.

Hendee, J. C., E. Mueller, C. Humphrey, and T. Moore, A data-driven expert system for producing coral bleaching alerts at Sombrero Reef in the Florida Keys. *Bulletin of Marine Science,* 69(2): 673-684, 2001.

Hoegh-Guldberg, H. and O. Hoegh-Guldberg, The Implications of Climate Change for Australia's Great Barrier Reef: People and Industries at Risk. A joint report between WWF Australia and the Queensland Tourism Industry Association: 356 pp, 2004.

Hoegh-Guldberg, O., Climate change, coral bleaching and the future of the world's coral reefs. *Marine and Freshwater Research*, 50: 839-866, 1999.

Hoegh-Guldberg, O. and B. Salvat, Periodic mass-bleaching and elevated sea temperatures: bleaching of outer reef slope communities in Moorea, French Polynesia. *Marine Ecology Progress Series*, 121: 181-190, 1995.

Hoeksema, B. W., Control of bleaching in mushroom coral populations (Scleractinia: Fungiidae) in the Java Sea: stress tolerance and interference by life history strategy. *Marine Ecology Progress Series*, 74: 225-37, 1991.

Hughes, T., A. H. Baird, D. R. Bellwood, M. Card, S. R. Connolly, C. Folke, R. Grosberg, O. Hoegh-Guldberg, J. B. C. Jackson, J. Kleypas, J. M. Lough, P. Marshall, M. Nystrom, S. R. Palumbi, J. M. Pandolfi, B. Rosen, and J. Roughgarden, Climate Change, Human Impacts, and the Resilience of Coral Reefs. *Science*, 301: 929-933, 2003.

Hughes, T. P, Herbivory on coral reefs: community structure following mass mortalities of sea urchins. *Journal of Expt. Mar. Biol. Ecol.* 113: 39-59, 1987.

IPCC, Climate Change 2001: The Scientific Basis. Contribution of Working Group I to the Third Assessment Report of the Intergovernmental Panel on Climate Change Houghton, J. T.,Y. Ding, D. J. Griggs, M. Noguer, P.J. van der Linden, X. Dai, K. Maskell, and C. A. Johnson (eds.). Cambridge University Press, Cambridge, United Kingdom and New York, NY, USA: 881 pp., 2001.

Jokiel, P. L. and S. L Coles, Response of Hawaiian and other Indo-Pacific reef corals to elevated temperatures associated with global warming. *Coral Reefs*, 9: 155-162, 1990.

Jones, G. P, M. I. McCormick, M. Srinivasan, and J. V. Eagle, Coral decline threatens fish biodiversity in marine reserves. *PNAS*, 101(21): 8251-8253, 2004.

Kassem, K., M. Toscano, G. Llewellyn, and K. Casey. Where Do Coral Reefs Feel the Heat? A Global Analysis of HotSpot Frequencies and the Consequences for Tropical Marine Biodiversity Conservation Planning. American Geophysical Union/American Society of Limnology and Oceanography Ocean Sciences Meeting. Honolulu, Hawaii, 2002.

Kelleher, G., Guidelines for Marine Protected Areas. IUCN, Gland, Switzerland and Cambridge, UK: xxiv +107 pp, 1999.

Kleypas, J., R. Buddemeier, and J. Gattuso, The future of coral reefs in an age of global change. *Int. J. Earth Sci.*, 90: 426-437, 2001.

Kleypas, J. A., R. Buddemeier, D. Archer, J. P. Gattuso, C. Langdon, and B. N. Opdyke, Geochemical consequences of increased atmospheric CO_2 on corals and coral reefs. *Science*, 284: 118-120, 1999.

Langdon, C., T. Takahashi, F. Marubini, M. Atkinson, C. Sweeney, H. Aceves, H. Barnett, D. Chipman, and J. Goddard, Effect of calcium carbonate saturation state on the calcification rate of an experimental coral reef. *Global Biogeochemical Cycles*, 14: 639-654, 2000.

Langdon, C., W. S. Broecker, D. E. Hammond, E. Glenn, K. Fitzsimmons, S. G. Nelson, T. S. Peng, I. Hajdas, and G. Bonani, Effect of elevated CO_2 on the community metabolism of an experimental coral reef. *Global Biogeochemical Cycles,* 17: 10, 2003.

Leclercq, N., J. P. Gattuso, and J. Jaubert, CO_2 partial pressure controls the calcification rate of a coral community. *Global Change Biology*, 6: 329-334, 2000.

Little A. F., M. J. H. Van Oppen, and B. L. Willis, Flexibility in algal endosymbioses shapes growth in reef corals. *Science*, 304: 1492-1494 2004.

Littler, M. M. and D. S. Littler, Epizoic red alga allelopathic to a Caribbean coral. *Coral Reefs*, 16: 168, 1997.

Lough, J. M., Sea surface temperature variations on coral reefs: 1903-1998. AIMS Report No. 31. Australian Institute of Marine Science, Townsville, 109 pp, 2000.

Loya, Y., K. Sakai, K. Yamazoto, Y. Nakano, H. Sembali, and R. van Woesik., Coral bleaching: the winners and losers. *Ecology Letters*, 4: 122-131.2001.

Lubchenco, J., S. R. Palumbi, S. D. Gaines, and S. Andelman, Plugging a Hole in the Ocean: The Emerging Science of Marine Reserves. *Ecological Applications*, 13: S3-S7, 2003.

Marshall, P. and A. Baird, Bleaching of corals on the Great Barrier Reef: differential susceptibilities among taxa. *Coral Reefs*, 19: 155-163, 2000.

Marubini, F., H. Barnett, C. Langdon, and M. J. Atkinson, Dependence of calcification on light and carbonate ion concentration for the hermatypic coral *Porites compressa*. *Marine Ecology Progress Series*, 220: 153-162, 2001.

McClanahan, T., N. Polunin, and T. Done, Ecological states and the resilience of coral reefs. Conservation Ecology 6(2): 18. [online] URL: http://www.consecol.org/vol6/iss2/art18, 2002.

McClanahan, T. R., Primary succession of coral-reef algae: Differing patterns on fished versus unfished reefs. *Journal of Experimental Marine Biology and Ecology*, 218: 77-102, 1997.

McCook, L. J., Macroalgae, nutrients and phase shifts on coral reefs: scientific issues and management consequences for the Great Barrier Reef. *Coral Reefs*, 18: 357-367, 1999.

Nakamura, T. and R. Van Woesik, Differential survival of corals during the 1998-bleaching event is partially explained by water-flow rates and passive diffusion. *Marine Ecology Progress Series*, 212: 301-304, 2001.

Palumbi, S. R., Molecular biogeography of the Pacific. *Coral Reefs*, 16: S47-S52, 1997.

Palumbi, S. R. Population Genetics, Demographic Connectivity, and the Design of Marine Reserves. *Ecological Applications.* 13: S146-S158, 2003.

Pandolfi, J. M., R. H. Bradbury, E. Sala, T. P. Hughes, K. A. Bjorndal, R. G. Cooke, D. McArdle, L. McClenachan, M. J. Newman, G. Paredes, R. R. Warner, and J. B. Jackson, Global trajectories of the long-term decline of coral reef ecosystems. *Science*, 301: 955-958, 2003.

Pauly D., V. Christensen, J. Dalsgaard, R. Froese and F. Torres, Fishing down marine food webs. *Science*, 279: 860-863, 1998.

Pittock, A. B., Coral reefs and environmental change: Adaptation to what? *Am. Zool*, 39(1): 10-29, 1999.

Pomeroy, R. S., J. E. Parks, and L. M. Watson, How is your MPA doing? A guidebook of natural ands social indicators for evaluating marine protected area management effectiveness. IUCN Gland. Switzerland and Cambridge, UK. xvi +216 pp, 2004.

Rowan, R., Thermal adaptation in reef coral symbionts. *Nature*, 430:742, 2004.

Rowan, R. and D. A. Powers, Molecular genetic identification of symbiotic dinoflagellates (zooxanthellae). *Marine Ecology Progress Series*, 71: 65-73 1991.

Rowan, R., N. Knowlton, A. C. Baker and J. Jara, Landscape ecology of algal symbiont communities explains variation in episodes of coral bleaching. *Nature,* 388: 265-269, 1997.

Russ, G., Yet another review of marine reserves as reef fishery management tools. pp. 421-443 in Sale, P. F. (ed.). Coral Reef Fishes. Dynamics and diversity in a complex ecosystem. *Academic Press*, *San Diego*, *2002*.

Salm, R.V., S. E. Smith and G. Llewellyn, Mitigating the impact of coral bleaching through marine protected area design. pp. 81-88 in Schuttenberg, H.Z. (ed.). Coral Bleaching: Causes, Consequences and Response. Selected papers presented at the 9th International Coral Reef Symposium on “Coral Bleaching: Assessing and Linking Ecological and Socioeconomic Impacts, Future Trends and Mitigation Planning.” Coastal Management Report #2230, Coastal Resources Center, University of Rhode Island: 102 pp, 2001.

Sammarco, P. W., *Diadema* and its relationship to coral spat mortality: grazing, competition, and biological disturbance. *Journal of Experimental Marine Biology and Ecology*, 45: 245-272, 1980.

Siebek, O., Experimental investigation of UV tolerance in hermatypic corals (Scleractinia). *Marine Ecology Progress Series*, 43: 95-103, 1988.

Smith, S. V. and R. W. Buddemeier, Global change and coral reef ecosystems. *Annual Review of Ecological Systems*, 23: 89-118, 1992.

Smith, S. V., W. J. Kimmerer, E. A. Laws, R. E. Brock, and T. W. Walsh, Kaneohe Bay sewage diversion experiment: perspectives on ecosystem responses to nutritional perturbation. *Pacific Science*, 35: 279-402, 1981.

Soto, C., The potential impacts of global climate change on marine protected areas. *Reviews in Fish Biology and Fisheries*, 11: 181-195, 2002.

Spalding, M., K. Teleki, and T. Spencer. Biodiversity and Climate Change: Climate Change and Coral Bleaching Report. UNEP-World Conservation Monitoring Centre, 2000.

Stafford-Smith, M. G. and R. F. G. Ormond, Sediment rejection mechanisms of 42 species of Australian scleractinian corals. *Australian Journal of Marine and Freshwater Research*, 43: 683-705, 1992.

Steneck, R. S. and M. N. Dethier, A functional group approach to the structure of algal-dominated communities. *Oikos*, 69: 476-498, 1994.

Van Oppen, M. J. H., A. J. Mahiny and T.J. Done, Geographic distribution of zooxanthellae types in three coral species on the Great Barrier Reef sampled after the 2002 bleaching event. *Coral Reefs*, 24: 482.2005.

Veron, J. E. N., Corals in space and time: biogeography and evolution of the Scleractinia. Ithica, Comstock, *Cornell*, 321 pp, 1995.

Wagner, G. M., Y. D. Mgaya, F. D. Akwilapo, R. G. Ngowo, B. D. Sekadende, A. Allen, N. Price, E. A. Zollet, N. Mackentley, Restoration of coral reef and mangrove ecosystems at Kunduchi and Mbweni, Dar es Salaam, with community participation. Western Indian Ocean Marine Science Association, Zanzibar (Tanzania); Institute of Marine Sciences, Zanzibar (Tanzania) University of Dar es Salaam. Marine Science Development in Tanzania and Eastern Africa. 467-488, 2001.

Wellington, G. M., P. W. Glynn, A. E. Strong, S. A. Navarrete, E. Wieters, and D. Hubbard, Crisis on Coral Reefs Linked to Climate Change. EOS, Transactions, American Geophysical Union 82(1): 1-7, 2001.

West, J. M. and R. V. Salm, Resistance and resilience to coral bleaching: Implications for coral reef conservation and management. *Conservation Biology*, 17(4): 956-967, 2003.

Williams, I. D., N. V. C. Polunin, and V. J. Hendrick, Limits to grazing by herbivorous fishes and the impact of low coral cover on macroalgal abundance on a coral reef in Belize. *Marine Ecology Progress Series*, 222: 187-296, 2001.

Wooldridge, S. and T. J. Done, Learning to predict large-scale coral bleaching from past events: A Bayesian approach using remotely sensed data, in-situ data, and environmental proxies. *Coral Reefs*, 23: 96-108, 2004.

World Wide Fund for Nature (WWF). Buying Time: A User's Manual for Building Resistance and Resilience to Climate Change in Natural Systems. Hansen, L. J., J. L. Biringer, and J. R. Hoffmann (eds.). Published by Martin Hiller, 2003.

12

Adapting Coral Reef Management in the Face of Climate Change

Paul Marshall and Heidi Schuttenberg

Abstract

Climate change is now recognized as one of the most significant threats to coral reefs worldwide. The effects of increased sea temperatures on coral populations are the main cause for concern for the future of reefs, due to predicted increases in the frequency and severity of mass coral bleaching events over the next century. Additionally, acidification of the oceans, changes in current patterns, increased incidence of disease and physical changes at the coastal margins all have implications for coral reef ecosystems and associated human communities. The urgency and magnitude of the threat posed by climate change requires coral reef managers to develop meaningful response strategies and to adapt their management practices in the face of climate change. Here we recommend a two-part approach for adapting coral reef management in the face of climate change that consists of: (1) strengthening efforts to mitigate the impacts of climate change by raising awareness about the implications of climate change for coral reefs, and (2) managing coral reef ecosystems for resilience. Managing for resilience involves taking actions that support the capacity of coral reef ecosystems to successfully reorganize during stress and change rather than shifting to a predominantly algal state. This capacity is largely determined by ecosystem condition, biological diversity, connectivity between reefs, and local environmental factors. Two key strategies for supporting reef resilience are: adapting marine protected area design and management; and reviewing and revising management targets for recreation, water quality and fishing.

1. Introduction

Global climate change is already affecting the earth's ecosystems, with impacts observed on phenology, species ranges, reproductive success, migrations and species interactions in a broad range of habitats and environments [Parmesan and Yohe, 2003; Walther et al., 2002]. Coral reefs appear to be particularly sensitive, and are recognised as among the ecosystems most vulnerable to climate change [IPCC, 2001a]. In combination with the more localised stresses such as overfishing and degraded water quality, climate change has lead to concerns that coral reefs are in "crisis" [Bellwood et al., 2004].

The principle basis for the alarming prognoses for coral is the critical tolerances of corals to anomalous water temperatures [Buddemeier et al., 2004; Hoegh-Guldberg, 1999;

Coral Reefs and Climate Change: Science and Management
Coastal and Estuarine Studies 61
Copyright 2006 by the American Geophysical Union.
10.1029/61CE13

McClanahan, 2002; McClanahan et al., 2002b]. Temperatures that exceed normal summer maxima by only 1.5-2.0°C are enough to cause coral bleaching, and prolonged high temperatures over large areas can lead to "mass" coral bleaching, with the potential to cause extensive mortality [Brown, 1997]. With a warming climate, mass coral bleaching events are predicted to increase in frequency and severity, threatening the ecological foundations of reef ecosystems [Hoegh-Guldberg, 1999]. Unusually high temperatures are estimated to have caused the loss of 16% of the world's coral reefs during the 1998 mass bleaching event alone, with some regions such as the Maldives, Palau and Seychelles suffering more than 50% coral mortality over large areas [Wilkinson, 1998; Wilkinson, 2002a].

Climate change is also expected to have a range of other implications for coral reefs. Most significant among these is the affect on ocean chemistry of increased concentrations of CO_2 [Gattuso and Buddemeier, 2000]. Measurable increases in acidity suggest that the availability of aragonite for the building of skeletons by corals and other calcifying organisms will be marginal by the middle of the century in many parts of the tropical seas [Kleypas et al., 1999]. Changes in ocean currents are also predicted, with potential implications for patterns of connectivity, location and strength of upwelling, and timing and persistence of productivity zones [Buddemeier et al., 2004]. Diseases of corals and other reef organisms are also expected to increase as a result of climate change, as the virulence of many disease organisms increase with temperature [Harvell et al., 2002].

Furthermore, significant physical changes at the coastal margins are predicted that have the potential to impact a range of coral reef organisms and processes [Buddemeier et al., 2004; Pittock, 2003]. Increased sea levels may exceed the vertical growth rate of some reefs, while inundation of coastal areas has the potential to change shoreline morphology, reduce the availability of habitat for shore-dependent species (such as nesting turtles) and mobilise new sources of sediments. Changes in terrestrial climate also have the potential for downstream effects on coral reefs, with increased drought severity and greater rainfall intensity expected to deliver more runoff and higher loads of sediments, nutrients and contaminants to many nearshore reef areas [Pittock, 2003].

Through a combination of direct and indirect effects, climate change has the potential to threaten the existence of species and habitats [Thomas et al., 2004]. At a minimum, climate change can be expected to change the relative abundances of key species and degrade the condition of many coral reef habitats [Done et al., 2003; Hughes et al., 2003]. Additionally, and significantly, climate change is almost certain to undermine the sustainability of human uses of coral reef resources under current management regimes [Hughes et al., 2003]. As a consequence, climate change requires managers to assess the implications of climate change for coral reefs in their jurisdiction, and to adapt their management approaches to take account of this emergent threat in relation to other issues [Buddemeier et al., 2004; Hughes et al., 2003; Hughes, 2005; Wooldridge et al., 2005].

The aim of this paper is to provide an overview of the range of consequences of climate change for coral reefs, and to outline the types of strategies available to managers to respond to climate change. This should be seen as a prelude to more detailed knowledge and guidance that will continue to emerge as scientists and management practitioners collaborate in an effort to meaningfully tackle the challenge of climate change over the coming decade.

2. Implications of Climate Change for Coral Reef Ecosystems

Climate change has far-reaching implications for coral reef ecosystems. Changes in physico-chemical variables that define the environmental niche of coral reefs, such as temperature and the carbonate saturation state of seawater, are predicted to accelerate in the course of this century [Buddemeier et al., 2004]. Other effects of climate change on coral reefs are also beginning to be examined, such as shifts in the strength, location or timing of ocean currents, increases in disease risk, rising sea levels and altered physical processes at the coastal margins. The majority of our knowledge about climate change impacts on coral reef ecosystems is centred on corals, yet studies of the implications for other coral reef organisms, such as fish [Spalding and Jarvis, 2002], algae [Flanagan et al., 2003; McClanahan, 2002], seabirds [Smithers et al., 2003] and turtles [Fish et al., 2005] are beginning to accumulate. The following section summarises current knowledge of the major implications of climate change for coral reefs. This provides a foundation for the discussion of management responses to climate change in Section 3.

2.1. Increased Sea Temperatures

The dramatic effects of increased temperatures are already evident in the mass bleaching events and associated mortality of corals that have been observed at many locations since 1998 [Wilkinson, 2004a]. Corals are highly sensitive to elevated temperatures outside their normal range. While contemporary coral reefs span a temperature gradient of 9°C (i.e., 34°C in the Arabian Gulf; 25°C at Easter Island [Coles and Brown, 2003; Wellington et al., 2001]), corals are strongly adapted to local thermal regimes [Coles and Brown, 2003]. Temperatures need only exceed the long-term summer maximum of the local environment by 1.5-2.0°C for a short period (6-10 weeks) to cause stressful conditions that lead to coral bleaching [Berkelmans, 2002]. The worldwide mass bleaching event of 1998 demonstrated the potential for elevated sea temperatures to dramatically affect the abundance and composition of coral reefs on a global scale [Wilkinson, 2002b]. Conservative scenarios for climate change project an increase in temperatures of 1.4-5.8°C relative to 1990 by the end of the century [IPCC, 2001a], driving an increase in the frequency and severity of coral bleaching events [Hoegh-Guldberg, 1999], and prompting concerns that mass bleaching will substantially alter the abundance and composition of coral communities at a global scale [Hughes et al., 2003; Wilkinson, 2004a]. Increased sea temperatures also have implications for disease for corals and a diversity of other benthic and mobile reef species [Harvell et al., 1999]. Furthermore, the strong links between metabolic processes and temperature suggests that increased sea temperatures will have a wide range of sub-lethal implications for organisms on coral reefs. The various implications of elevated sea temperature are examined below.

2.1.1. Mass coral bleaching

Coral bleaching is a stress response in corals that can be triggered by various stressors, including changes in temperature or salinity and exposure to toxins [Brown, 1997]. However, *mass* coral bleaching, which occurs when the majority of corals are affected near-simultaneously over spatial scales of kilometres, is primarily triggered by anomalously high water temperatures [Brown, 1997]. Thermally-induced bleaching results when corals expel the symbiotic dinoflagellates (zooxanthellae) hosted within their tissues in response to dysfunction in the light-processing mechanism in the algae [Jones et al., 1998]. The tissue of bleached corals can be repopulated with zooxanthellae if thermal stress abates soon enough, and the coral will survive, albeit with various sub-lethal impacts such as reduced growth and reproductive output [Baird and Marshall, 2002; Michalek-Wagner and Willis, 2001a]. However, prolonged thermal stress readily leads to mortality, with long-lasting ecological consequences for coral reef ecosystems [Brown, 1997; Hoegh-Guldberg, 1999; McClanahan, 2002].

Bleaching of corals is not a new phenomenon, and low levels of bleaching can be observed on many reefs during average summer conditions. However, mass bleaching, and particularly mass bleaching events that result in extensive coral mortality, appear to be a relatively recent occurrence that is increasing in frequency [Buddemeier et al., 2004; Coles and Brown, 2003; Hoegh-Guldberg, 1999]. The last decade has seen reports of mass bleaching events that are unprecedented in extent and severity, resulting in impacts on the abundance and community composition of coral reefs at a global scale [Wilkinson, 2002a; Wilkinson, 2004b].

A future in which bleaching events occur more frequently, or with greater severity, is likely to drive a general decline in the abundance of corals on reefs [Done et al., 2003; Hoegh-Guldberg, 1999; Hughes et al., 2003; Wooldridge et al., 2005]. While the long term effects of bleaching on coral abundance are difficult to predict with confidence, future scenarios have been explored in recent modelling studies [Done et al., 2003; Wooldridge et al., 2005; WWF, 2004]. These projections demonstrate the plausibility of massive degradation of coral reefs due to warming seas. Even under relatively conservative warming scenarios, many reefs previously dominated by a diverse assemblage of hard corals may give way to low abundances and depauperate communities [Sheppard, 2003]. In turn, this could lead to algal-dominated reefs with low habitat complexity in as little as 50 years [Bellwood et al., 2004; Hoegh-Guldberg, 2000]. Should this occur throughout the ecosystem, the capacity for recovery of hard corals will also decrease, exacerbating the rate of decline. Although these model results may appear overly pessimistic, they do not rely on catastrophic change. Rather they assume, very conservatively, that reefs can recover between bleaching events as long as there are fewer than three massive mortality events per decade [Hoegh-Guldberg, 2000]. Related modelling studies predict, however, that severe bleaching events will become annual events by 2020 [Hoegh-Guldberg, 1999]. Additionally, these models do not include the cumulative or synergistic effects of other stresses such as water pollution or destructive fishing practices [Hughes et al., 2003; Wooldridge et al., 2005].

In addition to the impacts of coral bleaching on live coral cover, the composition of coral communities is likely to change as a result of a warming climate [Buddemeier et al., 2004; Coles and Brown, 2003; Hughes et al., 2003; Wooldridge et al., 2005]. Among corals there is substantial variation in the susceptibility to bleaching of different coral taxa [Marshall and Baird, 2000]. The coral species most likely to show declines in abundance immediately after a severe bleaching event include those that tend to be relatively fast growing and visually dominant, such as staghorn and tabular *Acropora* [Brown et al., 2000; Marshall and Baird, 2000]. The loss of these species is likely to lead to an initial shift toward the slower-growing, massive growth forms such as *Porites* and Faviids. While this can have a dramatic visual impact on many reefs, it will also result in decreased habitat for many reef-dependent species [Jones et al., 2004]. These bleaching-susceptible coral species may also be among the quickest to recover by way of larval recruitment and rapid growth if the return intervals of mass bleaching events are large enough [Brown et al., 2000; Buddemeier et al., 2004; Coles and Brown, 2003]. However, modelling studies suggest that this last condition is increasingly unlikely to be met as climate change progresses, even under conservative scenarios [Hoegh-Guldberg, 1999]. Once temperatures increase to the point that bleaching events are frequent and severe enough that even bleaching-resistant species are killed [Hoegh-Guldberg, 1999], reef communities are likely to become dominated by more ruderal (non-coral) species as the coral community is forced into constant recovery [Wooldridge et al., 2005]. These changes in coral abundance and community composition suggest that coral bleaching has substantial implications not just for corals, but also for the diverse array of species that rely on coral-dominated reefs [Hughes et al., 2003; Jones et al., 2004; McClanahan, 2002], including humans [Moberg and Folke, 1999].

A critical factor in determining the fate of coral reefs under climate change is the potential for corals to adapt to increasing temperatures, and thereby mitigate the cumulative impacts of coral bleaching over coming decades [Coles and Brown, 2003]. Research demonstrating the role of zooxanthellae in the temperature sensitivity of corals, and the potential flexibility of the symbiosis [Little et al., 2004], suggests there exists the potential for adaptation of the holobiont in response to selective pressures during mass bleaching events [Baker, 2001]. However, switching between the two most common clades of zooxanthellae (clades c and d) appears unlikely to increase bleaching resistance beyond predicted temperature change, as well as imposing a growth penalty [Little et al., 2004].

2.1.2. Changes in physiology, behaviour and distribution

In addition to coral bleaching, elevated water temperatures can be expected to have physiological implications for a wide variety of organisms in coral reef ecosystems [Buddemeier et al., 2004; Kennedy, 2002]. Many metabolic processes are temperature regulated [Buddemeier et al., 2004; Hochachka and Somero, 2002], and increases in ambient temperatures can be expected to contribute to physiological, behavioural and distributional changes for a variety of organisms. Observed effects that are indicative of future impacts of climate change include sub-lethal impacts to sessile marine species such as corals [Baird and Marshall, 2002; Michalek-Wagner and Willis, 2001a] and significant and relatively rapid changes in distribution of key fish species [Perry et al., 2005].

Detecting and attributing the effects of global warming is a major challenge generally [IPCC, 2001b]. It can be particularly difficult for complex ecosystems such as coral reefs, or for species subject to a diversity of other influences on individual behaviour and population dynamics [Parmesan and Yohe, 2003]. In these cases, knowledge of temperature sensitivity can help evaluate risks from climate change. As an example, the sensitivity to temperatures of the early life history in sea turtles can be used to gain insight into their vulnerability to climate change, even though detecting and attributing changes in these long-lived and far-ranging species is likely to be very difficult [Fish et al., 2005]. The temperature-dependence of sex-determination of sea turtles during egg incubation suggests that an increase in global sand temperatures could affect sex ratios in future cohorts [Hays et al., 2003]. Should temperatures increase on average across all nesting sites of a population, a shift in the sex ratio toward a predominance of females could result [Limpus, 1993]. A comprehensive review of the temperature sensitivities of coral reef species is beyond the scope of this paper, but such an approach can provide a useful foundation for assessing risks associated with climate change where there are few long-term studies of climate impacts.

2.1.3. Increases in coral disease

Many coral diseases increase in virulence at higher temperatures, suggesting greater prevalence of disease outbreaks as average sea temperatures increase [Harvell et al., 2002]. Furthermore, there appears to be a link between coral bleaching and coral disease outbreaks, with observations of significant and sometimes dramatic increases in disease incidence following bleaching events, particularly in the Caribbean [Harvell et al., 2001; Rosenberg and Ben-Haim, 2002; Santavy, 2005]. Diseases have already caused chronic coral mortality in many reefs in areas such as the Florida Keys and the Caribbean, and reports of coral disease are increasing in the Great Barrier Reef and other Pacific locations

[Dinsdale, 2000; Rosenberg and Ben-Haim, 2002]. Coral disease is now recognised as a serious management issue in many reef locations and is increasingly the subject of targeted research programs in areas such as the Florida Keys and the Great Barrier Reef (personal observations).

2.2. Acidification of the Oceans

In contrast to temperature effects, the implications of changes in ocean chemistry are likely to be much less obvious in the short term. However, the chronic effects of reduced calcification rates in corals and other calcifying organisms are emerging as serious concerns for the long term future of reefs and the functions they provide human societies, such as shoreline protection [Buddemeier et al., 2004; Kleypas et al., 1999]. Dramatic increases in the levels of carbon dioxide (CO_2) in the earth's atmosphere are leading to a reduction in the pH of seawater, which in turn is decreasing the availability of carbonate ions [Gattuso and Buddemeier, 2000; Kleypas et al., 1999]. Reduced calcium carbonate saturation of seawater is expected to significantly reduce the rates of calcification in key reef-building organisms such as corals and coralline algae, resulting in slower growth and/or more brittle skeletons [Buddemeier et al., 2004; Gattuso et al., 1998; Kleypas et al., 1999]. The implications of this for the ability of coral reefs to withstand storms and to maintain their role in shoreline protection are still being examined, but early indications are that these changes will be important, even if they manifest themselves only slowly or subtly [Buddemeier and Fautin, 1993]. In particular, lowered rates of skeletal extension may reduce the ability of shallow-water corals in some locations (particularly in higher latitudes) to keep pace with sea level rise, even under moderate scenarios [Riegl, 2003].

2.3. Changes in Current Patterns

The potential for climate change to have dramatic impacts on oceanic circulation has been well established [Stocker, 2000], with support emerging from observational studies [Gille, 2002]. In particular, the possibility of major changes in the thermo-hyaline circulation that drives ocean-wide currents and the weather patterns of entire continents has been a focal issue for climate science over the recent decade [Stocker, 2000]. However, the effects of climate change on circulation patterns at regional and sub-regional scales are much less studied [Harley et al., 2006], including for the tropics. Yet critical oceanographic processes such as current strength and intrusion, location and persistence of convergence zones, and the timing and frequency of upwelling are all strongly driven by water temperature and/or density differences [Siedler et al., 2001]. It is logical, therefore, to assume that climate change will result in alterations to the oceanography of coral reef areas, even though the nature (and in most cases, even the direction) of these changes is not well known.

In some locations, such as the Great Barrier Reef, observations of ecological changes are the first evidence of changes in ocean processes. Seabirds have been observed to suffer provisioning failures during periods of anomalously high sea surface temperatures [Smithers et al., 2003]. These events, resulting in over 50% mortality of fledgling chicks, appear to be linked to changes in location of prey-rich convergence zones, which in turn are correlated with anomalous sea surface temperature patterns [Smithers et al., 2003]. The potential for temperature increases to cause changes in ocean currents, and the strong dependence of at least some species on particular oceanographic processes and features, suggests that the ecological implications of changes in current patterns may prove to be an important impact of climate change.

2.4. Physical Changes at the Coastal Margins

The expected rate of sea level rise is not expected to pose a significant challenge to vertical reef growth in healthy coral reef ecosystems [Smith and Buddemeier, 1992]. However, reductions in rates of skeletal accretion of hard corals due to ocean acidification [Kleypas et al., 1999] could compromise the ability of reefs to keep pace with sea level rise. Furthermore, there has been recent speculation [De Angelis and Skvarca, 2003] that the rates of sea level rise could be significantly different from those presented in the Third Assessment Report of the IPCC [IPCC, 2001a].

Apart from the direct risk of reefs being "drowned", rising sea levels have other implications for coral reef ecosystems. Shoreline inundation as sea levels rise is likely to increase the export of sediments, nutrients and pollutants from newly-flooded coastal areas [Buddemeier et al., 2004]. Increased inputs of these materials are known to add stress to coral reef ecosystems [Fabricius, 2005]. In particular, increased sediment loads and nutrient concentrations can significantly reduce the resilience of coral communities by compromising critical recovery processes such as larval settlement and survival [Anthony and Connolly, 2004; Fabricius, 2005]. In addition, animals that rely on the low-lying habitat provided by coral reef islands and cays, such as sea turtles and sea birds, are likely to be significantly affected by rising water levels [Fish et al., 2005].

3. Managing Coral Reefs in a Changing Climate

Current knowledge about the effects of climate change on coral reefs suggests that their future condition will depend largely on two factors: (1) the rate and extent of increased thermal stress, and (2) the resilience of coral reef ecosystems. This section outlines a two-part strategy for managing coral reefs in a changing climate based on these two factors. Section 3.1 discusses the importance of using climate change impacts on coral reefs as a catalyst for efforts to mitigate the rate and severity of global temperature increase. Section 3.2 outlines various approaches that are emerging from scientific and management forums to guide reef managers toward positive action to support coral reef resilience in the face of climate change [Marshall and Schuttenberg, 2006; Obura, 2005; Salm and West, this volume; Schuttenberg et al., 2000; The Nature Conservancy, 2005; Westmacott et al., 2000].

3.1. Increasing the Imperative for Mitigating Climate Change

Further deterioration of the world's coral reefs due to climate change is inevitable [Buddemeier et al., 2004; Hoegh-Guldberg, 1999; Hughes et al., 2003; Pandolfi et al., 2003]. The challenge for coral reef managers, then, is to identify strategies and implement activities that will minimise the extent of this deterioration. This challenge is most likely to be met through a combination of strategies that increase the resilience of the system (next section), and reduce the rate and extent of anthropogenic climate change [Buddemeier et al., 2004; Hansen et al., 2003; Hughes et al., 2003; West and Salm, 2003]. Human activities are significant contributors to observed and projected changes to the earth's climate [G8 Summit, 2005; IPCC, 2001b], creating opportunities for mitigation through increased management of these activities.

Coral reef managers and scientists can play an influential role in efforts to mitigate climate change, even though they are unlikely to be directly involved in greenhouse reduction activities or climate policy discussions. In partnership with scientists and other relevant experts, managers are well placed to measure, understand and communicate the implications of climate change for coral reefs and the people who depend on them. This information is critical for a holistic evaluation of the costs of mitigation activities against the costs of damage from unbridled climate change. While ascribing values to observed and potential impacts to coral reef ecosystems and the goods and services that they provide humanity remains a challenge for natural resource managers and scientists [Folke et al., 1993; Moberg and Folke, 1999], the vulnerability of coral reef ecosystems to climate change is increasingly being recognised in both science [Hughes et al., 2003] and policy forums [Allen Consulting Group, 2005]. By taking a proactive role in communicating the implications of climate change for coral reefs and those who depend on them, coral reef managers can increase the imperative for concerted action to mitigate climate change. Even small reductions in the rate and extent of climate change may buy important time for ecosystems to adjust and thereby minimise the impact of changing conditions [Hansen et al., 2003].

3.2. Managing Coral Reefs for Resilience

Supporting the resilience of the ecosystem will serve as a critical insurance policy against unpredictable or unmanageable stresses, such as climate change and coral bleaching [West and Salm, 2003]. Coral reef ecosystems are highly dynamic systems that have evolved to cope with a wide range of disturbances [Connell et al., 1997]. While a resilient system will have the best chances of coping with future threats, human influences have eroded the natural resilience of many coral reef systems, reducing their capacity to cope with disturbance [Nyström et al., 2000]. Strategies aimed at rebuilding and supporting the resilience of the system are the best investment for ensuring that reefs can continue to provide the goods and services upon which humans depend [Bellwood et al., 2004; Salm and West, this volume]. Contemporary thinking on resilience includes consideration of the larger socio-ecological system, including the institutions that are responsible for conservation and natural resource management [Gunderson, 2000]. In this context, resilience-based management is likely to require significant changes in the way management organisations approach their business [Hughes, 2005]. However, even within current institutional settings, there is much that can be done to support the resilience of the ecosystem in the face of climate change [Hughes et al., 2003; Marshall and Schuttenberg, 2006; McClanahan et al., 2003; Nyström et al., 2000; Obura, 2005]. This section introduces the concept of ecosystem resilience and the ideas behind resilience-based management.

Defining resilience Ecosystem resilience relates to the ability of a system to undergo change and yet retain the processes and dynamics that maintain function and structure [Hughes et al., 2003; Nyström and Folke, 2001]. For coral reef ecosystems, this includes maintaining coral dominance rather than shifting to a predominantly algal state. It also includes the potential of the system to reorganise and build its capacity to adapt to change [Holling et al., 1995; Nyström and Folke, 2001]. In essence, resilience is the quality that determines the magnitude of disturbance that can be tolerated without coral reefs shifting into another, undesirable state.

In a mass bleaching context, resilience can be considered as the ability to resist, survive, or recover after recurrent bleaching events [Marshall and Schuttenberg, 2006; Obura, 2005]. Individual corals may resist bleaching or survive the bleaching process, and if significant mortality of corals on a reef occurs, a reef system may still maintain key system characteristics (structure and function) through rapid recovery and reorganisation, relative to less resilient reefs. The capacity of coral reefs to recover from disturbances will become increasingly important as the frequency and severity of bleaching events increases [Hoegh-Guldberg, 2004; McClanahan, 2002]. Reefs with lowered resilience are more susceptible to serious and lasting impacts from coral bleaching events [Nyström and Folke, 2001]. In a broader context, the cumulative effects of global and local stressors will determine the long-term resilience of coral reef ecosystems. While both global and local stressors influence resilience of reef ecosystems, local stressors are much easier to manage in the short term, and provide opportunities for coral reef managers to play an active role in supporting the resilience of reefs within their jurisdiction.

Characteristics of resilient coral reef areas Patterns of past bleaching, mortality and reef recovery provide insights into an area's resilience to mass coral bleaching. Based on evidence from the literature and systematically compiled observations from researchers in the field, a number of local factors that correlate with resistance and resilience to coral bleaching have been identified [Gunderson, 2000; Obura, 2005; West and Salm, 2003]. Resistance and resilience to bleaching are associated with features that:

- Reduce temperature stress, e.g., localised upwelling and proximity to deep or cooler water.
- Increase water movement and flush harmful toxins, e.g., topographic features such as narrow channels and strong currents.
- Screen corals from damaging radiation, e.g., high island shading, reef shelf shading, aspect relative to the sun, or water turbidity.
- Indicate potential pre-adaptation to temperature and other stressors, e.g., highly variable temperature regimes, regular exposure at low tides, history of corals surviving bleaching events.
- Indicate strong recovery potential, e.g., abundance of coral larvae or strong recruitment.
- Improve coral larval transport to the site, e.g., connectivity with source reefs.
- Maintain a favourable substrate for coral larval recruitment, e.g., diverse community structure present or effective management regime in place.

These characteristics can act as criteria in site selection processes for MPA planning activities where the goal is to maximise the resilience of the ecosystem [West and Salm, 2003]. However, a range of other considerations should be made when using marine protected areas as part of a resilience-based management program, as discussed below.

3.2.1. Ecological goals to guide managing for resilience

Factors that influence the resilience of coral reef ecosystems can be grouped into four categories: (1) ecosystem condition, (2) biological diversity, (3) connectivity, and (4) local environment. Each of these categories includes attributes that can strengthen resistance, survival, and recovery from mass bleaching, as well as recovery from other types of disturbances. These factors can be the basis for goals for management strategies discussed in sections 3.2.2 and 3.2.3. Management efforts can be designed to take advantage of and protect areas that are naturally resilient due to local environmental variation and connectivity between reefs. They can also aim to maintain and strengthen coral reef condition and biological diversity as a way of enhancing coral reef resilience. These four factors are discussed further below.

Ecosystem condition There is ample evidence that coral ecosystems in good condition will recover from mortality more successfully than degraded ecosystems [Buddemeier et al., 2004; Done, 2001; Hughes et al., 2003; McClanahan et al., 2002a]. Key influences on coral reef ecosystem condition include coral condition, coral cover, water quality, and fish abundance [West and Salm, 2003]. These attributes influence survivorship during mass bleaching events and recovery after mass bleaching events or other disturbances. Healthy reef ecosystems are better able to provide the conditions required for the recruitment, survival and growth of corals to replace those killed by bleaching. Recovery potential is especially dependant on ecosystem conditions, as it requires a source for new coral recruits, suitable substrate for settlement and survival of larval corals, good water quality, and an abundant and diverse community of herbivores [McClanahan et al., 2003; McCook, 1999]. Management efforts that are effective in restoring or maintaining ecosystem condition are likely to be influential in determining the resilience of reefs in the face of climate change.

Biological diversity Biological diversity includes genetic diversity within species, and species diversity within ecosystem functions. These attributes influence resistance to bleaching, survivorship during bleaching, and recovery after mass bleaching events or other disturbances. In particular, variation in the temperature sensitivity in different genetic groups (clades) of zooxanthellae has been shown to be an important source of variation in the thermal tolerance of corals [Baker, 2001; Little et al., 2004]. Genetic differences between coral species will also strongly influence the future reef seascape. Notably, species vary in their ability to resist bleaching and in the speed with which they can recover from disturbance [Baird and Marshall, 2002]. This variation will compound over time as reef ecosystems are repeatedly exposed to thermal stresses, resulting in a shift in species composition. More diverse ecosystems will have greater scope for remaining in a coral-dominated state during such an ecosystem reshuffle [Done, 2001; McClanahan, 2002].

Biological diversity has a very practical function in protecting reef ecosystems from future threats [Holling et al., 1995]. A system is less prone to failure when key functions are performed by multiple species, because there is less chance that any one disturbance will eliminate all organisms that perform that key function. The importance of this property, 'functional redundancy' [Nyström et al., 2000], is well illustrated by the role of herbivores in coral reef ecosystems. In a case study from Jamaica, overfishing had prevented herbivorous fishes from playing a significant role in controlling algal growth [Hughes, 1994]. As a result, this herbivory function, which is essential for allowing coral populations to recover from mortality events, was largely dependent on only one species of sea urchin. Subsequently, a disease epidemic killed most of the urchin population, leaving too few herbivores in the system to perform the function of algal removal. When a major storm caused widespread damage to coral communities, the unchecked algal growth prevented substantial recovery of corals. These reefs have remained algal-dominated for over a decade. This lack of functional redundancy made the system very susceptible to disturbances and led to a phase shift to an algal-dominated system, which has substantially lower value in providing ecosystem services to humans [Hughes, 1994; Moberg and Folke, 1999].

Connectivity Connectivity between damaged populations and refugia is critical to recovery processes following a disturbance. From the perspective of a coral reef manager, however, connectivity needs to be considered as more than larvae drifting in largely unmanageable ocean currents. Many habitats and habitat patches are critical for different life-stages of organisms, or are sites of sub-populations that can act as ecological links between larger populations [Cappo and Kelley, 2001]. These linking habitats, which can be important for maintenance and regeneration of populations, may become increasingly critical to ecosystem resilience as reefs spend greater time in recovery mode due to larger and more frequent disturbance events. Management efforts that provide effective protection of adequate areas of all of these habitats are likely to be essential to the coral reef system's capacity to successfully adapt to climate change [Done, 2001; McClanahan, 2002].

Local environment Variation in the local environment can determine exposure to heat stress or current speed, which are factors that influence resistance of corals to bleaching, and their survival when bleached [Nakamura and Van Woesik, 2001; West and Salm, 2003]. For example, exposure to heat stress will vary depending on location within the reef (such as reef flat v. reef slope) or, at a larger scale, a reef's orientation with respect to upwelling or coastal topography. Topographic complexity and variability in aspect and surface rugosity are all likely to play an important role in conferring resistance to coral bleaching.

3.2.2. Using marine protected areas to increase resilience

Marine Protected Areas (MPAs) are a key strategy in resilience-based management. MPAs can help build coral reef resilience by supporting and enhancing the local factors that confer resilience: good coral reef condition, biological diversity, connectivity, and favourable local conditions. Traditionally, principles of MPA selection, design and management have not specifically addressed the threat of mass coral bleaching. However, with the inclusion of the following considerations, the value of MPAs in a resilience-based management system can be optimised:

- *MPA site selection*: the basis for identifying candidate MPA sites should include resilience principles, especially as they relate to coral bleaching.
- *Representation and replication*: MPA network design should aim to replicate a range of reef types and related habitats in order to maximise biodiversity protection.
- *Refugia*: sites with natural resistance to coral bleaching should be considered for inclusion in a MPA network, as these can serve as source reefs for less resistant areas.
- *Connectivity*: knowledge of connectivity between reefs should guide the selection and spatial arrangement of MPAs in a reserve network.
- *Effective management*: MPA site selection should give priority to sites where management is likely to be most effective in restoring and maintaining resilience of reef areas.

A detailed examination of these issues is presented in Salm et al. (this volume).

3.3. Managing MPAs in the Context of Mass Coral Bleaching

Once sites are selected for inclusion in a MPA network, managers must decide on the management objectives and management regime for each protected area. In the context of mass bleaching, one goal for management is to strengthen the factors that confer resilience [West and Salm, 2003; Westmacott et al., 2000]. MPAs are particularly suited to management of direct threats to coral reefs, such as those from over-fishing and recreational overuse or misuse. MPAs can also assist in addressing larger scale or indirect threats, such as land-based pollution, although usually only as part of a broader management initiative within an integrated coastal zone management (ICZM) approach.

A principle objective for MPA management in the context of climate change should be to maintain healthy populations of herbivorous fishes. The role of herbivores in maintaining conditions that are conducive for coral recruitment and survival makes their protection critical for reefs subject to increasing sea temperatures [Bellwood et al., 2004; McClanahan et al., 2003]. While some level of harvest may be sustainable, the importance of herbivores to future reef resilience means that managers should carefully manage fishing activity to ensure adequate levels of herbivory are sustained (a conservation objective), and not merely to ensure a sustainable or maximum harvest (a fisheries objective). Ecological modelling supports this focus for reef management by illustrating the importance of maintaining coral cover and herbivory if future reef condition is to be optimised under scenarios of repeated mass bleaching [Wooldridge et al., 2005].

Marine Protected Areas can also play a role in reducing damage associated with recreational activities. Resilience of a reef can be eroded through a range of human activities, including physical damage from diving and boat anchoring, and from release of nutrients and combustion products from vessels [Ailen, 1992; Saphier, 2005; Westmacott et al., 2000]. Where MPAs have been established to protect important refugia, even localised stresses associated with recreational activities may pose a significant threat to ecosystem resilience. MPA managers may already have regulations and best-practice guidelines in place, but further measures to ensure users avoid imposing additional stresses during periods of temperature stress are worthy of consideration.

3.3.1. Reviewing and revising management targets

While effectively managed MPAs can play a key role in resilience-based management, reef managers will also need to identify and implement broader management strategies in order to maximise the resilience of the ecosystem. Management efforts beyond MPA boundaries will play an increasingly critical role in protecting coral reefs as climate stress and other pressures continue to increase [Bellwood et al., 2004; Buddemeier et al., 2004].

Many of the broader strategies required for effective restoration and maintenance of ecosystem resilience are not unique to climate change responses. However, with the additional threat of climate change, management of localized stressors, such as water quality, over-fishing, or recreational misuse, may need to become more conservative if levels of ecological condition and services are to be maintained. Drawing from recent examinations of the cumulative and synergistic impacts between local stressors and climate change [Anthony, 2004; Buddemeier et al., 2004; Coles and Brown, 2003] this section reviews the relationships between local and global stressors, notably mass coral bleaching, and highlights ways in which management targets for local stressors can be reviewed and revised to promote coral reef ecosystem resilience [Marshall and Schuttenberg, 2006; West and Salm, 2003; Westmacott et al., 2000].

Recreational activities: snorkelling, diving and boat anchoring The temperature anomalies that trigger coral bleaching events impose stress on coral colonies, even before there are any visible signs of bleaching [Jones, 1997]. Once a coral is bleached it is severely stressed with reduced energy input and a consequential drop in capacity for essential physiological functions, such as injury repair and resistance to pathogens [Meesters and Bak, 1993]. Snorkelling, diving, and boat anchoring are all activities that can cause physical injuries to corals if not carefully managed [Davis and Tisdell, 1995; Hawkins et al., 1999; Rouphael and Inglis, 2002; Zakai and Chadwick-Furman, 2002]. While recreational activities alone can be important sources of stress to a coral reef community in high-use situations, the potential for physical damage to exacerbate mortality risk during periods of thermal stress suggests that a resilience-based approach to management should include a review of relevant policies that influence activities at heavily used, high value tourism sites.

Water quality Degraded water quality affects various life stages of corals, increasing the vulnerability of corals to other stresses such as anomalous temperatures [Fabricius, 2005; Gilmour, 1999; McClanahan, 2002; Nugues and Roberts, 2003]. Acute increases in sediment and pollutants associated with coastal development or dredging deliver additional stress to corals, which must clear sediment from colony surfaces [Gilmour, 1999; Nugues and Roberts, 2003; Riegl, 1995; Sakai and Nishihira, 1991]. Corals stressed from mass bleaching are likely to be less effective at defending against invasion by microalgae or resisting competition from macroalgae [Meesters and Bak, 1993]. Additionally, nutrient inputs can favour growth of macroalgae, providing a competitive disadvantage to coral recruits attempting to recolonise after bleaching-induced mortality [McCook, 1999].

In light of these implications, managers are beginning to consider restricting the timing of coastal activities such as dredging or coastal development projects during periods of increased temperature stress [Orpin et al., 2004]. Strategic timing of particular coastal activities could reduce the risk of damage to coral communities while also reducing the risk that developers and others involved in the activities will be blamed for any coral mortality that could be due to coral bleaching.

Fishing activities Herbivores play a critical role in facilitating recovery of coral reefs after major disturbances [Bellwood et al., 2003; McClanahan et al., 2003]. In many locations, the grazing activity of herbivores is essential to the maintenance of substrate suitable for coral recruitment, so maintaining the functional link between herbivores and coral recovery should be a strategic goal in management efforts to support reef resilience [Belliveau and Paul, 2002; Bellwood et al., 2003]. Reef managers may consider strategies for implementing initiatives to protect the herbivory function that is necessary for reef recovery, such as policies restricting or preventing harvesting of herbivores on a temporary or long-term basis [Bellwood et al., 2004; McClanahan et al., 2001; McCook, 1997]. These could involve general, spatial or temporary limits on herbivorous fishes, or other restrictions thought to be effective in reducing harvest of key herbivore species. These initiatives would need to be maintained until significant recovery is evident or until there is other evidence that adequate settlement substrate can be maintained despite fishing pressures.

4. Conclusions

Predicted increases in the frequency and severity of coral bleaching events over the next century are the basis for grave concerns about the future of coral reefs in a changing climate [Hoegh-Guldberg, 1999; Hughes et al., 2003]. Yet climate change also poses other issues for coral reef ecosystems, many of which are only just beginning to be understood [Buddemeier et al., 2004]. Responding to these threats requires a dual management strategy. First, mitigation efforts are required to minimize the rate and extent of climate change. Coral reef scientists and managers can increase the imperative for mitigation by raising awareness about the impacts of climate change on coral reef ecosystems and the services they provide to people. Second, managers can enhance and focus efforts to more effectively manage local stressors in order to support ecosystem resilience. With the emergent threat of climate change, these efforts are becoming critically important to "buy time" for coral reefs [Hansen et al., 2003]. Despite uncertainty in the details, the principles behind managing for resilience are scientifically sound, and they offer a proactive, "no regrets" strategy for addressing this most challenging issue [Marshall and Schuttenberg, 2006; West and Salm, 2003; Westmacott et al., 2000]. Adapting coral reef management in the face of climate change will best be most effective if associated with deep changes to management approaches and structures [Hughes, 2005]. However, strategies can be developed now, using current scientific understanding and emerging resilience principles, to enable reef managers to improve the prospects of coral reefs in the face of climate change.

Acknowledgments. We wish to thank David Obura, Jordan West, Ove Hoegh-Guldberg, Billy Causey, Terry Hughes, Jo Johnson, David Wachenfeld, Naneng Setiasih, Nadine Marshall and Lara Hansen for stimulating and challenging discussions about how coral reef managers might meaningfully respond to the threat of climate change. In particular, Rod Salm and Stephanie Wear have facilitated important learning networks for coral reef managers attempting to tackle climate change implications, which have provided many insights that have influenced our thinking. Two anonymous reviewers helped to substantially improve the manuscript.

References

Ailen, W. H., Increased dangers to Caribbean marine ecosystems: Cruise ship anchors and intensified tourism threaten reefs, *Biosci.*, 42, 330-335, 1992.

Allen Consulting Group, Climate change: risk and vulnerability, promoting an efficient adaptation response in Australia, *Final Report, March* 2005, p. 159, Australian Greenhouse Office, Department of the Environment and Heritage, 2005.

Anthony, K. R. N., and S. R. Connolly, Environmental limits to growth: physiological niche boundaries of corals along turbidity-light gradients, *Oecologia*, 141, 373-384, 2004.

Baird, A. H., and P. Marshall, Mortality, growth and reproduction in scleractinian corals following bleaching on the Great Barrier Reef, *Mar. Ecol. Prog. Ser.*, 237, 133-141, 2002.

Baker, A. C., Reef corals bleach to survive change, *Nature*, 411, 765-766, 2001.

Belliveau, S. A., and V. J. Paul, Effects of herbivory and nutrients on the early colonization of crustose coralline and fleshy algae, *Mar. Ecol. Prog. Ser.*, 232, 105-114, 2002.

Bellwood, D. R., A. H. Hoey, and J. H. Choat, Limited functional redundancy in high diversity systems: resilience and ecosystem function on coral reefs, *Ecol. Lett.*, 6, 281-285, 2003.

Bellwood, D. R., T. P. Hughes, C. Folke, and M. Nystrom, Confronting the coral reef crisis, *Nature*, 429, 827-833, 2004.

Berkelmans, R., Time-intergrated thermal bleaching thresholds of reefs and their variation on the Great Barrier Reef, *Mar. Ecol. Prog. Ser.*, 229, 73-82, 2002.

Brown, B. E., Coral bleaching: causes and consequences, *Coral Reefs*, 16, S129-S138, 1997.

Brown, B. E., R. P. Dunne, M. S. Goodson, and A. E. Douglas, Bleaching patterns in reef corals, *Nature*, 404, 142-143, 2000.

Buddemeier, R. W., and D. G. Fautin, Coral bleaching as an adaptive mechanism, *Biosci.*, 43, 320-325,1993.

Buddemeier, R. W., J. A. Kleypas, and R. Aronson, Coral Reefs and Global Climate Change: potential contributions of climate change to stresses on coral reef ecosystems, Prepared for the Pew Centre on Global Climate Change, 2004.

Cappo, M., and R. Kelley, Connectivity in the Great Barrier Reef World Heritage Area—An Overview of Pathways and Processes, In *Oceanographic Processes of Coral Reefs: Physical and Biological Links in the Great Barrier Reef*, edited by E. Wolanski, p. 161-185, CRC Press, Boca Raton, Florida, 2001.

Coles, S. L., and B. E. Brown, Coral bleaching-capacity for acclimatization and adaptation, *Ad. Mar. Biol.*, 46, 183-223, 2003.

Connell, J. H., T. P. Hughes, and C. C. Wallace, A 30-year study of coral abundance, recruitment, and disturbance at several scales in space and time, *Ecol. Monogr.*, 67, 461-488, 1997.

Davis, D., and C. Tisdell, Recreational scuba-diving and carrying capacity in marine protected areas, *Ocean and Coastal Management*, 26, 19-40, 1995.

De Angelis, H., and P. Skvarca, Glacier surge after ice shelf collapse, *Science*, 299, 1560-1562, 2003.

Dinsdale, E. A., Abundance of black-band disease on corals from one location on the Great Barrier Reef: a comparison with abundance in the Caribbean region, Proceedings 9th International Coral Reef Symposium, Bali, *Indonesia*, 2, 1239-1243, 2000.

Done, T., Scientific Principles for Establishing MPAs to Alleviate Coral Bleaching and Promote Recovery, In *Coral Bleaching and Marine Protected Areas, Proceedings of the Workshop on Mitigating Coral Bleaching Impact Through MPA Design*, edited by R. V. Salm and S. L. Coles, pp. 53-59, Bishop Museum, Honolulu, Hawaii, Asia Pacific Coastal Marine Program Report # 0102, The Nature Conservancy, Honolulu, Hawaii, USA, 2001.

Done, T. J., P. Whetton, R. Jones, R. Berkelmans, J. Lough, W. Skirving, and S. Wooldridge, Global climate change and coral bleaching on the Great Barrier Reef, *Final report to the State of Queensland Greenhouse Taskforce*, Department of Natural Resources and Mining, *Townsville*, 2003.

Fabricius, K. E., Effects of terrestrial runoff on the ecology of corals and coral reefs: review and synthesis, *Mar. Pollut. Bull.*, 50, 125-146, 2005.

Fish, M. R., I. M. Cote, J. A. Gill, A. P. Jones, S. Renshoff, and A. R. Watkinson, Predicting the impact of sea-level rise on Caribbean Sea Turtle nesting habitat, *Conserv. Biol*, 19, 482-491, 2005.

Flanagan, K. M., E. McCauley, F. Wrona, and T. Prowse, Climate change: the potential for latitudinal effects on algal biomass in aquatic ecosystems, *Can. J. Fish. Aquat. Sci.*, 60, 635-639, 2003.

Folke, C., C. Perrings, J. A. McNeely, and N. Myers, Biodiversity conservation with a human face: ecology, economics and policy, *Ambio*, 22, 62-63, 1993.

G8 Summit (Last Update 2005). *Climate change, clean energy and sustainable development*. Date accessed: 11 July 2006. http://www.fco.gov.uk/Files/kfile/PostG8_Gleneagles_CCChapeau.pdf.

Gattuso, J. P., and R. W. Buddemeier, Ocean Biogeochemistry: Calcification and CO_2, *Nature*, 407, 312-313, 2000.

Gattuso, J. P., M. Frankignoulle, I. Bourge, S. Romaine, and R.W. Buddemeier, Effect of calcium carbonate saturation of seawater on coral calcification, *Global Planetary Change*, 18, 37-47, 1998.

Gille, S. T., Warming of the Southern Ocean since the 1950s, *Science*, 295, 1275-1277, 2002.

Gilmour, J., Experimental investigation into the effects of suspended sediment on fertilisation, larval survival and settlement in a scleractinian coral, *Mar. Biol.*, 135, 451-462, 1999.

Gunderson, L. H., Resilience in theory and practice, *Annu. Rev. Ecol. Syst.*, 31, 425-439, 2000.

Hansen, L. J., J. L. Biringer, and J. R. Hoffmann, *BuyingTime: A User's Manual for Building Resistance and Resilience to Climate Change in Natural Systems*, World Wildlife Fund, Washington, DC, 2003.

Harley, C. D. G., A. Randall Hughes, K. M. Hultgren, B. G. Miner, C. J. B. Sorte, C. S. Thornber, L. F. Rodriguez, L. Tomanek, and S. L. Williams, The impacts of climate change in coastal marine systems, *Ecol. Lett.*, 9, 228-241, 2006.

Harvell, C. D., K. Kim, J. M. Burkholder, R. R. Colwell, P. R. Epstein, D. J. Grimes, E. E. Hofmann, E. K. Lipp, A. D. M. E. Osterhaus, R. M. Overstreet, J. W. Porter, G. W. Smith, and G. R. Vasta, Emerging marine diseases—climate links and anthropogenic factors, *Science*, 285, 1505-1510, 1999.

Harvell, C. D., C. E. Mitchell, J. R. Ward, S. Altizer, A. P. Dobson, R. S. Ostfeld, and M. D. Samuel, Climate warming and disease risks for terrestrial and marine biota, *Science*, 296, 2158-2162, 2002.

Harvell, D., K. Kim, C. Quirolo, J. Weir and G. Smith, Coral bleaching and disease: contributors to 1998 mass mortality in *Briareum asbestinum* (Octocorallia, Gorgonacea), *Hydrobiologia*, 460, 97-104, 2001.

Hawkins, J. P., C. M. Roberts, T. Van'T Hof, K. De Meyer, J. Tratalos, and C. Aldam, Effects of recreational scuba diving on Caribbean coral and fish communities, *Conserv. Biol.*, 13, 888-897, 1999.

Hays, G. C., A. C. Broderick, F. Glen, and B. J. Godley, Climate change and sea turtles: a 150-year reconstruction of incubation temperatures at a major marine turtle rookery, *Global Change Biology*, 9, 642-646, 2003.

Hochachka, P. W., and G. N. Somero, *Biochemical Adaptation: mechanism and process in physiological evolution*, Oxford Univ. Press, New York, 2002.

Hoegh-Guldberg, O., Climate change, coral bleaching and the future of the world's coral reefs, *Mar. Freshw. Res.*, 50, 839-866, 1999.

Hoegh-Guldberg, O., The future of coral reefs: Integrating climate model projections and the recent behaviour of corals and their dinoflagellates, Proceedings of the 9th International Coral Reef Symposium, Bali, Indonesia, Vol. 2, 1105-1110, 2000.

Hoegh-Guldberg, O., Coral reefs in a century of rapid environmental change, *Symbiosis*, 37, 1-31, 2004.

Holling, C. S., D. W. Schindler, B. W. Walker, and J. Roughgarden, Biodiversity in the functioning of ecosystems: an ecological synthesis, In *Biodiversity loss, ecological and economical issues*, edited by C. A. Perrings et al., Cambridge University Press, Cambridge, UK, 1995.

Hughes, T. P., Catastrophes, phase shifts and large-scale degredation of a Caribbean coral reef, *Science*, 265, 1547-1551, 1994.

Hughes, T. P., A. H. Baird, D. R. Bellwood, M. Card, S. R. Connolly, C. Folke, R. Grosberg, O. Hoegh-Guldberg, J. B. C. Jackson, J. Kleypas, J. M. Lough, P. Marshall, M. Nystroem, S. R., Palumbi, J. M., Pandolfi, B. Rosen and J. Roughgarden, Climate change, human impacts, and the resilience of coral reefs, *Science*, 301, 929-933, 2003.

Hughes, T. P., D. R. Bellwood, C. Folke, R. S. Steneck, and J. Wilson, New paradigms for supporting the resilience of marine ecosystems, *Trends Ecol. Evol.*, 20, 380-386, 2005.

IPCC, Climate Change 2001: Synthesis Report. A Contribution of Working Groups I, II, and III to the Third Assessment Report of the Intergovernmental Panel on Climate Change, in *IPCC 3rd Assessment Report*, edited by R.T. Watson and the Core Writing Team, p. 398, Cambridge University Press, New York, NY, USA, 2001a.

IPCC, Climate Change 2001: The Scientific Basis. Contribution of Working Group I to the Third Assessment Report of the Intergovernmental Panel on Climate Change. J. T. Houghton et al. eds.). (Cambridge University Press, Cambridge, United Kingdom and New York, NY, USA), pp. 881, 2001b.

Jones, G. P., M. I. McCormick, M. Srinivasan, and J. V. Eagle, Coral decline threatens fish biodiversity in marine reserves, PNAS, 101, 8251-8253, 2004.

Jones, R. J., Zooxanthellae loss as a bioassay for assessing stress in corals, *Mar. Ecol. Prog. Ser.*, 149, 1-3, 1997.

Jones, R. J., O, Hoegh-Guldberg, A. W. D. Larkum, and U. Schreiber, Temperature-induced bleaching of corals begins with impairment of the CO_2 fixation mechanism in zooxanthella, *Plant Cell Environ.*, 21, 1219-1230, 1998.

Kennedy, V. S., R. R. Twilley, J. A. Kleypas, J. H. Cowan, S. R. Hare. Coastal and marine ecosystems and global climate change: Potential effects on US resources. (Pew Center on Global Climate Change, Arlington, Va), pp. 52, 2002.

Kleypas, J. A., R. W. Buddemeier, D. Archer, J. P. Gattuso, C. Langdon, and B. N. Opdyke, Geochemical consequences of increased atmospheric carbon dioxide on coral reefs, *Science*, 284, 118-120, 1999.

Limpus, C. J., A Marine Resource Case Study: Climate change and sea level rise probable impacts on marine turtles, Proceedings of the 2nd SPREP Meeting, p. 157, Fiji, 1993.

Little, A. F., M. J. H.v. Oppen, and B. L. Willis, Flexibility in algal endosymbioses shapes growth in reef corals, *Science*, 304, 1492-1494, 2004.

Marshall P., and H. Z. Schuttenberg, *A Reef Manager's Guide to Coral Bleaching*, Great Barrier Reef Marine Park Authority, Townsville, Australia, 2006.

Marshall, P. A., and A. H. Baird, Bleaching of corals on the Great Barrier Reef: differential susceptibilities among taxa, *Coral Reefs*, 19, 155-163, 2000.

McClanahan, T., T. Done, and N. C. Polunin, Resiliency of coral reefs, in *Resilience and the Behaviour of Large Scale Systems*, edited by L. H. Gunderson and L. Pritchard, p. 240, Island Press: *Washington DC*, USA, 2003.

McClanahan, T., N. Polunin, and T. Done, Ecological states and the resilience of coral reefs, *Conserv. Ecol.*, 6, 18, 2002a.

McClanahan, T. R., The near future of coral reefs, *Environ. Conserv.*, 29, 460-483, 2002.

McClanahan, T. R., J. Maina, and L. Pet Soede, Effects of the 1998 coral mortality event on Kenyan coral reefs and fisheries, *Ambio*, 31, 543-550, 2002b.

McClanahan, T. R., M. McField, M. Huitric, K. Bergman, E. Sala, M. Nystrom, I. Nordemar, T. Elfwing, and N. A. Muthiga, Response of algae, corals and fish to the reduction of macroalgae in fished and unfished patch reefs of Glovers Reef Atoll, Belize, *Coral Reefs*, 19, 367-379, 2001.

McCook, L. J., Effects of herbivory on zonation of *Sargassum* spp. within fringing reefs of the central Great Barrier Reef, *Mar. Biol.*, 129, 713-722, 1997.

McCook, L. J., Macroalgae, nutrients and phase shifts on coral reefs: scientific issues and management consequences for the Great Barrier Reef, *Coral Reefs*, 18, 357-367, 1999.

Meesters, E. H., and R. P. M. Bak, Effects of coral bleaching on tissue regeneration potential and colony survival, *Mar. Ecol. Prog. Ser.*, 96, 189-198, 1993.

Michalek-Wagner, K., and B. L. Willis, Impacts of bleaching on the soft coral *Lobophytum compactum*. I. Fecundity, fertilization and offspring viability, *Coral Reefs*, 19, 231-239, 2001a.

Moberg, F., and C. Folke, Ecological goods and services of coral reef ecosystems, *Ecological Economics*, 29, 215-233, 1999.

Nakamura, T., and R. Van Woesik, Water-flow rates and passive diffusion partially explain differential survival of corals during the 1998 bleaching event, *Mar. Ecol. Prog. Ser.*, 212, 301-304, 2001.

Nugues, M. M., and C. M. Roberts, Partial mortality in massive reef corals as an indicator of sediment stress on coral reefs, *Mar. Pollut. Bull.*, 46, 314-323, 2003.

Nyström, M., and C. Folke, Spatial resilience of coral reefs, *Ecosystems*, 4, 406-417, 2001.

Nyström, M., C. Folke, and F. Moberg, Coral-reef disturbance and resilience in a human-dominated environment, *Trends Ecol. Evol.*, 15, 413-417, 2000.

Obura, D. O. Resilience and climate change: lessons from coral reefs and bleaching in the Western Indian Ocean. Estuarine, *Coastal and Shelf Science*, 63: 353-372, 2005.

Orpin, A. R., P. V. Ridd, S. Thomas, K. R. N. Anthony, P. Marshall, and J. Oliver, Natural turbidity variability and weather forecasts in risk management of anthropogenic sediment discharge near sensitive environments, *Mar. Pollut. Bull.*, 49, 602-612, 2004.

Pandolfi, J. M., R. H. Bradbury, E. Sala, T. P. Hughes, K. A. Bjorndal, R. G. Cooke, D. McArdle, L. McClenachan, M. J. H. Newman, G. Paredes, R. R. Warner, and J. B. C. Jackson, Global trajectories of the long-term decline of coral reef ecosystems, *Science*, 301, 955-957, 2003.

Parmesan, C., and G. Yohe, A globally coherent fingerprint of climate change impacts across natural systems, *Nature*, 421, 37-42, 2003.

Perry, A. L., P. J. Low, J. R. Ellis, J. D. Reynolds. Climate change and distribution shifts in marine fishes. *Science*, 308: 1912-1915, 2005.

Pittock, B., *Climate Change: an Australian guide to the science and potential impacts.* B. Pittock (ed.). (Australian Greenhouse Office, Canberra), pp. 239, 2003.

Riegl, B., Effects of sand deposition on scleractinian and alcyonacean corals, *Mar. Biol.*, 121, 517-526, 1995.

Riegl, B., Climate change and coral reefs: different effects in two high-latitude areas (Arabian Gulf, South Africa), *Coral Reefs*, 22, 433-446, 2003.

Rosenberg, E., and Y. Ben-Haim, Microbial diseases of corals and global warming, *Environmental Microbiology*, 4, 318-326, 2002.

Rouphael, A. B., and G. J. Inglis, Increased spatial and temporal variability in coral damage caused by recreational scuba diving, *Ecol. Appl.*, 12, 427-440, 2002.

Sakai, K., and M. Nishihira, Immediate effect of terrestrail runoff on a coral community near a river mouth in Okinawa, *Galaxea*, 10, 125-134, 1991.

Salm, R., and J. West, this volume.

Santavy, D. L., J. K. Summers, V. D. Engle, and L. C. Harwell, The condition of coral reefs in southern florida using coral disease and bleaching as indicators, *Environ. Monit. Assess.*, 100, 129-152, 2000.

Saphier, A. D., and T. C. Hoffman, Forecasting models to quantify three anthropogenic stresses on coral reefs from marine recreation: Anchor damage, diver contact and copper emission from antifouling paint, *Mar. Pollut. Bull.*, 51, 590, 2005.

Schuttenberg, H. Z., E. D. Gomez, S. Westmacott, and H. Cesar, Coral bleaching causes, consequences and response, Proceedings of the 9th International Coral Reef Symposium, Bali, *Indonesia*, 1-11, 2000.

Sheppard, C. R. C., Predicted recurrences of mass coral mortality in the Indian Ocean, *Nature*, 425, 294-296, 2003.

Siedler, G., J. Church, and J. Gould, Ocean circulation and climate: observing and modelling the global ocean, *International Geophysics*, *77*: Elsevier, Oxford, UK, 2001.

Smith, S. V., and R. W. Buddemeier, Global change and coral reef ecosytems: *Annu. Rev. Ecol. Syst.*, 23, 89-118, 1992.

Smithers, B. V., D. R. Peck, A. K. Krockenberger, and B. C. Congdon, Elevated sea-surface temperature, reduced provisioning and reproductive failure of wedge-tailed shearwaters (*Puffinus pacificus*) in the southern Great Barrier Reef, Australia, *Mar. Freshw. Res.*, 54, 973-977, 2003.

Spalding, M. D., and G. E. Jarvis, The impact of the 1998 coral mortality on reef fish communities on the Seychelles, *Mar. Pollut. Bull.*, 44, 309-321, 2002.

Stocker, T. F., Past and future reorganisations in the climate system: *Quaternary Science Reviews*, 19, 301-319, 2000.

The Nature Conservancy, R2—Reef Resilience—Toolkit. http://www.outsidetheboxdesign.com/r2coral/start. Htm, 2005.

Thomas, C. D., A. Cameron, R. E. Green, M. Bakkenes, L. J. Beaumont, Y. C. Collingham, B. F. N. Erasmus, M. F. De Siqueira, A. Grainger, L. Hannah, L. Hughes, B. Huntley, A. S. Van Jaarsveld, G. F. Midgley, L. Miles, M. A. Ortega-Huerta, A. Townsend-Peterson, O. L. Phillips, and S. E. Williams, Extinction risk form climate change, *Nature*, 427, 145-148, 2004.

Walther, G., E. Post, P. Convey, A. Menzel, C. Parmesan, T. J. C. Beebee, J. Fromentin, O. Hoegh-Guldberg, and F. Bairlein, Ecological responses to recent climate change, *Nature*, 416, 389-395, 2002.

Wellington, G. M., P.W. Glynn, A. Strong, S. A. Nauarrete, E. Wieters, and D. Hubbard, Crisis on coral reefs linked to climate change, EOS, *Transactions*, *American Geophysical Union*, 82, 1-7, 2001.

West, J. M., and R. V. Salm, Resistance and resilience to coral bleaching: implications for coral reef conservation and management, *Conserv. Biol.*, 17, 956-967, 2003.

Westmacott, S., K. Teleki, S. Wells, and J. West, *Management of bleached and severely damaged coral reefs*, IUCN, The World Conservation Union, Washington, DC, USA, 2000.

Wilkinson, C., *Status of Coral Reefs of the World: 1998*, pp. 184, Australian Institute of Marine Science, Townsville, Australia, 1998.

Wilkinson, C., Coral bleaching and mortality—the 1998 event 4 years later and bleaching to 2002, In *Status of Coral Reefs of the World 2002*, edited by C. Wilkinson, p. 33-44, Australian Institute of Marine Science, Townsville, Australia, 2002a.

Wilkinson, C., *Status of Coral Reefs of the World: 2002*, pp. 378. Australian Institute of Marine Science, Townsville, Australia, 2002b.

Wilkinson, C., Status of coral reefs of the world: 2004, in *Status of Coral Reefs of the World*, edited by C. Wilkinson, pp. 301, Australian Institute of Marine Science, Townsville, Australia, 2004a.

Wilkinson, C., Status of coral Reefs of the world: 2004 Summary, Australian Institute of Marine Science, Townsville, Australia, 2004b.

Wooldridge, S., T. Done, R. Berkelmans, R. Jones, and P. Marshall, Precursors for resilience in coral communities in a warming climate: a belief network approach, *Mar. Ecol. Prog. Ser.*, 295, 157-169, 2005.

WWF, *The Implications of Climate Change for Australia's Great Barrier Reef*, pp. 345, World Wildlife Fund, Sydney, Australia, 2004.

Zakai, D., and N. E. Chadwick-Furman, Impacts of intensive recreational diving on reef corals at Eilat, northern Red Sea, *Biol. Conserv.*, 105, 179-187, 2002.

List of Contributors

Felipe Arzayus
NOAA Coral Reef Watch
1335 East-West Highway RA31
Silver Spring, MD 20910

John Bruno
Department of Marine Sciences
12-7 Venable Hall, CB# 3300
The University of North Carolina
Chapel Hill, NC 27599

Kenneth S. Casey
NOAA Coral Reef Watch
1335 East-West Highway RA31
Silver Spring, MD 20910

Billy Causey
Florida Keys National Marine Sanctuary
P.O. Box 500368
Marathon, FL 33050

Julie Church
IUCN- The World Conservation Union
IUCN Regional Office for Eastern Africa
PO Box 68200
Nairobi KENYA

Terry Done
Australian Institute of Marine Science
Townsville MC
Townsville, QLD 4810 AUSTRALIA

Sophie G. Dove
University of Queensland
Centre for Marine Studies
Gehemann Laboratories
Research Rd.
St. Lucia, QLD 4072 AUSTRALIA

E. Mark Eakin
NOAA Coral Reef Watch
1335 East-West Highway RA31
Silver Spring, MD 20910

Andrea Grottoli
Ohio State University
Dept of Geological Sciences
125 South Oval Mall
Columbus, OH 43210

Drew Harvell
Department of Ecology and Evolutionary Biology
Cornell University
Ithaca, NY 14853

Malcolm Heron
Marine Geophysical Laboratory
School of Mathematical and Physical Sciences
James Cook University
Townsville, QLD 4811 AUSTRALIA

Scott Heron
NOAA Coral Reef Watch
1335 East-West Highway
Silver Spring, MD 20910

Ove Hoegh-Guldberg
Centre for Marine Studies
University of Queensland
St. Lucia, 4072 QLD AUSTRALIA

Roger Jones
CSIRO Marine and
Atmospheric Research
Private Bag No 1
Aspendale, 3195 AUSTRALIA

Joan A. Kleypas
National Center for Atmospheric Research
1850 Table Mesa Drive
Boulder, CO 80305

Semen Koksal
Department of Mathematical Sciences
Florida Institute of Technology
150 West University Boulevard
Melbourne, FL 32901-6988

Chris Langdon
Rosenstiel School of Marine and Atmospheric Science (RSMAS)
Room E208 Grosvenor East
4600 Rickenbacker Cauesway
Miami, FL 33149

Paul Marshall
Great Barrier Reef Marine Park Authority
2-68 Flinders Street
4810 Townsville
Queensland AUSTRALIA

Elizabeth McLeod
The Nature Conservancy
923 Nu'uanu Avenue
Honolulu, HI 96817

David Obura
CORDIO
#8/9 Kibaki Flats
Kenyatta Beach
Box 10135-80101
Mombasa KENYA

Cathie A. Page
School of Marine Biology and Aquaculture
James Cook University
Townsville, QLD 4811 AUSTRALIA

Jonathan T. Phinney
Maryland Sea Grant College
4321 Hartwick Road, Suite 300
University of Maryland
College Park, Maryland 20740

Rodney Salm
The Nature Conservancy
923 Nu'uanu Avenue
Honolulu, HI 96817

Elizabeth R. Selig
University of North Carolina
Campus Box 3275
216 Miller Hall
Chapel Hill, NC 27599

William Skirving
NOAA Science Center
5200 Auth Road
Camp Springs, MD 20746

Heidi Schuttenberg
School of Tropical Environment Studies and Geography
James Cook University
Townsville, QLD 4811 AUSTRALIA

Hugh Sweatman
Australian Institute of Marine Science
Townsville MC
Townsville 4810
Queensland AUSTRALIA

Bette L. Willis
School of Marine Biology and Aquaculture,
James Cook University of North Queensland
Townsville, QLD 4811 AUSTRALIA

Alan E. Strong
NOAA Science Center
5200 Auth Road
Camp Springs, MD 20746

Robert van Woesik
Department of Biological Sciences
Florida Institute of Technology
150 West University Boulevard
Melbourne, FL 32901-6988

John (Charlie) Veron
Australian Institute of Marine Science
Townsville MC
Townsville 4810
Queensland AUSTRALIA